W0043809

ALLE·ZEIT·WACH
1842
SJ

ILSI Human Nutrition Reviews

Series Editor: Ian Macdonald

*Already published:*

**Sweetness**
Edited by John Dobbing

*Forthcoming titles in the series:*

**Calcium in Human Biology**
Edited by B. E. C. Nordin

**Zinc in Human Biology**
Edited by C. F. Mills

# Foreword

This volume in the International Life Sciences Institute series *Human Nutrition Reviews* is unusual in that it is largely an up-to-date compilation of references. Many pronouncements are made from various sources which offer advice on the wisdom or otherwise of having sugar in our diet and it was felt therefore that there must be many people who would wish to confirm the validity of these pronouncements. To this end this monograph on sugars – mainly sucrose – will enable the reader to seek out the appropriate scientific references with minimal effort.

Students interested in or studying or doing research on sugar will, with this book, have an easy access to that part of the scientific literature appropriate to their needs. This volume could also be of value to those in the media who may wish to confirm, at first hand, reports that they would like to present to the public.

In the main, the style adopted in this monograph is that of a brief outline of the contents of a scientific paper which is followed by its reference. Although the monograph like others in the series, is a compilation from several authors, the contributors to various sections have not been identified because the volume editors have had the original drafts checked by others and the result is therefore a "refereed book of references".

The International Life Sciences Institute (ILSI) is a non-profit scientific foundation established to encourage and support research and educational programmes in nutrition, toxicology and food safety and to encourage cooperation in these programmes among scientists from universities, industry and government in order to facilitate the resolution of health and safety issues world-wide.

Ian Macdonald
Series Editor

# Sucrose

## Nutritional and Safety Aspects

Edited by Gaston Vettorazzi and
Ian Macdonald

Springer-Verlag
London Berlin Heidelberg New York
Paris Tokyo

Gaston Vettorazzi, MD, MS, PhD
Director, International Toxicology Information Centre, Miraconcha 8, 1-B, E-200007 San Sebastian, Spain (Formerly Senior Toxicologist, International Programme on Chemical Safety, WHO, Geneva, Switzerland)

Ian Macdonald, MD, DSc, FIBiol
Head, Department of Physiology, Guy's Hospital Medical and Dental School, St. Thomas' Street, London, SE1 9RT, UK

*Series Editor*
Ian Macdonald, MD, DSc, FIBiol (address as above)

ISBN-13: 978-1-4471-1651-6 e-ISBN-13: 978-1-4471-1649-3
DOI: 10.1007/978-1-4471-1649-3

British Library Cataloguing in Publication Data
Sucrose nutritional and safety aspects. — (ILSI human nutrition reviews)
1. Man. Diseases. Role of sugar
I. Vettorazzi, Gaston II. Macdonald, Ian III. Series
616.07'1

Library of Congress Cataloging-in-Publication Data
Sucrose : nutritional and safety aspects / edited by Gaston Vettorazzi and Ian Macdonald.
p. cm. — (ILSI human nutrition views)
Includes bibliographies and indexes.

1. Sucrose—Physiological effect. 2. /Sucrose—Health aspects.
I. Vettorazzi, Gaston. II. Macdonald, Ian, 1921– . III. Series. QP702.S8S83 1988 613.2'8—dc19


Softcover reprint of the hardcover 1st edition 1988



Filmset by Photo-graphics, Honiton, Devon

2128/3916–543210

# Preface

This book was designed to fill the void for a synopsis or reference book on sucrose, and it is presented as a user-friendly thesaurus serving multiple purposes. These will be contingent on whether the reader is an individual interested in the full range of scientific work on sucrose, or wishes to be informed of the major expert opinions on the subject (Section I), or is searching for a particular author or published paper in the comprehensive literature (Section II). All three aspects are represented in the book.

Section I brings out in monographic style (using decimally numbered sections) a series of summaries of scientific work reflecting the various aspects of current interest. This section follows an outline that facilitates the location of a particular topic. The criteria adhered to in writing the summaries and in selecting the appropriate literature are described in the introduction to the monograph. This section represents a collaborative effort among several contributors and also includes major conclusions reached and recommendations made by authoritative scientific and evaluative expert committees on aspects related to the subject. It concludes with a summary by the editor.

Section II is a bibliography of practically all the most significant scientific works related to the nutritional and safety aspects of sucrose. It represents an impressive array of literature citations, which required considerable time, effort, and investment in collection assembly, and organization. This is then followed by a list of books on sucrose that may be considered useful additional reading.

The editors would like to record their appreciation to the scientists whose names and addresses appear in the list of contributors to the monograph; their eminent collaboration makes it possible for readers to benefit from an informative monograph that outlines the current position of research on nutritional and safety issues involving sucrose. Their valuable comments on the draft and their advice on the inclusion of significant work are gratefully acknowledged.

Thanks are also due to the International Life Science Institute for accepting this volume as part of their series on *Human Nutrition Reviews*.

Geneva, Switzerland
August 1987

Gaston Vettorazzi

## *Acknowledgement*

The Editors would like to acknowledge the assistance of Diane Rahim in proof-reading, reference work and typing.

# Contents

# Contributors to the Monograph

Prof. Joseph F. Borzelleca
Department of Pharmacology and Toxicology, Medical College of Virginia, Virginia Commonwealth University, Box 613 – MCV Station, Richmond, Virginia 23298-001, USA

Prof. Corrado L. Galli
Institute of Pharmacological Sciences, University of Milan, 20133 Milan, Italy

Prof. Franciso Grande (Emeritus)
Department of Biochemistry, University of Zaragoza, 50009 Zaragoza, Spain

Prof. Giuseppe Grippaudo
Department of Dental Sciences, University of Rome "La Sapienza", 00161 Rome, Italy

Prof. Sidney H. Ingbar
Beth Israel Hospital, Harvard Medical School, 330 Brookline Avenue, Boston, Massachusetts 02215, USA

Prof. Ancel Keys (Emeritus)
Division of Epidemiology, School of Public Health, University of Minnesota, 611 Beacon Street, SE Stadium Gate 27, Minneapolis, Minnesota 55455, USA

Dr. Harris R. Lieberman
Department of Brain and Cognitive Sciences, Massachusetts Institute of Technology, Cambridge, Massachusetts 02139, USA

Dr. Armstrong D. Lowe
Technical Terminology Service, World Health Organization, 1211 Geneva 27, Switzerland

Prof. Ian Macdonald
Department of Physiology, Guy's Hospital Medical and Dental Schools, London SE1 9RT, England

Prof. Felix G. Reyes
Department of Food Science, State University of Campinas, 13081 Campinas, SP, Brazil

Prof. Gerard Slama
Diabetology Service, Hotel-Dieu Public Hospital, 1, Place du Parvis-Notre-Dame, 75181 Paris Cedex 04, France

Dr. Edward Smith
International Programme on Chemical Safety, World Health Organization, 1211 Geneva 27, Switzerland

Prof. Gianni Tomassi
Department of Experimental Nutrition, National Institute of Nutrition, Via Ardeatina 546, 0000179 Rome, Italy

# Abbreviations

| | |
|---|---|
| AD | alcohol dehydrogenase (EC 1.1.1.1) |
| ADP | adenosine diphosphate |
| AMP | adenosine monophosphate |
| AOAC | Association of Official Analytical Chemists |
| ATP | adenosine triphosphate |
| BNF | British Nutrition Foundation |
| °C | degree Celsius (centigrade) |
| CAC | Codex Alimentarius Commission |
| CAS | Chemical Abstract Service |
| CHD | coronary heart disease |
| DNA | deoxyribonucleic acid |
| E | enolase (EC 4.2.1.11) |
| EC | Enzyme Commission |
| FAO | Food and Agriculture Organization of the United Nations |
| FASEB | Federation of American Societies for Experimental Biology |
| FDP | fructose (hexose) diphosphatase (EC 3.1.3.11) |
| FK | fructokinase (EC 2.7.1.4) |
| g | gram |
| GD | glyceraldehyde dehydrogenase (EC 1.2.1.3) |
| GIP | gastrointestinal polypeptide |
| GK | D-glycerate kinase |
| GPD | glucose-6-phosphate dehydrogenase (EC 1.1.1.49) |
| GPD | glyceraldehyde phosphate dehydrogenase (EC 1.2.1.9) |
| GPI | glucose phosphate isomerase (EC 5.3.1.9) |
| GTP | guanosine triphosphate |
| GRAS | generally recognized as safe |
| HbA | haemoglobin A |

| | |
|---|---|
| HDL | high-density lipoprotein |
| HGS | hydrogenated glucose syrup |
| HK | hexokinase (EC 2.7.1.1) |
| HPLC | high-pressure liquid chromatography |
| 5HT | 5-hydroxytryptamine |
| ICUMSA | International Commission for Uniform Methods of Sugar Analysis |
| IDDM | insulin-dependent diabetes mellitus (Type I diabetes) |
| ILO | International Labour Office |
| INCAP | Institute of Nutrition for Central America and Panama |
| IPCS | ILO/UNEP/WHO International Programme on Chemical Safety |
| ITP | inosine triphosphate |
| IUPAC | International Union for Pure and Applied Chemistry |
| kg | kilogram |
| $K_m$ | Michaelis constant |
| LA | liver aldolase (EC 4.1.2.13) |
| LC | lactonase (EC 3.1.1.17) |
| LD | lactate dehydrogenase (EC 1.1.1.17) |
| $LD_{50}$ | lethal dose (50%) |
| LSRO | Life Sciences Research Office |
| mg | milligram |
| mM | millimole |
| μM | micromole |
| min | minute |
| MMT | million metric ton |
| NAD | nicotinamide adenine dinucleotide |
| NADP | nicotinamide adenine dinucleotide phosphate |
| NADPH | dihydronicotinamide adenine dinucleotide phosphate |
| NAG | *N*-acetyl-β-glucosaminidase |
| NIDDM | non-insulin-dependent diabetes mellitus (Type II diabetes) |
| NRC | National Research Council of the US National Academy of Sciences |
| ORS | oral rehydration salts |
| PAHO | Pan American Health Organization |
| PCT | proximal convoluted tubules |
| PGD | 6-phosphogluconate dehydrogenase (EC 1.1.1.44) |
| PGK | phosphoglycerate kinase (EC 2.7.2.3) |
| PGM | phosphoglyceromutase (EC 2.7.5.3) |
| Pi | phosphorus (inorganic) |

| | |
|---|---|
| PK | pyruvate kinase (EC 2.7.1.40) |
| PPi | diphosphate (inorganic) |
| PSCC | Panel on Simple and Complex Carbohydrates |
| RI | ribose phosphate isomerase (EC 5.3.1.6) |
| SCOGS | Selected Committee on GRAS Substances |
| SIDA | Swedish International Development Authority |
| STF | Sugar Task Force (US FDA) |
| TFSS | Task Force on Sugars and Syrups (BNF) |
| TI | triophosphate isomerase (EC 5.3.1.1) |
| TK | triokinase (EC 2.7.1.28) |
| UDP | uridine diphosphate |
| UNEP | United Nations Environment Programme |
| UNICEF | United Nations Children's Fund |
| US FDA | United States Food and Drug Administration |
| UTP | uridine 5′-triphosphate |
| $V_{max}$(or $V$) | maximum velocity (i.e., rate of formation of product of an enzyme catalysed reaction) |
| VLDL | very low density lipoprotein |
| WARF | Wisconsin Alumni Research Foundation |
| WHO | World Health Organization |

Section I

# The Monograph

Chapter 1

# Introduction

## 1.1 Objective and Scope

This monograph on the nutritional and safety aspects of sucrose has been written with the aim of supplying information and highlighting contemporary research work. Its objective is to convey to readers the essential information on the subject, thus giving them a comprehensive insight into the state of the art of the topics of interest. This has been accomplished by means of brief summaries and references to relevant scientific reports from the current literature.

## 1.2 Criteria for Literature Selection

Only articles from refereed journals have been given preferential consideration. Reports of expert committees, reviews, and books that contain references to the scientific literature are also included in this monograph.

There are a considerable number of papers related to the nutritional and safety aspects of sucrose, especially the former. Though an extensive coverage is included in this monograph, older, and perhaps superseded, papers have not been included or summarized in detail. However, by using the references given both in the monograph and in Section II of this book, it is possible to gain access to the entire literature on a specific aspect of interest. No judgement has been made in deciding the references to be included beyond those mentioned above.

## 1.3 Criteria for Assessing the Validity of Data

The scientific soundness of an experimental study or epidemiological survey must be ascertained before its results can be accepted as valid. Several approaches can be used. Peer reviews and the use of refereed journals are widely used by scientists and others. The specialized, as well as non-specialized, reader may now profit from some recently published international guidelines (WHO 1987); these guidelines may provide alternative means when peer reviews are not yet available for a particular study. The most important guidelines are summarized below.

1. *Sources of Differences Between Treated and Control Groups.* An objective of many studies is to determine whether a treatment elicits a response. However, the observation of a difference in response between a treated and a control (untreated) group does not necessarily mean that the difference is a result of a treatment. There are two other potential causes of a difference: bias and chance. Bias implies systematic differences other than treatment between the groups; in other words, failure to compare like with like. Properly conducted studies analysed appropriately can eliminate bias.

Chance factors cannot be wholly excluded, because identically treated animals will not all respond identically, however carefully the study is conducted. While it is impossible to be absolutely certain that even very extreme differences in response are not due to chance, appropriate statistical analysis will allow the experimentalist to assess the probability of a "false positive", that is, the probability of the observed difference having occurred had there been no effect of treatment at all. The smaller the probability, the greater the confidence of having found a real effect. To improve the likelihood of detecting a true effect with confidence, it is necessary to try to minimize the role of chance by seeking to ensure that the "signal" can be recognised above the "noise" by the use of appropriate statistical analysis.

2. *Choice of Species and Strain.* While maximizing the "signal" means avoiding a species in which the response of interest is very rare, the use of an over-responsive species may also be problematical. Furthermore, it is not certain that an increased incidence of an effect that is a common spontaneous finding in the animal species used provides biological evidence of an effect that can be extrapolated to other species. Other considerations relate to the choice of species, whether they are practical (short life span, small size, availability, existence of detailed knowledge of the species) or theoretical (biochemical, physiological, or anatomical similarity to man). The application of these observations to older papers reporting on effects caused by the administration of sucrose to experimental animals whose species and strain were not clearly specified may invalidate them.

3. *Dose Levels.* Dose selection is an important and controversial element in the development of an experimental protocol for testing a substance. On biological grounds, it would be reasonable to test only at dose levels comparable with those to which human beings are exposed. On statistical and economic grounds this is not usually practical, because the effect will be too small to detect without a very large number of animals. To avoid the possibility of missing an effect that would occur in a small proportion of millions of exposed human beings in a study on hundreds or even thousands of animals, test animals are often dosed at levels many times higher than the maximum human exposure level. Then, assuming any effect that exists is dose related, the demonstration of a non-significant increase in response at a high dose level, though not providing evidence of absolute safety (an impossible goal), can give

reasonable grounds for believing that any effects that might occur at a very much lower dose level would be, at most, very slight.

A particular problem with this procedure is to decide how high the dose level should be. If the substance seems relatively non-toxic, the high dose may represent about 5% of the diet, or even more for substances, such as some nutritive food ingredients.

It is more important to have more than one dose level for a number of reasons. One is to compensate for the possibility that a misjudgement has occurred and that the highest dose may prove to cause undesirable effects. A second is that the metabolic pathways may differ at the various dose levels. A third reason is that the purpose of the study may be to obtain dose–response information. Finally, it may be necessary to ensure that the effect does not occur at dose levels in the range to be used by man.

The application of these guidelines to many of the experiments involving sucrose may permit the reader to draw important conclusions as to their validity and scientific soundness.

4. *Number of Animals.* The number of animals to be used is clearly an important determinant of the precision of the findings and must be large enough to permit statistical analysis.

5. *Duration of the Study.* The duration of the study can also markedly affect the sensitivity of tests. While studies should not be terminated too early, it is also important that they do not go on too long. This is because the last few weeks or months may produce relatively little data. Disease and old age may be of little interest in themselves, but may render it more difficult to detect conditions that are of interest.

Observation time following treatment is an important feature often overlooked in many studies including those involving sucrose. Such a time allows one to establish the reversibility of the effects observed.

6. *Accuracy of Determinations.* Accuracy of observations is clearly important in minimizing error. The advent of good laboratory practice and quality control units has done much to improve the quality of recording observations, but the quality of the observations and, hence, the study, still depends on interested and diligent personnel.

7. *Test Substance.* The test substance should be clearly identified and characterized chemically. In many experimental studies, including those involving sucrose, the origin and purity of the test substance are not reported. This omission completely invalidates the study.

8. *Adequacy of Control Groups.* The principle of comparing like with like implies that control groups should be randomly allocated from the same source as the treatment groups. While historical control data can be of value in the interpretation of rare findings in treated animals, there is sufficient evidence of quite large systematic differences in response between apparently identical untreated control groups that it is often impossible to be sure whether a difference seen between a treated group and a historic group is really due to treatment at all.

It is also essential to be sure that the treated group differs from the control group only with respect to the treatment of interest. Thus, if a treatment is applied in a solvent, an untreated control is not a proper basis for comparison, as one cannot be sure whether observed differences are the result of the treatment or the solvent. In this case, the appropriate control group would be one in which animals are given only the solvent.

9. *Animal Placement.* The general underlying requirement to avoid systematic differences between groups other than their treatment also demands that attention be given to the question of animal placement. If all treated animals are placed on the highest racks or are at one end of the room, differences in heating, lighting, or ventilation may produce effects that are erroneously attributed to treatment. Such systematic differences should be avoided; randomization of cage positions is desirable.

In recent years, a number of books have appeared in North America and the United Kingdom that purported to describe "the evil of sugar" (Starenkyj 1981; Dufty 1975; Du Ruisseau 1973; Yudkin 1972; Abrahamson and Pezet 1971; Cleave et al. 1969; Fredericks and Goodman 1969) (data that did not support this concept were often not considered).

In the absence of valid data to support this position, opinions were presented as scientific conclusions. For example, "the difference between sugar addiction and narcotic addiction is largely one of degree" (Dufty 1975). Since there are no data to support addiction to sucrose, this comparison cannot be made. Consequently, the reader is invited to exercise judgement when using the guidelines indicated above. These guidelines will certainly help in separating science from pseudoscience.

## 1.4 The Outline

The outline of the monograph follows the decimally numbered system of presentation. This has been done in order to facilitate the rapid location of a topic of interest. Such a system assists the reader in finding a particular subject, since it moves gradually down from general issues to more detailed topics in descending order.

The selection of the main chapter titles was guided equally by logical considerations and findings from searches of literature. For instance, it would be logical that a monograph on nutritional and safety aspects of a substance should contain a chapter on the production, use, and occurrence of that substance that gives an idea of the level of exposure. However, a subtitle dealing with "wound healing' may not be expected, unless this aspect has been found in the literature.

References are given in alphabetical order at the end of each chapter. Also, at the end of many sections relevant review articles are cited. References and presentation of work summaries in the text have been arranged in chronological order, the most recent works listed first.

The content of, and rationale for, each main chapter title is briefly given below.

*Chemical and Physical Data.* The data included in Chapter 2 on chemical and physical data are limited to the properties that are considered relevant to the appreciation of exposure and to the understanding of the effects. The Chemical Abstracts Service Registry Number, the Chemical Abstract Primary

Name – Ninth Collective Index – (CAS 1978), and the IUPAC Systematic Name (IUPAC 1965) are recorded for sucrose and its products of hydrolysis, glucose, and fructose. Other synonyms are given, but the list is not necessarily comprehensive. Trade names have not been recorded, because they present difficulty and may represent mixtures in which sucrose, glucose, and fructose are only a few of the ingredients.

The structural and molecular formulae, molecular weight, and the most important chemical and physical properties are given for the three compounds. The properties listed refer to the pure substance, unless otherwise specified, and include data that might be relevant to identification, fate, human exposure, and to the biological effects. The specifications and standards for sucrose established by the Codex Alimentarius Commission have also been included.

Analytical techniques are discussed only to the extent needed to understand data on levels and biological standards. These may not necessarily apply to all the research work cited in the parts of the monograph dealing with metabolic, biochemical, health, and safety aspects. The purpose of this section is to give readers an overview, rather than a complete list, of current methods cited in the literature. No critical evaluation or recommendation of any of the methods is meant or implied.

*Production, Uses and Occurrence.* Data on production and use are included in Chapter 3 to provide indications of the extent of past and present human exposure to sucrose. They are not meant to represent an exhaustive or critical review. Production and foreign trade data are derived mostly from governments, international organizations, and trade publications. Information on uses, which are drawn from public sources, identify uses that may not be current or major applications, and the coverage is not necessarily comprehensive.

Statements concerning regulations, standards, and guidelines (e.g., maximum levels permitted in foods, use of sucrose as a vehicle for iron supplementation) in specific countries may not reflect the most recent situation, since such standards are continuously reviewed and modified.

*Metabolism and Kinetics and Biochemical Aspects.* Under these headings (in Chapters 4 and 5), data on sucrose, glucose, and fructose are presented. Current and/or widely accepted theories are presented. Data derived from both experiments on laboratory animals and observations in man are reported. Comparison between data from experimental animals and anecdotal or epidemiological observations in human beings helps determine the relevance of the effects observed in animals, which is likely to result in a more rational assessment of their applicability to the human situation.

Sugar is consumed for its taste, but also – and principally, in many instances – because of its nutritional properties to serve as a food that supplies ready energy to the body.

*Health Aspects.* Chapter 6 deals with the impact of sucrose consumption on human health. In the past, a number of unfavourable effects have been attributed to the ingestion of sucrose as food or food ingredient, and this has led some authors to advise that its intake be sharply restricted. Data are reported that address all these issues. A critical appraisal of many of these data is found in Chapter 8 of this monograph.

*Safety Aspects.* Scientific work relating to the safety of sucrose is reported in Chapter 7.

This monograph is the result of a collaborative effort among several authors who take the responsibility for accuracy in reporting the data, but who make no allowance or express judgement on the validity of the data. Some of the authors provided their input by drafting one or more chapters; others have contributed comments and suggestions for improving the texts. Individual contributions were then pooled and systematized by the editors.

## 1.5 References

Abrahamson EM, Pezet AW (1971) Body, mind, and sugar. Pyramid, New York

CAS (1978) Chemical abstracts ninth collective index (9Cl), 1972–76. Chemical Abstracts Services, Columbus, Ohio, vols 76–85

Cleave TL, Campbell GC, Painters NS (1969) Diabetes, coronary thrombosis and the saccharine disease. John Wright, Bristol

Dufty W (1975) Sugar blues. Warner Books, New York

Du Ruisseau J-P (1973) La mort lente par le sucre. Editions du Jour, Montreal

Fredericks C, Goodman H (1969) Low blood sugar and you. Constellation International, New York

IUPAC (1965) Nomenclature of organic chemistry, section C. International Union of Pure and Applied Chemistry, Butterworths, London

Starenkyj D (1981) Le mal du sucre. Orion, Québec

WHO (1987) EHC 70: Principles for the safety assessment of food additives and contaminants in food. World Health Organization, Geneva

Yudkin J (1972) Sweet and dangerous. Bantam Books, New York

Chapter 2

# Chemical and Physical Data

## 2.1 General Observations and Terminology

1. All data listed apply to the form with the D configuration, i.e., D-glucose, D-fructose.
2. The names "glucose" and "fructose" are ambiguous if the stereochemistry is not specified.

Sugars are a group of sweet-tasting substances which occur widely in natural foodstuffs. They form part of a larger group of substances called *carbohydrates*.

Carbohydrates, in general, are chemical compounds mainly constituted of carbon, hydrogen, and oxygen which may occur together to form a single unit like the monosaccharide glucose, or long chains of saccharide units, known as polysaccharides, such as amylose and cellulose.

The most commonly occurring single unit sugars are glucose and fructose. These simple sugars are called *monosaccharides* and occur in nature largely as free sugars but also as molecules in which two or more units either of the same type or of different structure, are combined. A molecule consisting of two sugar units is referred to as a *disaccharide*, while carbohydrates containing short chains of sugar units are called *oligosaccharides*.

Oligosaccharides are intermediate in complexity between simple sugars (e.g., glucose and maltose) and *polysaccharides* (e.g., amylose and amylopectin). They are also intermediate in sweetness. The so-called "high-conversion" glucose syrup, in which glucose predominates, is nearly as sweet as glucose itself. Lower conversion syrups are progressively less sweet as the ratio of oligosaccharides to glucose and maltose increases. Polysaccharides are not sweet.

Syrups are solutions of mixtures of sugars in water. Firstly, they may be made by dissolving a sugar or sugars in water as is typical of the syrup used in canning fruit. Secondly, they may be an end-product or by-product of some process. For example, glucose syrups which are widely used in the food industry are the end-products of a process which begins with starch whereas golden syrup is a by-product of sucrose-refining. Thirdly, they may occur naturally like honey or maple syrup which is mainly a solution of sucrose and is a concentrate of natural exudate from the sugar maple tree.

## 2.2 Chemical Names and Synonyms

### 2.2.1 Sucrose

| | |
|---|---|
| CAS registry number: | 57-50-1 |
| CAS systematic name: | β-D-fructofuranosyl-α-D-glucopyranoside |
| IUPAC systematic name: | β-D-fructofuranosyl-α-D-glucopyranoside |
| Synonyms: | saccharose, sugar (in common usage) |
| Trade names: | none |

### 2.2.2 D-Glucose

| | |
|---|---|
| CAS registry number: | 50-99-7 |
| CAS preferred name: | D-glucose |
| IUPAC preferred name: | D-glucose (*Note*: the CAS/IUPAC systematic name is D-*gluco*-hexose, but, in both cases, the preferred name is D-glucose) |
| Synonyms: | dextrose, D(+)-glucopyranose, D(+)-glucose, *d*-glucose, (+)-glucose, grape sugar |
| Trade names: | none |

### 2.2.3 D-Fructose

| | |
|---|---|
| CAS registry number: | 57-48-7 |
| CAS preferred name: | D-fructose |
| IUPAC preferred name: | D-fructose (*Note*: the CAS/IUPAC systematic name is D-*arabino*-2-hexulose, but, in both cases, the preferred name is D-fructose) |
| Synonyms: | diabetin, D(−)-fructopyranose, D(−)-fructose, fruit sugar, laevulose, levulose |
| Trade names: | none |

## 2.3 Structural and Molecular Formulae and Molecular Weight

For glucose and fructose, the structural formulas given are those of the α-D-forms.

### 2.3.1 Sucrose

Molecular formula: $C_{12}H_{22}O_{11}$
Molecular weight: 342.30

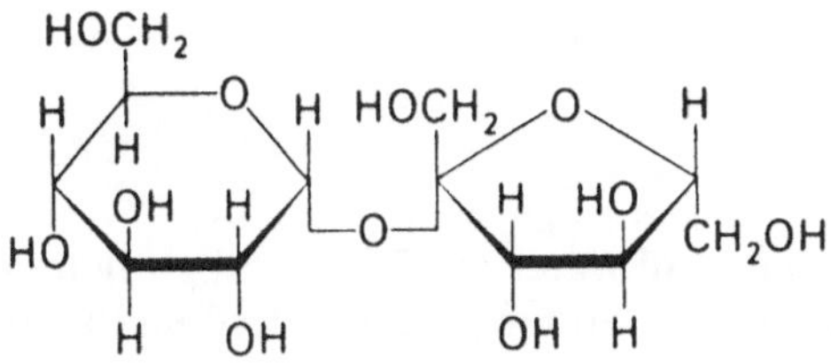

### 2.3.2 Glucose

Molecular formula: $C_6H_{12}O_6$
Molecular weight: 180.16

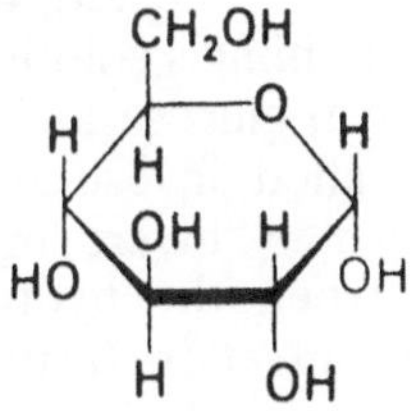

### 2.3.3 Fructose

Molecular formula: $C_6H_{12}O_6$
Molecular weight: 180.16

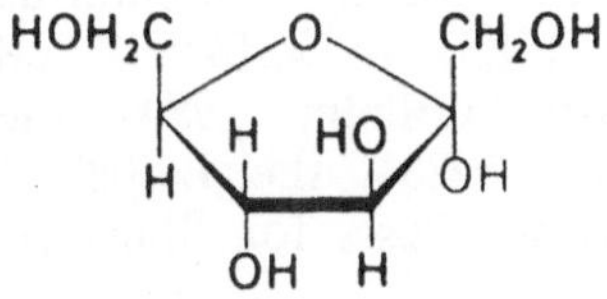

## 2.4 Chemical and Physical Properties of the Pure Substance

### 2.4.1 Sucrose

#### *2.4.1.1 Description*

Monoclinic sphenoidal, colourless, odourless crystals, crystalline masses, blocks of white crystalline powder, with sweet taste, obtained from the juice of sugar cane *Saccharum officinarum* L. (Family: Gramineae) and of white-rooted varieties of sugar beet *Beta vulgaris* L. (Family: Chenopodiaceae) and other sources (Windholz et al., 1983; Reynolds and Prasad 1982). Because sucrose is used as a standard for relative sweetness, its sweetness is chosen to be 1.0.

#### 2.4.1.2 *Solubility*

One gram dissolves in 0.5 ml water at 25 °C, in slightly more than 0.2 ml water at 100 °C, in 170 ml ethanol, and in about 100 ml methanol; it is freely soluble in 70% ethanol and sparingly soluble in chloroform, diethyl ether, glycerol, and pyridine (Windholz et al. 1983; Reynolds and Prasad 1982).

#### 2.4.1.3 *Stability*

Sucrose melts at 185–186 °C; when heated to about 200 °C it chars and caramelizes. It does not reduce Fehling's solution or form an osazone or show mutarotation. It is hydrolysed to D-glucose and D-fructose by diluted acid and by β-fructofuranosidase (also called invertase or sucrase). The mixture of D-glucose and D-fructose is known as "invert sugar". On hydrolysis, the optical rotation falls and is negative when the hydrolysis is complete. Sucrose is fermentable; when in high concentration (usually above 65% w/w) it resists decomposition by bacteria; at this strength, crystallization may occur (Weast et al. 1986; Windholz et al. 1983; Reynolds and Prasad 1982).

#### 2.4.1.4 *Identity and Specification Test*

Such tests are given in the Codex Alimentarius Commission (1981), the European Pharmacopoeia (1983), the Official Methods of the Association of Official Analytical Chemists (Williams 1984), the US Pharmacopoeia (US Pharmacopoeial Convention 1985), the British Pharmacopoeia (Medicines Commission 1986), and Basic Tests for Pharmaceutical Substances (World Health Organization 1986).

### 2.4.2 Glucose

#### 2.4.2.1 *Description*

Colourless crystals or a white odourless crystalline or granular powder with a sweet taste; it is 0.74 times as sweet as sucrose. It occurs naturally and in the free state in fruits and other parts of plants or combined in glucosides, such as di-, oligo-, and polysaccharides. It is a product of hydrolysis of sucrose and a main source of energy for living organisms (Windholz et al. 1983).

### 2.4.2.2 *Solubility*

One gram dissolves in 1.1 ml water at 25 °C, in 0.8 ml water at 30 °C, in 0.41 ml water at 50 °C, in 0.28 ml water at 70 °C, in 0.18 ml water at 90 °C, and in 120 ml methanol at 20 °C. It is very sparingly soluble in absolute ethanol, ether, acetone; it is soluble in hot glacial acetic acid, pyridine, and aniline (Windholz et al. 1983).

### 2.4.2.3 *Stability*

Below 50 °C, α-D-glucose hydrate is the stable crystalline form, above 50 °C, the anhydrous form is obtained, and, at still higher temperatures, the β-D-glucose is formed (Windholz et al. 1983).

### 2.4.2.4 *Identity and Specification Test*

Such tests are given in the Codex Alimentarius Commission (1981), the Food Chemical Codex (Committee on Codex Specifications NRC 1981), the Martindale Extra Pharmacopoeia (Reynolds and Prasad 1982), the European Pharmacopoeia (1983), the US Pharmacopoeia (US Pharmacopoeial Convention 1985), the British Pharmacopoeia (Medicines Commission 1986), and Basic Tests for Pharmaceutical Substances (World Health Organization 1986).

## 2.4.3 Fructose

### 2.4.3.1 *Description*

Odourless, colourless crystals or a white crystalline or granular powder with a very sweet taste inversely related to temperature. It occurs in a large number of fruits, honey, and as the sole sugar in bull and human semen. It occurs in both the furanose and pyranose forms (Windholz et al. 1983).

### 2.4.3.2 *Solubility*

One gram dissolves in 0.3 ml water, 15 ml ethanol, and in 14 ml methanol; it is practically insoluble in chloroform and ether. It is slightly soluble in cold acetone, freely soluble in hot acetone, and soluble in pyridine, ethylamine, and methylamine (Windholz et al. 1983; Reynolds and Prasad 1982).

### 2.4.3.3 *Stability*

It shows mutarotation. Rapid and anomalous mutarotation involves pyranose–furanose interconversion. Precipitated from ethanol in orthorhombic, bisphenoidal prisms, it decomposes at 103–105 °C.

#### 2.4.3.4 *Identity and Specification Test*

Such tests are given in Rosin (1955), the Codex Alimentarius Commission (1981), the Food Chemical Codex (Committee on Codex Specifications NRC 1981), the Martindale Extra Pharmacopoeia (Reynolds and Prasad 1982), the European Pharmacopoeia (1983), and the US Pharmacopoeia (US Pharmacopoeial Convention 1985).

## 2.5 Codex Alimentarius Specifications

Recommended International Standard for White Sugar

1. Scope
   This standard applies to white sugar, except that paragraph 3.1.4 (loss on drying) does not apply to white sugar in lump or cube forms or to crystal candy sugar (crystal korizato) or rock sugar (korizato).
2. Description
   White sugar is purified and crystallized sucrose (saccharose).
3. Essential composition and quality factors

| | |
|---|---|
| 3.1 Specification A | |
| 3.1.1 Polarization | not less than 99.7 °S |
| 3.1.2 Invert sugar content | not more than 0.04% m/m |
| 3.1.3 Conductivity ash | not more than 0.04% m/m |
| 3.1.4 Loss on drying (3 h at 105 °C) | not more than 0.2% m/m |
| 3.1.5 Colour | not more than 60 ICUMSA units |
| 3.2 Specification B | |
| 3.2.1 Polarization | not less than 99.5 °S |
| 3.2.2 Invert sugar content | not more than 0.1% m/m |
| 3.2.3 Conductivity ash | not more than 0.1% m/m |
| 3.2.4 Loss on drying (3 h at 105 °C) | not more than 0.1% m/m |
| 3.2.5 Colour | not more than 150 ICUMSA units |

4. Food Additives

| | |
|---|---|
| 4.1 Sulphur dioxide | Maximum level |
| Specification A | 20 mg/kg |
| Specification B | 70 mg/kg |

5. Contaminants

| | |
|---|---|
| Specifications A and B | Maximum level |
| 5.1 Arsenic (As) | 1 mg/kg |
| 5.2 Copper (Cu) | 2 mg/kg |
| 5.3 Lead (Pb) | see Sect. 2.6.3 |

From: Codex Alimentarius Commission (1981).

## 2.6 Identity and Purity Aspects

Sucrose has a purity matched by hardly any other commercial product. A maker factory will produce a raw sugar of 98%–99% purity, which, by agreement, is sold to refineries on the basis of its 96° sucrose, a standardized unit based on calculated direct polarization, roughly equal to 96% purity.

### 2.6.1 Degradation Products

Sucrose is readily attacked by acids, oxidizing agents, and also by alkalis and catalytic hydrogenation. Because of the sensitivity of the molecule, sucrose is partially or wholly degraded by these reagents to compounds of lower molecular weight. The products thus formed include acids, glycols, higher polyols, and cyclic compounds. The action of a number of microorganisms – fungi, moulds, yeast, and bacteria – also produces these materials. Degradation products of interest include lactic acid, levulinic acid, mannitol, piperazine, and imidazole derivatives (Kollonitsch 1970).

### 2.6.2 Additives

Except for hydrated lime ($CaO{\cdot}H_2O$) and occasionally magnesium oxide and small amounts of phosphoric acid used in the neutralization of sugar juices, few chemicals are widely used in the processing of sugar. A partially hydrolysed polyacrylamide, Separon AP-30 (Dow Chemical Company), is approved in the USA at no more than 5 mg/kg on juice weight as an aid in flocculating and settling the heated limed colloids. The residue data for this product were largely based on experiments with $^{14}C$-labelled compounds.

A slime control agent has been approved for use against *Dextranicum* organisms in sugar-cane mills. The product, Busan 881, may be added to sugar cane as it enters the mill, and consists of 2.5 mg/kg of disodium cyanodithioimidocarbonate, 1 mg/kg of ethylenediamine, and 3.5 mg/kg of potassium *N*-methyldithiocarbamate. Liquid sugars are perishable and require considerable care in storage. In general, they are protected by ultraviolet radiation, for which federal approval has been granted in the United States (Hilton 1966).

Icing sugar may contain up to 5% starch as long as no other anti-caking agent is used. If starch is not present, it may contain any of the following, singly or in combination, up to a maximum of 1.5%: tribasic calcium phosphate, magnesium carbonate, magnesium stearate, silicon dioxide, amorphous (dehydrated) silica gel, calcium silicate, magnesium trisilicate, sodium calcium alumino-silicate. The brown sugar with a characteristic taste known as Demerara may have had the colour stabilized with stannous chloride, titanium chloride, or artificial colour (FAO 1986, 1979).

### 2.6.3 Extraneous Materials

The extraneous water-insoluble material in good quality white sugar is usually less than 25 mg/kg. Lead content will usually be very considerably lower than the limit of 2 mg/kg (FAO 1986, 1979).

The level of contaminants established by the Codex Alimentarius Commission reported in Sect. 2.5 has been recently modified as far as lead is concerned. The Codex Alimentarius Commission had temporarily endorsed a level for lead of 1 mg/kg for white sugar and 0.5 mg/kg for fructose (Codex Alimentarius Commission 1984).

### 2.6.4 Pesticide Residues

Pesticides are extensively used in growing sugar cane and sugar beets. A list of commonly used compounds is found in Gleason et al. (1969, p 29). These comprise classes such as dormant stage herbicides, fumigants, insecticides, seed treatment fungicides and insecticides, and soil treatment fungicides, herbicides, and insecticides. The residue data found in the raw commodities and in the final product have been reviewed by Hilton (1966). Processing to raw sugar consists of a series of purification steps involving lime and heat precipitation of many impurities in the extracted juice, vacuum concentration to remove water, repeated crystallization, centrifugation, and reboiling of the liquid to produce raw sugar and final molasses. Refining of raw sugar for market involves washing the raw sugar, dissolving it in water, filtration through clay, percolation through animal bone charcoal, or vegetable carbon to remove colour and impurities, and finally repeated crystallization from vacuum-concentrated syrups (see also Sect. 3.2.2.1 and 3.2.2.2). Most of the published and available information on pesticide residues concerns the fate of herbicides, which tend to decrease rapidly with time, partly by the rapid growth dilution and partly by the abscission of leaves. Processing further reduces the pesticide residue so that the refined products are practically free of pesticide contamination.

However, considerations on the purity aspects of sucrose are particularly important in the interpretation of results obtained in experimental animal studies.

A recent review published by FAO and prepared with the support of the Swedish International Development Authority (SIDA) deals with composition, routine analysis, identification tests, adulteration and other laboratory aspects or sucrose and sugar products (FAO 1986).

## 2.7 References

Codex Alimentarius Commission (1981) Code standards for sugars, 1st edn. Food and Agriculture Organization of the United States, World Health Organization, Rome (CAC, vol 3)

Codex Alimentarius Commission (1984) Codex alimentarius. XVII. Contaminants, 1st edn. Secretariat of the Joint FAO/WHO Food Standards Programme, Food and Agriculture Organization of the United Nations, Rome

Committee on Codex Specifications NRC (1981) Food chemicals codex, 3rd edn. National Academy Press, Washington DC
European Pharmacopoeia (1983) 2nd edn. Maisonneuve SA, Sainte-Ruffine (sucrose pp 204/1–204/4; glucose pp 177/1–177/4; fructose pp 188/1–188/3)
FAO (1979) Manuals of food quality control. III. Commodities. Food and Agriculture Organization of the United Nations, Rome (FAO Nutrition Paper No 14/3)
FAO (1986) Manuals of food quality control. VIII. Food analysis: quality, adulteration, and tests of identity. Food and Agriculture Organization of the United Nations, Rome (FAO Nutrition Paper No 14/8)
Gleason MN, Gosselin RE, Hodge HC, Smith RP (1969) Clinical toxicology and commercial products, 3rd edn. Williams and Wilkins, Baltimore, Maryland
Hilton HW (1966) Pesticides and food additives in sugar cane and sugar products. Residue Rev 15:1–30
Kollonitsch V (1970) Sucrose chemicals. A critical review of a quarter-century of research by the Sugar Research Foundation. The International Sugar Research Foundation, Washington DC
Medicines Commission (1986) British pharmacopoeia 1980. Addendum 1986. Her Majesty's Stationery Office, London
Reynolds JEF, Prasad AB (1982) The Martindale extra pharmacopoeia, 28th edn. The Pharmaceutical Press, London
Rosin J (1955) Reagent chemicals and standards, 3rd edn. Van Nostrand, Toronto
US Pharmacopoeial Convention (1985) The US pharmacopoeia, 21st rev. US Pharmacopoeial Convention, Rockville, Maryland
Weast RC, Astle MJ, Beyer WH (1986) CRC handbook of chemistry and physics 1986–1987, 67th edn. CRC Press, Boca Raton, Florida
Williams S (1984) Official methods of analysis of the Association of Official Analytical Chemists (AOAC), 14th edn. Association of Official Analytical Chemists, Arlington, Virginia
Windholz M, Budavari S, Blumetti RF, Otterbein ES (1983) The Merck index, 10th edn. Merck, Rahway, New Jersey
World Health Organization (1986) Basic tests for pharmaceutical substances. World Health Organization, Geneva

Chapter 3

# Production, Uses and Occurrence of Sucrose

## 3.1 Production and Uses

A product of photosynthesis, sucrose constitutes the main sugar in the sap of plants and is also present in honey, along with fructose and glucose. Major sources of commercial sucrose are sugar cane and sugar beets. Less important sources are maple trees, sugar palm, particularly date palms, and sorghum. Commercial grades range from large brown crystals to granulated sucrose. Powdered sucrose, or icing sugar, is pulverized with an additive to prevent caking. Lump sugar is moulded and pressed from granulated sucrose using sugar syrups for bonding the crystals.

Sucrose and related compounds, belonging to the same group of carbohydrates, are termed sugars, i.e., corn sugar (glucose, dextrose), fruit sugar (fructose, levulose), milk sugar (lactose), and malt sugar (maltose) (Goetz 1985).

Sugar cane, the giant, thick perennial grass, cultivated in tropical and subtropical regions world-wide, is a major source of surcrose and molasses. It probably originated in what is now New Guinea, its cultivation spreading along with migration routes to Southeast Asia, India, and Polynesia. The original sugar cane, so-called noble canes, have 2n=80 chromosomes, though there are exceptions. The noble canes probably evolved from a wild ancestor *Saccharum robustum* Brandes and Jesweit ex Grassl, with 2n=60−194 chromosomes. Most current commercial varieties are interspecific hybrids involving primarily two or more of the following species: *S. officinarum, S. robustum,* and *S. spontaneum* (Dean 1982; Goetz 1985).

Sugar beet is second to sugar cane as the major source of the world's sugar. Sugar beets are grown in temperate or cold climates and in populated, well-developed areas where much of the product is consumed; this contrasts with sugar cane, which is grown in tropical regions. Sugar is stored in the root of the beet, instead of being present in the stalk, as in sugar cane. Beet culture is a comparatively recent development, whereas sugar-cane culture dates from antiquity. The sugar beet was grown as a garden vegetable for fodder before it was valued for its sugar content. Sugar was produced experimentally from beets in Germany in 1747 by the chemist Andreas Marggraf, but the first beet factory was built in Silesia in 1802. Napoleon became interested in the process in 1811, because the British blockade had cut off the supply of raw sugar from the West Indies, and, under his influence, several factories were established in France. The industry collapsed temporarily after Napoleon's fall, but recovered in the 1840s. Beet-sugar production then increased rapidly throughout Europe; by 1880, the tonnage had overtaken that of cane sugar, and, before World War I, the annual quantity of sugar produced from beets just about equalled that produced from cane (9 million tons each); since then, the production of cane sugar has exceeded that of beet sugar. Today, 60% of the total sugar is cane sugar (Eridania, personal communication 1987). Currently, the leading producers of beet sugar are France, the United State of America, and the Soviet Union (Goetz 1985).

Several publications on the history of origin and use of sucrose are available: *The history of sugar* is a detailed and authentic account of early history of sugar published in two volumes (Deer 1949, 1950): *The beet sugar story* (US Beet Sugar Association 1959) and *The sugar cane* (Barnes 1964) are detailed descriptions of world-wide cane and beet-growing techniques. Other important accounts include those of Hugill (1949) and Aykroyd (1967, 1974).

### 3.1.1 Production

From the nutritional and safety points of view, production data are important for assessing the exposure of the population to sucrose. These data are generally presented as estimates, and it should be realized that estimates are only approximations from which tendencies and trends may be established; however, it would be inappropriate to use these estimates to draw conclusions on individual situations pertaining to nutrition and safety.

There are many publications that contain data on the trade and production of sucrose on the world-wide scale. The most important are the *International trade statistics yearbook* of the United Nations, which lists combined trade figures in US dollars of sugar, honey, and sugar candies (non-chocolate), world-wide and by country (United Nations 1986); the *Food and Agriculture Organization trade yearbook*, which presents quantities and value of trade for different agricultural commodities (including sugar cane, sugar beet, refined and non-refined sugar) as well as agricultural requisites (FAO 1982); the *Food and Agriculture Organization production yearbook*, gives detailed accounts of area and production data on sugar cane and sugar beets, generally covering all crops harvested, except crops grown explicitly for food (FAO 1986). Most of the crop is used for the production of centrifugal and non-centrifugal sugar; however, in several countries, important quantities of sugar cane are also used for seed, feed, fresh consumption, the manufacturing of alcohol, and other uses. Some sugar beet is used for feed and alcohol production. Under the term centrifugal sugar, data include both cane and beet sugar and are shown, as far as possible, in terms of raw value as reported by individual countries. It is not certain, however, whether all countries report raw sugar in terms of 95° polarization, as requested in the survey questionnaires. Data reported in Table 3.1 as refined sugar are converted to a raw basis using the conversion factor of 1.087 (or 108.7%) (FAO 1986); the *FAO monthly bulletin of statistics* is a source of current facts and data on world food and agricultural conditions

**Table 3.1.** World sugar production (1000 MT)

| | 1979–81 | 1983 | 1984 | 1986 |
|---|---|---|---|---|
| Sugar cane | 757 281 | 805 050 | 912 141 | 940 911 |
| Sugar beets | 272 654 | 273 755 | 295 684 | 282 925 |
| Centrifugal sugar | 88 585 | 97 929 | 100 043 | 97 867 |
| Non-centrifugal sugar | 13 419 | 13 707 | 15 109 | 14 889 |

Data from: FAO (1986)

(including sugar and sugar products), with an analysis of the factors that influence them.

Another world-wide source of information on sugar production is the International Sugar Organization's yearbook. The 1984 *Sugar yearbook* (the latest issue) continues the sequence of 21 sugar yearbooks published by the International Sugar Council, and nine sugar yearbooks published by the International Sugar Organization under the International Sugar agreement of 1968. As in the Organization's *Monthly statistical bulletin*, the statistics relate to centrifugal sugar only. It reports statistics submitted by member countries of the organization. The book contains data on production, import, consumption, and stock by country, as well as general tables covering statistics on world-wide production, exports, imports, consumption, and other factors (International Sugar Organization 1985).

Because of the widespread interest in the world production and trade of food and agricultural commodities, a vast number of pertinent publications are available. Some of them have been mentioned earlier. Many are contributed by national surveys to international political or trade organizations. Various factors are used when national surveys are undertaken: exports, imports, consumption, stock, and others. Furthermore, international organizations may also use results from diverse national sources, such as government departments, local trade associations, manufacturer, producer, or consumer associations, unions, or corporations. Thus, estimates for a specific commodity proceeding from the same country may be at variance when presented in international surveys, raising the question of reliability of the sources. For this reason, statistics aimed at defining trends are more generally useful than individual data.

Misuse and unwarranted reference to world sugar production figures may lead to gross misinterpretations, when they are applied to nutritional and safety aspects of sucrose. This possibility has been exemplified by Jukes (1986) in pointing out the need to correlate increasing trends in sugar production with population growth. World-wide production of cane plus beet sugar was estimated, in 1983, as 94.7 million metric tons (MMT). This represents a 28% increase compared with the 1972 world production, estimated as 74 MMT in the Life Sciences Research Office's Report (SCOGS/FASEB 1976), and perhaps shows how the production of sugar is keeping pace with the world's increase in population. According to a University of California Survey (University of California Food Task Force 1974), of the world's food crop output, 5% of total calories produced was ascribed to sugar, and the net food demand for sugar was expected to increase by 50% between 1970 and 1985, because of population growth. According to this, the increase in sugar production between 1974 and 1983 was somewhat behind target to cope with the increase in population.

Another possible error in generalizing human consumption values from production figures may be incurred when the amounts destined to industrial uses are disregarded; for instance, Bollenback (1982) estimated that about 300 000 tons of sugar may be consumed by yeast in the baking industry. Therefore, it is important that, in calculating per capita sugar consumption figures, care is exercised not to derive such figures from the gross production of sugar cane and beets, but only from the quantity of sugar produced from these sources, and destined for human consumption as such.

### 3.1.2 Uses

In general, sugar is not eaten alone; rather it is added as a component to many foods in the home, in the food beverage, and in the pharmaceutical industries. It is used for its sweet taste and as a source of energy, but it is also important for its physical properties. Because of its stickiness and viscosity in solution, it binds food elements together; these properties are widely used by the food-manufacturing industry, and also by the drug industry. Sugar in solution is an important food preservative, because, at high concentration, it inhibits the growth of spoilage microorganisms by raising osmotic pressure. The most important functions of sugar in food technology (sweetener, preservative, texture modifier, fermentation substitute, and bulking agent) have been described by Koivistoinen and Hyvonen (1985).

Another important use of sugar, as well as glucose, is its life-saving role in oral rehydration therapy, when used in combination with salt and boiled water. While the concept of an oral rehydration solution containing common salt and sugar is not new, the scientific basis for its use was not known until the 1960s, when in vitro and in vivo studies showed that glucose could mediate sodium transport across the mucosa of the small intestine. Evidence that the absorption of glucose, glucose-mediated sodium chloride, and water remained largely intact during diarrhoea was first developed in cholera patients in 1964. In 1968, further studies in patients with cholera revealed that fluid and electrolyte losses due to watery diarrhoea could be adequately replaced by oral electrolyte solution of optimum composition (World Health Organization/Pan American Health Organization 1983; Mahalanabis et al. 1981; Chatterjee et al. 1977; Pierce and Hirschhorn 1977). The management of diarrhoea by oral rehydration therapy using "salt and sugar" solution is presently a major programme of the World Health Organization (WHO) and the United Nations Children's Fund (UNICEF) in their joint effort to combat an illness responsible, in the poorest areas of the world, for 20%–50% of all deaths in children less than 4 years of age, and affecting 500 million children annually (Mahalanabis 1986; World Health Organization 1986; World Health Organization/United Nations Children's Fund 1985).

An interesting application of sucrose that is worthy of mention relates to its use as a vehicle for vitamin A. In Central America, nutrition surveys have revealed a high prevalence of hypovitaminosis of vitamin A. The diets were generally low in foods containing this vitamin, resulting in inadequate intakes. Serum-retinol values in rural children were low, and evidence of xerophthalmia was present.

A programme was undertaken in South America (INCAP report 1975) of adding vitamin A to sucrose. Crystalline sucrose was found to be the most suitable carrier. In this preparation of sugar, a vitamin A/sugar premix is first made. Separation of the vitamin A (vitamin A palmitate, type 250-CWS, Roche) particles from the sugar crystals in the premix is prevented by providing an edible bonding agent when blending the dry vitamin A with the crystalline sugar. At an appropriate processing stage, by means of a mechanical feeder, this premix is added to the flow of sugar at a set ratio, followed by adequate mixing to ensure uniformity of the final product. Biological effectiveness of the added vitamin A in sugar was demonstrated by animal assays and by increased vitamin A levels in blood-serum of human beings. As a consequence of these findings, various countries in South America have passed laws requiring sugar used at home to be so treated (Araujo et al. 1978a,b; Aguilar et al. 1977; Arroyave et al. 1975, 1979, 1981; Simmons and de Mello 1975; Varela et al. 1972).

Layrisse et al. (1976a,b) investigated iron absorption from a number of foods, including sugar. They concluded that sucrose was a better vehicle than bread and wheat products for iron fortification.

The effect of ascorbic acid on iron absorption was determined in 116 Indian housewives of low socio-economic status living near Durban, South Africa. They were given a cereal meal, with and without sucrose fortified with ascorbic acid. It was concluded that the fortification of the diet of these peoples with ascorbic acid, so as to enhance iron absorption, might be more effective than adding iron to the diet. Sucrose appeared to be a suitable vehicle for ascorbic acid (Derman et al. 1977).

Therapeutic uses of sucrose include management of severe hypoglycaemia, wound healing (Quatraro et al. 1985; Selwyn 1983; Knuston et al. 1981), control of odour from malignant ulcers of the breast (Thomlinson 1980), and cessation of prolonged hiccups (Engleman et al. 1971). Sucrose is an ingredient in most formulae used in enteral nutrition (Foppiani 1986). However, its intake is contraindicated in patients with the glucose–galactose malabsorption syndrome, fructose intolerance, or sucrase deficiency (Reynolds and Prasad 1982) (see also Sect. 6.1.7 on wound healing).

In nutritional trials, sucrose had a positive influence on acceptability of instant CSM (corn, soy flour, skimmed milk), a food supplement for young children in developing areas of the world (Grewal et al. 1973).

Extreme exhaustion and mountain sickness have been greatly reduced in recent years by the frequent consumption of sugar at high altitude by mountain climbers (Jukes 1975).

The use of sucrose, as a chemical raw material, for non-food production of agricultural chemicals and pharmaceuticals is being explored (Chang 1974; Parker 1974). Considerable success has been achieved, as some sucrose derivatives have already found applications as emulsifiers, denaturants, fibres, adhesives, viscosity modifiers, antioxidants, chelating agents, surfactants, detergents, and resin intermediates (Vlitos 1977).

## 3.2 Occurrence

Sucrose occurs as a natural constituent in fruits and plants. It is an extraction product from sugar cane and sugar beet.

The synthesis of sucrose from D-glucopyranose-1-phosphate and fructose by means of an enzyme from the microorganism *Pseudomonas saccharophile* was accomplished in 1943 (Noller 1966). Many early attempts to synthesize sucrose by chemical methods (e.g., Pictet and Vogel 1928) failed because of the difficulty in obtaining the correct configuration of the glycosidic linkage (Noller 1966). In 1953, sucrose octaacetate was obtained in 5.5% yield by heating of 1,2-anhydro-3,4,6-tri-*O*-acetyl-α-D-glucopyranose with 1,3,4,6-tetra-*O*-acetyl-D-fructofuranose. Purification and deacetylation yielded sucrose identical to the natural product (Lemieux and Huber 1956).

### 3.2.1 Natural Occurrence

Sucrose is produced in all green plants by photosynthesis, the process by which organic substances (carbon compounds) are produced from carbon dioxide and water by reactions that occur in the presence of chlorophyll, found in chloroplasts, on exposure to visible light, especially sunlight. The series of reactions that leads to sucrose ends when glucose and fructose, as phosphate compounds, join together in the juices of green plants.

Sucrose has been found in many fruits and vegetables (Dako et al. 1970; Lee et al. 1970; Davies 1966).

### 3.2.2 Extracted Sucrose

Industrial and commercial sucrose is the product of extraction of two major agricultural commodities, sugar cane and sugar beets. The only difference between the two extraction products is the nature of their impurities: impure cane sugar is flavoursome, whereas impure beet sugar has bitter undertones (Newton and Wardrip 1974). The final product is the same, 99.9% pure sucrose; however, the technical preparations are different.

#### *3.2.2.1 Technical Preparation of Sucrose from Sugar Cane*

In the extraction of sucrose from cane, the stalk is ground and the juice is squeezed out of it with rollers. The juice, which contains about 15% sucrose, is made slightly alkaline with lime to prevent hydrolysis. When heated, most of the impurities separate as a heavy scum and a precipitate, which are removed. This process is known as "defecation".

A better purification of the juice is obtained by adding small amounts of phosphoric acid or phosphate. This helps in filtering off the colloidal substances and other impurities, thus reducing the colour of the juice. Sometimes, sulphur dioxide is also added to obtain clearer juice. In alternative processes, more lime is added, which is then precipitated as calcium carbonate by means of carbon dioxide. The clear juice is then concentrated under reduced pressure and allowed to crystallize. The raw sugar is removed by centrifugation, and the centrifugate is reconcentrated. The operation is repeated until no more crystals can be obtained economically. The final mother liquor is a dark viscous liquid that contains about 50% fermentable sugars and is known as blackstrap. It is used for preparing cattle feeds and for the production of alcohol. The raw sugar is shipped to refineries where the brown colour is removed. Sugar is washed in centrifuges and dissolved in water; the syrup is purified by a lime-phosphoric acid or lime-carbonic acid treatment, then passed through decolourizing carbon. Concentration and crystallization in vacuum yield the pure sugar of commerce.

Brown sugars are mixtures of the purified sugar with various amounts of molasses. Demerara is a crude sugar originally from Guyana, which is popular in the United Kingdom. If the white crystalline sugar is ground with 3% starch to prevent caking, the product is known as powdered sugar, lump or moulded, or loaf sugar (Eridania, personal communication 1987; Noller 1966).

### *3.2.2.2 Technical Preparation of Sucrose from Sugar Beets*

The beets are sliced in thin sections (cossettes) by mechanical means. The sugar is extracted by a diffusion process involving continuous counter-current treatment with water. The extracted cossettes are pressed and used as animal fodder; frequently, they are dried and stored for the same purpose. The subsequent processing phase aims at purifying the juice by removing as many organic substances as possible. The presence of organic matter hinders the crystallization of sucrose, lowers the yield, and creates difficulties in all technical operations. To this end, the extracted juice is mixed with milk of lime, heated, and finally treated with carbon dioxide. By these means, the insoluble calcium salts are precipitated, proteins and pigments are coagulated, colouring and other substances are adsorbed. Muds are generally concentrated by sedimentation, separated by filtration, washed, and then removed. Sometimes, the purified juice is treated with sulphur dioxide to reduce the colour increment during the following processing phases.

To avoid decomposition, prior to crystallization in vacuum pans, the purified juice is concentrated under controlled temperature and pressure. Once this operation is over, the mass of crystals (massecuite) is dissolved, and the mother liquor is removed by centrifugation. The crystals, as discharged from the centrifuge, constitute the raw sugar, and the dark brown run-off liquid is known as "run-off", "green syrup", or "green molasses".

In the case of beet sugar, after having been initially drained, the raw sugar is generally washed during the spinning in the centrifuge. After drying and conditioning, the white sugar (granulated sugar), thus obtained, is stored, ready to be marketed. This sugar is similar to refined sugar except that it still contains some pigments and ashes. By repeating the processes of crystallization and spinning several times, a mother liquor is obtained from which no more sucrose can be crystallized and recovered economically. This dark, viscous liquor is known as "final molasses", or simply, molasses.

Raw sugar is refined by bringing it into a solution, washing and filtering through activated charcoal, as well as taking it through a new crystallization cycle. Refined sugar contains 99.9% sucrose.

During the refining of sugar, an ion exchange process is sometimes used to remove salts.

Molasses contains about 20% water and 50% sugar. The sucrose content, however, cannot be recovered by further precipitation, because it is held in solution by impurities. It is necessary, therefore, to separate the sugar from the residual matter by special treatment, for which purpose a large number of processes are available, e.g., separation by means of strontium or calcium sucrates, or by ion exchange processes. The former process depends on the property that sugar possesses of yielding insoluble or sparingly soluble sucrates with lime or strontium hydroxide. When the diluted molasses are treated with either of the above hydroxides, a precipitate of the corresponding sucrate is obtained. Inorganic and organic impurities in the molasses remain dissolved and are removed by filtration. After the sucrate is washed with a small amount of water, it is decomposed with carbon dioxide. The sugar solution obtained is processed as before, i.e. evaporation in vacuum pans (Eridania, personal communication 1987).

### 3.2.3 Sucrose Content in Foods

Foods may generally contain naturally occurring or added sucrose. Data on the naturally occurring sucrose in fruits, vegetables, legumes, and other commodities may have some applicability on a world-wide scale. On the other hand, the concentrations of added sucrose in manufactured food products will depend largely on technical feasibility, economics, and government regulations in individual countries.

The sucrose content of the diet is difficult to estimate, because the concentration of naturally occurring sucrose in foods is not adequately known and the sugar content of fruits and vegetables is subject to natural variations, depending on the conditions under which they were grown, when they were harvested, and how they were stored. Even when sucrose is added intentionally to a food product, it does not necessarily remain in the form of sucrose; in certain beverages, such as cola drinks, the initial syrup used may be sucrose, but the prevailing acidity results in fairly rapid inversion of the sucrose to glucose and fructose (Lee 1981; Southgate et al. 1978).

Difficulties also arise from the state-of-the-art analytical methodology; in spite of important advances that have been made during the last few years in methods for sugar analysis, the task still appears to be a difficult one. Methods of analysis are available specifically to determine the concentration of sugar in fruit juices (Li and Schuhmann 1983), yoghurt (Li et al. 1983), candy bars (Hurst et al. 1983), granola cereals (Li and Schuhmann 1981), chocolate (Hurst and Martin 1980), molasses (Damon and Pettitt 1980), ready-to-eat breakfast cereals (Li and Schuhmann 1980), rice grain (Pascual et al. 1978), and strawberries (Reyes et al. 1982a) as well as general methods for sucrose and other sugars in different food commodities (Iverson and Bueno 1981; Richmond et al. 1981; Hurst and Martin 1980; DeVries et al. 1979; Hurst et al. 1979; Southgate et al. 1978; Shallenberger 1974).

Surveys of the concentrations of sucrose in selected classes of manufactured food products are available, for example, from the USA (SCOGS/FASEB 1976; Shannon 1978; US/FDA/STF 1986), Finland (Koivistoinen and Hyvonen 1985), New Zealand (Brake et al. 1980), Canada (Korsrud and Trick 1977, 1979), and the United Kingdom (BNF/TFSS 1987).

### 3.2.4 Browning Reaction of Sucrose

In food technology, sucrose is mainly used in baking and bakery products, confectionery and candy making, distilled liquors, jelly and jam products, meat curing, wine making, sugar coating, and soft beverages. Processing may cause reactions between inverted sugar and the other food components.

When heated, sucrose is hydrolysed, and the hydrolysis products undergo browning reactions, especially in the presence of amino acids and proteins. These reactions are very different when they occur in solution as opposed to the dry state. When a pure sugar browns, it is said to undergo caramelization. When it first reacts with amino acids or proteins and then browns, this process is called the "Maillard reaction". When browing is light, sugar degradation is minimal; when dark, the sugars may be totally destroyed (Shallenberger and Birch 1975). In certain manufactured food products, these reactions are essential for production of desired flavour and colour (baked goods, coffee roasting). In other foods (sterilized milk, dried potatoes), they cause unwanted colour and flavour, and decrease some of their nutritional value (FAO/WHO 1980). Maillard reactions have been recently studied in model systems designed to compare reactivities among glucose, fructose, and sucrose-glycine (Reyes et al. 1982b) and to determine the effect on colour formation and rate of hydrolysis of sucrose-glycine using citrate buffer (Reyes and Toledo 1982).

## 3.3 Consumption Data

The collection of valid data on the food consumption habits of a population poses the most difficult problem to be overcome before any assessment can be made of the nutritional, health, and toxicological consequences of the dietary intake of a food commodity. Patterns of food consumption vary considerably within individuals and groups of individuals. This variability affects the choice of the method of dietary intake assessment, as well as the interpretation of the data obtained. Some groups within the population may show patterns of food consumption that are widely different from those of the population as a whole, and these include, for example, ethnic and cultural minorities within a community; people living on a subsistence diet; people whose employment or situation provides them with free food or who produce their own; members of the population with extreme eating and drinking habits; infants and young children; the elderly; pregnant or lactating women; the sick (e.g., diabetics); people on restricted diets (low calorie, low sodium, vegetarian, etc.). Several methods for the assessment of patterns of food consumption are available (FAO/UNEP/WHO 1985). Table 3.2 gives a short account of some of the possible approaches.

**Table 3.2.** Methods for obtaining food consumption data: comparison of approaches (FAO/UNEP/WHO 1985)

| Method | Respondent literacy required | Data validity ranking[a] | Suitability of approach for | | Individual consumption data provided | Respondent burden | Field staff burden/cost | Processing dietary data | Comments |
|---|---|---|---|---|---|---|---|---|---|
| | | | Large-scale studies[b] | Small-scale studies[c] | | | | | |
| National food disappearance (food balance sheets) | No | 6 | Suitable | Unsuitable | No | None | None: assume data already available | None | Crude estimate: disappearance rather than consumption; waste and home-grown foods not usually considered |
| Household food disappearance | Yes | 5 | Suitable | Suitable | No | Heavy | Heavy; 2 home visits | Heavy[d] | Food waste not usually corrected for; not applicable to age-sex groups; participation required for householder only |
| Food diary | Yes | 3 | Suitable | Suitable | Yes | Heavy | Medium; home visit and instruction | Heavy[d] | Record keeping may cause subject to eat differently; one of the more reliable methods especially when carried out over 7 days; literate participants or enumerators necessary |

| | | | | | | | | | |
|---|---|---|---|---|---|---|---|---|---|
| Weighed intake | Yes | 2 | Unsuitable | Suitable | Yes | Heavy | Heavy; home visits | Heavy[d] | Same as food diaries; experience with food balances necessary |
| Dietary recall | No | 3 | Suitable | Suitable | Yes | Light | Heavy; individual interview | Heavy[d] | Honesty and memory of subject critical; skill of interviewer important factor; response rate for 24-hour recall good |
| Food frequency | No | 3 | Suitable | Suitable | Yes | Medium | Medium; individual interview of self-administered questionnaire | Heavy[d] | Food models for quantifying intakes are useful; honesty and memory of the respondent critical; skill of interviewer important factor; response rate good |
| Duplicate portion | No | 1 | Unsuitable | Suitable | Yes | Heavy | Heavy; 2 or more home visits; food storage instuctions | None | Subjects must be paid for cost of extra food; special modification necessary to collect data on individual foods or food commodity groups |

[a] Validity: how closely the method measures daily intake (1 = most valid; 6 = least valid).
[b] Large-scale – representative of the entire population.
[c] Small-scale – representative of individuals or groups of individuals including households.
[d] Computerized analysis will greatly facilitate staff burden.

The Food Balance Sheets (FAO 1984) published by the Food and Agriculture Organization of the United Nations are a follow-up of the Provisional Food Balance Sheets 1972–74 Average and Food Balance Sheets 1975–77 Average, Per Capita Food Supplies 1961–65 Average 1967 to 1977 published in 1977 and 1980, respectively, which were processed by the FAO's Agricultural Data Bank known as the Interlinked Computerized Storage and Processing System of Food and Agricultural Commodity Data (ICS). The current volume, unlike previous ones, shows food balance sheets in a standardized format presenting statistical information for processed commodities, where possible, in their primary commodity equivalent. Among other commodities, these balance sheets report data on production, import, export, processed trade, domestic supply, domestic utilization (feed, seed, manufacture, waste, food) and yearly and per day/per capita supply (grams, calories, protein, fat) for the following: sugar cane, sugar beets, raw sugar, sugar confectionery, glucose, non-centrifugal sugar, sugar and syrups, and honey for 146 countries and territories in the world.

Parallel to this undertaking, the FAO has also published information from direct food consumption surveys (sugar, honey, sugar products, others) as submitted by countries in Europe, North America, Oceania (FAO 1977) Africa, Latin America, Near East and Far East (FAO 1979).

Calculations of the world per capita consumption of centrifugal sugar (1978–84) values, published by the International Sugar Organization (1985), are the result of the application of the same disappearance principle. Therefore, data are not available on world-wide consumption of sugar and sugar products, other than those derived from the application of national food disappearance. The value of these data in predicting the nutritional, health and safety consequences of sugar intake by the general population needs to be further validated.

Limited validation studies on the accuracy obtained by comparing results from the application of different methods for obtaining food consumption data suggest that the actual sugar consumption per capita may be between 10% and 50% less than the amount indicated by the total disappearance figures (BNF/TFSS 1987; Dziezak 1986; US/FDA/STF 1986). For a more detailed breakdown of consumption patterns, see the above references.

## 3.4 References

Aguilar JR, Arroyave G, Gallardo C (1977) Manual de supervision y control del programa de fortificación de azucar con vitamina A. Instituto de Nutrición de Centro American y Panama, Guatemala City (Publ INCAP E-913)

Araujo RL, Borges EL, Silva JDB, Palhares RD, Vieira EC (1978a) Effect of the intake of vitamin A fortified sugar by pre-school children. Nutr Rep Int 18: 429

Araujo RL, Souza MSL, Mata-Machado AJ et al. (1978b) Response of retinol serum levels to the intake of vitamin A fortified sugar by pre-school children. Nutr Rep Int 17: 307

Arroyave G, Aguilar Jr, Portela IE (1975) Manual de operaciones para la fortificación de azucar con vitamina A. Instituto de Nutrición de Centro America y Panama, Guatemala City (Publ INCAP E-853)

Arroyave G, Aguilar JR, Flores M, Guzman MA (1979) Evaluation of sugar fortification with vitamin A at the national level. Pan American Health Organization, Washington DC (Sci Publ No 384)

Arroyave G, Mejia LA, Aguilar JR (1981) The effect of vitamin A fortification of sugar on the serum vitamin A levels of preschool Guatemalan children: a longitudinal evaluation. Am J Clin Nutr 34: 41–49

Aykroyd WR (1967) Sweet malefactor: sugar, slavery and human society. Heinemann, London

Aykroyd WR (1974) Sugar in history. In: Sipple HL, McNutt KW (eds) Sugar in nutrition. Academic Press, New York, pp. 3–9

Barnes AC (1964) The sugar cane. Leonard Hill, London

BNF/TFSS (1987) Sugars and syrups. The report of the British Nutrition Foundation's Task Force. The British Nutrition Foundation, London

Bollenback N (1982) Sucrose and health. In: Lineback DR, Inglett GE (eds) Food carbohydrates. AVI Publishing, Westport, Connecticut, pp 63–73

Brake PJ, Wright FAC, Melton LD (1980) Concentration of sugars in commercial infant foods in New Zealand. NZ Med J October 22: 320–323

Chang RH (1974) Sucrose chemicals and their industrial uses. In: Inglett GE (ed) Symposium on sweeteners. AVI Publishing, Westport, Connecticut, pp. 74–77.

Chatterjee A, Jalan KN, Agarwal SK et al. (1977) Evaluation of a sucrose electrolyte solution for oral rehydration in acute infantile diarrhoea. Lancet I: 1333–1335

Dako DY, Trautner K, Somogyi JC (1970) Studies on the glucose, fructose, and saccharose content of various fruits. Bibl Nutr Dieta 15: 184–198

Damon CE, Pettitt BC (1980) High performance liquid chromatographic determination of fructose, glucose, and sucrose in molasses. J Assoc Off Anal Chem 63: 476–480

Davies JN (1966) Occurrence of sucrose in the fruit of some species of *Lycopersion*. Nature (Lond) 209: 640–641

Dean JL (1982) Sugarcane. In: McGraw-Hill encyclopedia of science and technology, 5th edn. McGraw-Hill, New York

Deer N (1949) The history of sugar, vol 1. Chapman and Hall, London

Deer N (1950) The history of sugar, vol 2. Chapman and Hall, London

Derman D, Sayers M, Lynch SR, Charlton RW, Bothwell TH, Mayet F (1977) Iron absorption from a cereal based meal containing cane sugar fortified with ascorbic acid. Br J Nutr 38: 261–269

DeVries JW, Heroff JC, Egberg DC (1979) High pressure liquid chromatographic determination of carbohydrates in food products: evaluation of method. J Assoc Off Anal Chem 62: 1292–1296

Dziezak JD (1986) Special report: sweeteners and product development. J Food Technol 40(2): 11–130

Engleman EG, Lankton J, Lankton B (1971) Granulated sugar as treatment for hiccups in conscious patient. N Engl J Med 285: 1489

FAO (1977) Review of food consumption survey, vol 1. Europe, North America, Oceania. Food and Agriculture Organization of the United Nations, Rome (FAO Food and Nutrition Paper 1/1)

FAO (1979) Review of food consumption survey, vol 2. Africa, Latin America, Near East, Far East. Food and Agriculture Organization of the United Nations, Rome (FAO Food and Nutrition Paper 1/2)

FAO (1982) 1981 FAO trade yearbook, vol 35. Food and Agriculture Organization of the United Nations, Rome

FAO (1984) Food balance sheets 1979–81. Food and Agriculture Organization of the United Nations, Rome

FAO (1986) FAO production yearbook, vol 39. Food and Agriculture Organization of the United Nations, Rome

FAO/UNEP/WHO (1985) Guidelines for the study of dietary intakes of chemical contaminants. GEMS: Global Environmental Monitoring System. World Health Organization, Geneva

FAO/WHO (1980) Carbohydrates in human nutrition. Food and Agriculture Organization of the United Nations, Rome, p 27

Foppiani E (1986) Caratteristiche dei prodotti dietetici per nutrizione enterale. In: Alimentazione enterale specializzata. Milano Instituto Scotti Bassani per le Ricerche e l'Informazione Scientifica e Nutrizionale, pp 45–64

Goetz PW (1985) The new encyclopaedia Britannica, 15th edn. Encylopaedia Britannica, Chicago, vol 2, pp 355–356

Grewal T, Gopaldas T, Hartenberger P, Ramakrishnan I, Ramachandran G (1973) Influence of sugar and flavour on the acceptability of instant CSM. Trials on young children from an urban orphanage. J Food Sci Technol 10: 149–152

Hugill JAC (ed) (1949) Sugar. Cosmo Publications, London

Hurst WJ, Martin RA (1980) High performance liquid chromatographic determination of carbohydrates in chocolate. Collaborative study. J Assoc Off Anal Chem 63: 595–599

Hurst WJ, Martin RA, Zoumas BL (1979) Application of HPLC on characterization of individual carbohydrates in foods. J Food Sci 44: 892–895

Hurst WJ, Martin RA, Zoumas BL (1983) Carbohydrate composition of candy bars. J Am Diet Assoc 83: 53–54

INCAP report (1975) (English translation 1975) Fortification of sugar with vitamin A in Central America and Panama. Institute of Central America and Panama, Guatemala City (INCAP V-36)

International Sugar Organization (1985) Sugar yearbook 1984. International Sugar Organization, London

Iverson JL, Bueno MP (1981) Evaluation of high pressure liquid chromatography and gas-liquid chromatography for quantitative determination of sugars in foods. J Assoc Off Anal Chem 64: 139–143

Jukes TH (1975) Nutrition for mountaineers. Proc Yosemite Inst Mountain Med, pp 21–26

Jukes TH (1986) Sugar and health. World Rev Nutr Diet 48: 137–194

Knuston RA, Merbitz LA, Creekmore AA, Snipes HG (1981) Use of sugar and povidone iodine to enhance wound healing. Five years' experience. South Med J 74: 1329–1335

Koivistoinen P, Hyvonen L (1985) The use of sugar in foods. Intern Dent J 35: 175–179

Korsrud GO, Trick KD (1977) Sucrose, fructose, and glucose contents of Canadian breakfast cereals. Can Inst Food Sci Technol J 10: 134

Korsrud GO, Trick KD (1979) Sucrose, fructose, and glucose contents of infant cereals. J Can Diet Assoc 40: 56

Layrisse M, Martinez-Torres C, Renzi M (1976a) Sugar as a vehicle for iron fortification: further studies. Am J Clin Nutr 29: 274–279

Layrisse M, Martinez-Torres C, Renzi M, Velez F, Gonzalez M (1976b) Sugar as a vehicle for iron fortification. Am J Clin Nutr 29: 8–18

Lee CY, Shallenberger RS, Vittum MT (1970) Free sugars in fruits and vegetables. NY Food Life Sci Bull 1

Lee VA (1981) The nutritional significance of sucrose consumption 1970–1980. CRC Press, Boca Raton, Florida (CRC critical reviews in food science and nutrition)

Lemieux RU, Huber G (1956) A chemical synthesis of sucrose. A conformational analysis of the reactions of 1,2-anhydro-*d*-D-glucopyranose triacetate. J Am Chem Soc 20: 4117–4119

Li BW, Schuhmann PJ (1980) Gas liquid chromatographic analysis of sugars in ready to eat breakfast cereals. J Food Sci 45: 138–141

Li BW, Schuhmann PJ (1981) Gas chromatographic analysis of sugars in granola cereals. J Food Sci 46: 425–427

Li BW, Schuhmann PJ (1983) Sugar analysis of fruit juice content and method. J Food Sci 48: 633–635

Li BW, Schuhmann PJ, Holden JM (1983) Determination of sugar in yogurt by gas-liquid chromatography. J Agric Food Chem 31: 985–989

Mahalanabis D (1986) Development of an improved formulation of oral rehydration salts (ORS) with antidiarrhoeal and nutritional properties a "super ORS". In: Holmgren J, Lindberg A, Mollby R (eds) Development of vaccines and drugs against diarrhoea. 11th Nobel Conference, Stockholm 1985. Studenlitteratur, Lund, pp 240–256

Mahalanabis D, Merson MH, Barua D (1981) Oral rehydration therapy: recent advances. World Health Organization, Geneva, pp 245–249 (World Health Forum No 2)

Newton JM, Wardrip EK (1974) Sucrose in food systems. In: Inglett GE (ed) Symposium on sweeteners. AVI Publishing, Westport, Connecticut, pp 87–89

Noller CR (1966) Chemistry of organic compounds, 3rd edn. Saunders, Philadelphia, p 421

Parker KJ (1974) Sucrose as an industrial raw material. Sucrerie Belge 93: 15–27

Pascual CG, Singh R, Juliano BO (1978) Free sugars of rice grain. Carbohydr Res 62: 381–385

Pictet A, Vogel H (1928) Synthese du saccharose. Helv Chim Acta 11: 436–442

Pierce NF, Hirschhorn N (1977) Oral fluid: a simple weapon against dehydration in diarrhoea. How it works and how to use it. World Health Organization, Geneva, pp 87–93 (WHO Chronicle No 31)

Quatraro A, Minei A, Donzella C, Caretta F, Consoli G, Giugliano G (1985) Sugar and wound healing. Lancet II 1985: 664

Reyes FGR, Toledo MCF (1982) Maillard browning reaction of sucrose-glycine model system. Effect of citrate buffer on the color formation and on the hydrolysis of sucrose, Ciênc Technol Aliment (Brazil) 2: 134–142

Reyes FGR, Wrolstad RE, Cornwell CJ (1982a) Comparison of enzymic gas-liquid chromatographic and high-performance liquid chromatographic methods for determining sugars and organic acids in strawberries of three stages of maturity. J Assoc Off Anal Chem 65: 126–131

Reyes FGR, Poocharoen B, Wrolstad RE (1982b) Maillard browning of sugar-glycine model system: changes in sugar concentration, color, and appearance. J Food Sci 47: 1376–1377

Reynolds JEF, Prasad AB (1982) The Martindale extra pharmacopoeia, 28th edn. The Pharmaceutical Press, London

Richmond ML, Brandao SCC, Gray JI, Markakis P, Stine CM (1981) Analysis of simple sugars and sorbitol in fruit by high performance liquid chromatography. J Agric Food Chem 29: 4–7

SCOGS/FASEB (1976) Evaluation of the health aspects of sucrose as a food ingredient. Life Sciences Research Office (LSRO)/Federation of American Societies for Experimental Biology (FASEB), US Food and Drug Administration (Prepared for FDA under contract No. FDA 223–75–2004)

Selwyn S (1983) Evolution of antiseptics, Br J Clin Pract (suppl) 25: 1–3

Shallenberger RS (1974) Occurrence of various sugars in foods. In: Sipple HL, McNutt KW (eds) Sugars in nutrition. Academic Press, New York, pp 67–80

Shallenberger RS, Birch GG (1975) Sugar chemistry. AVI Publishing, Westport, Connecticut, pp 46–88

Shannon IL (1978) Concentration of sugar in commercial baby foods. J Dent Child 45: 19–22

Simmons WK, de Mello AV (1975) Blindness in the nine states of Northeast Brazil. Lett Am J Clin Nutr 28: 202

Southgate DAT, Paul AA, Dean AC, Christie AA, Fric C (1978) Free sugars in foods. J Hum Nutr 32: 335–347

Thomlinson RH (1980) Kitchen remedy for necrotic malignant breast ulcers. Lancet II: 707

United Nations (1986) 1984 International trade statistics yearbook, vol 2. Trade by commodity. United Nations, New York, pp 59–60

University of California Food Task Force (1974) A hungry world. The challenge to agriculture. University of California Press, Berkeley, California

US Beet Sugar Association (1959) The beet sugar story, 3rd edn. rev. US Sugar Beet Association, Washington DC

US/FDA/STF (1986) Evaluation of health aspects of sugars contained in carbohydrate sweeteners. In: Glinsmann WH, Irausquin H, Youngmee KP (ed) Report from FDA's sugar task force (reprinted from J Nutr 116: S1–S216)

Varela RM, Teixeira SG, Batista M (1972) Hypovitaminosis A in the sugar cane zone of southern Pernambuco State Northwest Brazil. Am J Clin Nutr 25: 800

Vlitos AJ (1977) Sweets for starters. Chem Br 13: 340–345

World Health Organization (1986) Oral rehydration therapy for treatment of diarrhoea in the home. World Health Organization, Geneva (WHO/CDD/SER/86.9)

World Health Organization/Pan American Health Organization (1983) Oral rehydration therapy: an annotated bibliography. Pan American Health Organization, Washington DC

World Health Organization/United Nations Children's Fund (1985) The management of diarrhoea and use of oral rehydration therapy. World Health Organization, Geneva

Chapter 4

# Metabolism and Kinetics

## 4.1 Sucrose

### 4.1.1 Digestion

Under normal physiological conditions, sucrose (like other disaccharides) must be hydrolysed before absorption can occur (Herman 1974). Although sucrose is sensitive to acid hydrolysis, no evidence has been found of sucrose digestion in the stomach (Dahlqvist and Borgström 1961).

Sucrose hydrolysis is carried out by disaccharidase, called sucrase or invertase (EC 3.2.1.26), which is present only in the small intestine. More precisely, sucrase is localized in the brush borders of mucosa epithelial cells where it exerts its physiological function (Dahlqvist and Borgström 1961; Crane 1960).

Histochemical staining methods indicate higher sucrase activity in the villi than in the crypts of the small intestinal mucosa (Dahlqvist 1974). More quantitative data on enzyme distribution were obtained by a microdissection technique (Nordstrom et al. 1968).

When given intravenously, sucrose is excreted unchanged in the urine, because there is no sucrase in the body other than that in the intestine wall (Macdonald 1981).

Sucrase develops during foetal life and is thus fully developed before birth (Dahlqvist and Lindberg 1966).

The feeding of diets containing high levels of sucrose appears to produce an adaptive increase in sucrase activity. Rats starved for 3 days and then refed a diet containing 70% sucrose for 24 hours showed greater levels of sucrase activity than rats refed a sucrose-free, high-casein diet (Blair and Tuba 1963). This increase in sucrase activity was subsequently shown to be specific for sucrose. Rats starved for 3 days and then refed a diet containing 68% sucrose for 24 hours had significantly greater levels of sucrase activity than those refed a diet different only in that maltose replaced sucrose (Deren et al. 1967). Sucrase activity in rats fed diets containing 70% carbohydrate was greater with sucrose than with either maltose or glucose (Reddy et al. 1968). Sucrase adaptation to sucrose feeding also appears to occur in human beings. Male volunteers consuming isocaloric diets differing in the nature and amount of dietary carbohydrate (40%–80% of the calories) exhibited increased sucrase activity when fed sucrose compared with glucose (Rosensweig and Herman 1970, 1968).

Intestinal absorption also appears to respond to dietary sucrose. Rats accustomed to a 65% sucrose diet showed increases in the intestinal absorption of sucrose and its constituent monosaccharides, compared with rats fed a stock diet (Reiser et al. 1975). The ad libitum or meal feeding of diets containing 54% sucrose, compared with 54% starch, to rats for 8–12 weeks produced similar increases in sugar absorption (Reiser and Hallfrisch 1977). After being adapted to a 75% sucrose diet for 9 weeks, baboons showed an increased absorptive ability for fructose and, to a lesser extent, glucose, following a sucrose meal (Rechcigl 1978; Crossley and Macdonald 1970).

### 4.1.2 Absorption and Transport

Intact sucrose may be absorbed across the intestine only if the concentration is high, as may occur in primary disaccharidase deficiency, or if there is damage to the mucosal cells with a secondary disaccharidase deficiency or injury to the cell membrane resulting in a loss of selective permeability.

After digestion of sucrose, the component monosaccharides (i.e., glucose and fructose) are absorbed in their usual way (Fridhandler and Quastel 1955; Wilson and Vincent 1955).

Fructose crosses the intestinal wall by "facilitated" transport, as will be discussed below.

Before glucose can be used by cells, it must be transported through the cell membrane into the cellular cytoplasm. However, glucose cannot diffuse through the pores of the cell membrane, because the maximum molecular weight of particles that can do that is about 100, while glucose has a molecular weight of 180. Fructose diffuses to the interior of the cells with a reasonable degree of freedom, but it is transported through the membrane by a mechanism of facilitated diffusion.

In the lipid matrix of the cell membrane, there are large numbers of protein carriers that can bind glucose. In this bound form, the glucose can be transported by the carrier from one side of the membrane to the other, and then be released. Therefore, if the concentration of glucose is greater on one side of the membrane than on the other, more glucose will be transported from the high concentration area than in the opposite direction.

It should be noted that this transport of glucose through the membranes of most tissue cells is quite different than that occurring through the gastrointestinal membrane or the epithelium of the renal tubules (Guyton 1981).

Glucose is transported against concentration gradients through the epithelial cell layer of the small intestine, as part of the process of absorption of sugars into the blood from the intestine. Glucose is also transported against a gradient by the epithelial cell layer of the kidney tubules, from the glomerular filtrate into the blood; thus, glucose is not normally excreted in the urine but is salvaged and kept in the blood-stream.

Glucose enters the cell by binding to a specific carrier, which can also bind $Na^+$, presumably at a second site on the carrier. The carrier molecule thus facilitates the simultaneous, compulsory transport of both $Na^+$ and glucose into the cell.

If the external $Na^+$ concentration is much higher than the internal concentration, $Na^+$ bound to this carrier will tend to move inward, down the $Na^+$ gradient. Since the carrier must also bind glucose in order to function, the inward transport of $Na^+$ down a gradient can drive glucose into the cell. In this way, glucose can be accumulated against a glucose gradient, so long as the inward gradient of $Na^+$ (generated by the pump mechanism) exceeds the outward gradient of glucose.

The energy required to transport glucose into the cell is furnished by the inward gradient of $Na^+$, generated by the pump mechanism ($Na^+$-$K^+$-ATPase = EC 3.6.1.3) (Lehninger 1976).

### 4.1.3 Sucrose as Energy Source

It was in 1931 that Rubner (1931) published a comparison of energy utilization in vertebrates and stated that the energy equivalent of various dietary carbohydrates was not the same, such that 234 g of cane sugar was equivalent to 256 g of glucose. The energy cost of storing fat from dietary carbohydrate is greater than from an iso-energetic intake of dietary fat (Wood and Reid 1975), and, in a study on baboons on high carbohydrate diets for 26 weeks, it was noted that the animals on sucrose gained more weight than those on a partial hydrolysate of starch, though the energy intake was similar (Brook and Noel 1969). When rats were offered ad libitum diets with 80% of the energy as carbohydrate, the greatest weight gain for 100 g food eaten over 20 weeks was with sucrose and the least with glucose (Allen and Leahy 1966). In rats on a reduced energy intake with approximately 75% of the dietary energy as sucrose or glucose, the body weight loss was greater on glucose compared with sucrose (Macdonald et al. 1981).

After the acute ingestion of various carbohydrates, the dietary induced thermogenesis in rats was greater after sucrose than after glucose and fructose given separately (Sharief and Macdonald 1982).

In man, a difference in heat increment after carbohydrate ingestion was first reported by Benedict and Carpenter (1918) who noted that the maximum heat increment was greater after sucrose than fructose, though overall the increment was similar. More recent work has confirmed this observation, and, in the 3-hour post-prandial period after ingestion of galactose, lactose, galactose + glucose, glucose, maltose, sucrose, or fructose + glucose, the metabolic rate was significantly greater after sucrose and fructose + glucose, with the other carbohydrates sharing similar responses (Macdonald 1984). Differences in the metabolic response to various dietary carbohydrates was less marked in the over-weight (Sharief and Macdonald 1982).

Over-weight men and women on a diet with 75% energy as sucrose or glucose with a reduced energy intake lost weight over a 28-day period more rapidly on glucose than sucrose, but a cross-over of diets at 28 days did not reverse this difference between carbohydrates (Macdonald 1986).

Several studies have measured the sugar intake in the over-weight or obese. In one study of short duration, the substitution of an intense sweetener for sucrose resulted in a 25% reduction in energy intake (Porikos et al. 1977). When studies were made on 4907 adolescents, the intake of sugar-containing foods was no higher in the over-weight compared with the normal weight individuals (Garn et al. 1980). In a questionnaire to 415 business men, an inverse relationship was found between sugar eaten and "fatness" (Richardson 1972). In South Wales, a study of 493 men aged 45–59 years revealed a sugar intake that was negatively associated with body mass index (Fehilly et al. 1984). The efficacy of sugar substitutes in the treatment of obesity and diabetes mellitus could not be demonstrated (Finer 1985).

The taste preference of the over-weight has been studied, and it was reported that they preferred food mixtures containing 34% fat and less than 5% sugar compared with normal weight persons, whose preference was for 20% fat and less than 10% sugar (Drewnowski et al. 1985). Obese individuals are not more sensitive to sweet taste (Grinker 1978), nor do they show a different in sweetness discrimination from normals (Rodin 1977).

The energy balance of eight volunteers fed diets supplemented with either lactitol or sucrose was measured. The energy contribution was clearly smaller from lactitol than from sucrose, possibly due to the effect of lactitol on digestion, and also probably because of the effect on the utilization of metabolizable energy (Es et al. 1986).

A review of obesity and sweet taste was published by Grinker (1978).

## 4.2 Glucose

### 4.2.1 Metabolism

To produce energy, glucose must enter the cells from blood and be phosphorylated in the cytoplasm to glucose-6-phosphate, which can be metabolized in a number of ways, depending on the needs and the biochemical versatility of the cell:

(a) breakdown to pyruvic acid/lactic acid;
(b) pentose phosphate cycle; and
(c) glycogen formation (Fig. 4.1).

#### *4.2.1.1 Breakdown to Pyruvic Acid/Lactic Acid*

The most important way by which energy is released from the glucose molecule is by the process of glycolysis and then oxidation of the end-products of glycolysis.

Glycolysis means splitting of the glucose molecule to form two molecules of pyruvic acid. This occurs by 10 successive steps of chemical reactions, as illustrated in Fig. 4.2.

Each step is catalysed by at least one specific enzyme. Glucose is first converted into fructose-1,6-diphosphate and then split into two 3-carbon atom

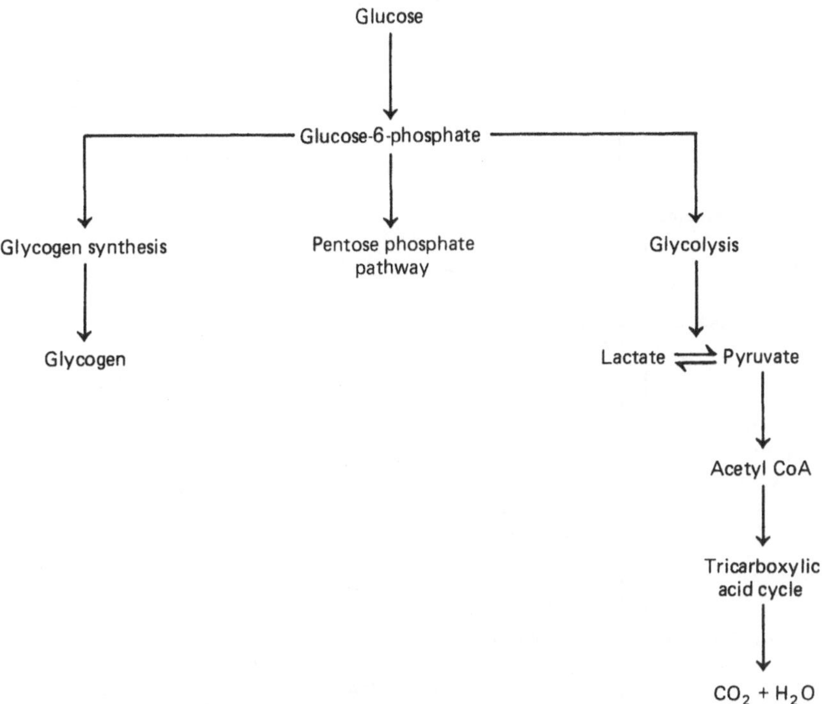


**Fig. 4.1.** The metabolic fate of glucose.

molecules, each of which is then converted through five successive steps into pyruvic acid.

Between the 1,3-diphosphoglyceric acid and the 3-phosphoglyceric acid stages, and again between the phosphopyruvic acid and the pyruvic acid stages, the packets of energy released are greater than 8000 calories per mole, the amount required to form ATP, and the reactions are completed in such a way that ATP is formed. Thus, a total of 4 moles of ATP are formed for each mole of fructose-1,6-diphosphate that is split into pyruvic acid.

Since 2 moles of ATP had been required to phosphorylate into the original glucose to form fructose-1,6-diphosphate, the net gain in ATP molecules by the entire glycolytic process is only 2 moles for each mole of glucose utilized. Finally, four hydrogen atoms are released when glyceraldehyde is converted to 1,3-diphosphoglyceric acid.

Pyruvic acid, the end-product of glycolysis, may enter the tricarboxylic acid cycle inside the mitochondria and be broken down completely to $CO_2$ and $H_2O$ with release of energy.

When pyruvic acid and hydrogen atom quantities begin to be excessive, these two end-products inhibit glycolysis and react with each other to form lactic acid in accordance with the following equation:

$$\text{pyruvic acid} + \text{NADH} + \text{H}^+ \longrightarrow \text{lactic acid} + \text{NAD}^+$$

The enzyme involved is lactic acid dehydrogenase (EC 1.1.1.27).

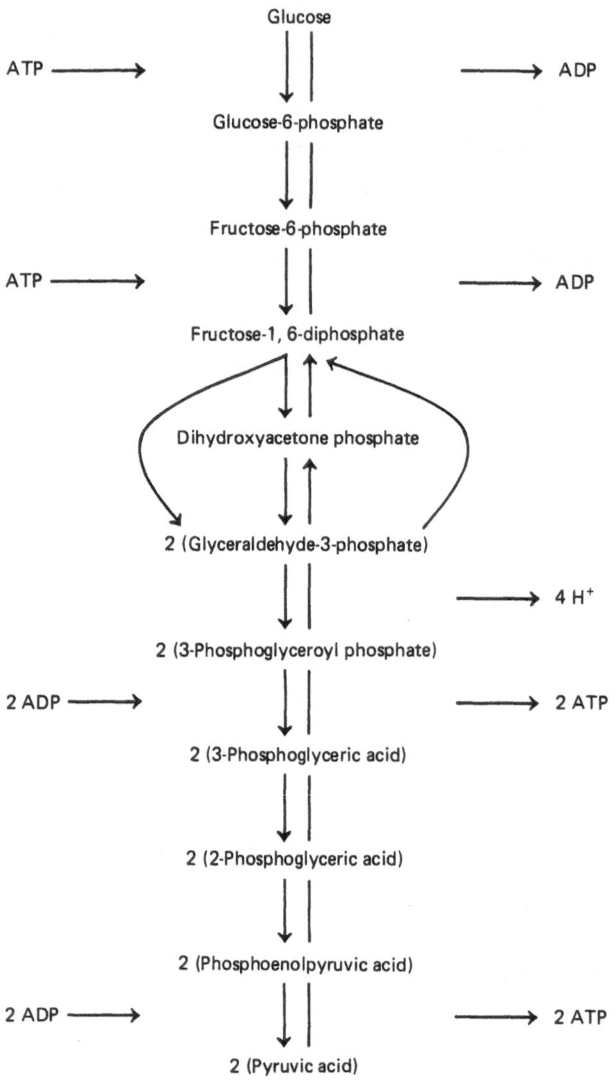


Net reaction:
Glucose + 2 ADP + 2 $PO_4^{3-}$ ⟶ 2 Pyruvic acid + 2 ATP + 4 $H^+$

**Fig. 4.2.** The sequence of chemical reactions responsible for glycolysis.

### *4.2.1.2 Pentose Phosphate Cycle (Phosphogluconate Pathway)*

The pentose cycle is responsible for as much as 30% of the glucose breakdown in the liver and even more than this in fat cells. Fig. 4.3 shows most of the basic chemical reactions in the phosphogluconate pathway.

The first reaction of the phosphogluconate pathway is the enzymatic dehydrogenation of glucose-6-phosphate to form 6-phosphoglucono-δ-lactone

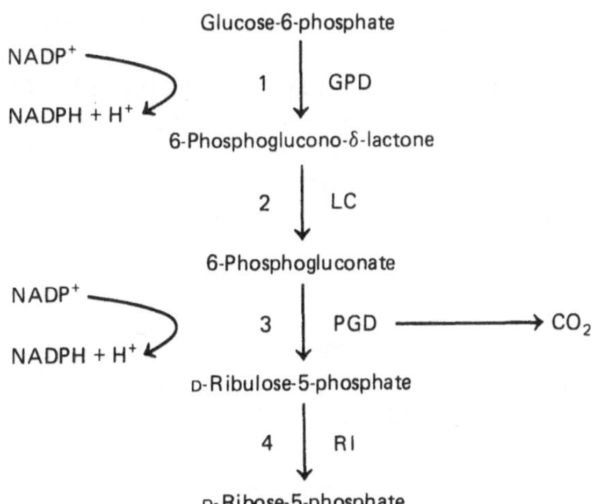


**Fig. 4.3.** The pentose phosphate pathway.

by glucose-6-phosphate dehydrogenase (GPD) (EC 11.1.1.49). The 6-phosphoglucono-δ-lactone is hydrolysed to produce 6-phosphogluconate by a specific lactonase (LC) (EC 3.1.1.17).

In the next step, 6-phosphogluconate undergoes oxidation and decarboxylation by 6-phosphogluconate dehydrogenase (PGD) (EC 1.1.1.44), a $Mg^{2+}$-dependent enzyme, to form D-ribulose-5-phosphate.

Then, by the action of ribose phosphate isomerase (RI) (EC 5.3.1.6), D-ribulose-5-phosphate is reversibly transformed into D-ribose-5-phosphate.

Under some metabolic circumstances, the phosphogluconate pathway ends at this point, and its overall equation may be written:

$$\text{glucose-6-phosphate} + 2\ NADP^+ + H_2O \longrightarrow$$
$$\text{D-ribose-5-phosphate} + CO_2 + 2\ NADPH + 2\ H^+$$

Under other circumstances, the phosphogluconate pathway may continue, and ribulose-5-phosphate can change progressively into several other 7-, 5-, 4-, and 3-carbon sugars.

Finally, various combinations of these sugars can resynthesize glucose. However, only five molecules of glucose are resynthesized for every six molecules of glucose that initially enter into the reactions. That is, the phosphogluconate pathway is a cyclic process in which one molecule of glucose is metabolized for each revolution of the cycle.

The net reaction is conversion of the single molecule of glucose plus six molecules of water into six carbon dioxide molecules and 24 hydrogen atoms:

$$6\ \text{glucose-6-phosphate} + 12\ NADP^+\ 6\ H_2O \longrightarrow$$
$$\text{5-glucose-6-phosphate} + 6\ CO_2 + 12\ NADPH + 12\ H^+ + Pi$$

Thus, by revolution of the cycle again and again, all the glucose can eventually be converted into $CO_2$ and hydrogen. The hydrogen, in turn, can enter the oxidative phosphorylation pathway to form ATP, or it is used for synthesis of fat.

The hydrogen released during the phosphogluconate cycle does not combine with $NAD^+$, but combines with $NADP^+$. This difference is extremely significant, because only hydrogen bound with $NADP^+$, in the form of NADPH, can be used for synthesis of fats from carbohydrates.

When glycolysis becomes slowed because of cellular inactivity, the phosphogluconate pathway still remains operative (mainly in the liver) to break down any excess glucose that continues to be transported into the cells. Since NADPH becomes abundant, acetyl radicals can be converted into long fatty acid chains.

#### *4.2.1.3 Glycogen Formation*

When glucose is not immediately required for energy, the extra glucose that continually enters the cells is either stored as glycogen or converted to fat. Glucose is preferentially stored as glycogen until the cells have stored as much glycogen as they can.

When the cells (mainly liver and muscle cells) approach saturation with glycogen, the additional glucose is then converted to fat (as described above in the pentose phosphate pathway) and stored in the fat cells. The synthesis of glycogen begins with the conversion of glucose-6-phosphate to glucose-1-phosphate, catalysed by the enzyme phosphoglucomutase (EC 2.7.5.1). The second step of glycogen synthesis:

$$\text{D-glucose-1-phosphate-UTP} \longrightarrow \text{UDP-D-glucose} + \text{PPi}$$

is catalysed by glucose-1-phosphate uridylytransferase (EC 2.7.7.9). Then, a different enzyme is involved, glycogen synthase (EC 2.4.1.11), which employs UDP-glucose as glycosyl donor.

$$\text{UDP-D-glucose} + (\text{glucose})_n \longrightarrow + (\text{glucose})_{n+1}$$

The glucosyl group of UDP-D-glucose is transferred to the terminal glucose residue at the non-reducing end of a glycogen chain to form an $\alpha$ (1→4) glycosidic linkage between carbon atom 1 of the added glucosyl residue and the 4-hydroxyl of the terminal glucose residue of the chain.

Glycogen synthase (EC 2.4.1.11) cannot make the $\alpha$(1→6) bonds found in the branch points of glycogen. However, the enzyme 1,4-$\alpha$-glucan branching enzyme (EC 2.4.1.18), which is present in many animal tissues, catalyses transfer of a terminal oligosaccharide fragment of 6 or 7 glucosyl residues from the end of the main glycogen chain to the 6-hydroxyl group of a glucose residue of the same or of another glycogen chain, so that an $\alpha$(1→6) linkage is formed and a new branch point is created.

## 4.3 Fructose

### 4.3.1 Absorption

Ingested fructose is absorbed by the small intestine (Fridhandler and Quastel 1955; Wilson and Vincent 1955). It disappears from the rat intestinal lumen at a rate intermediate between that of sugars (such as glucose or galactose), which are actively transported, and of sugars (such as mannose and pentoses), which are not (Wilson 1962; Cori 1925).

Most investigations support the existence of a specific system for fructose. This system seems to be distinct from that for glucose and operates more slowly; its energy dependence and its affinity for fructose have not been clearly established. A carrier-mediated diffusion process, as well as an active-transport mechanism, have been proposed (Van de Berghe 1986). In different species, fructose may be partially converted to another substance, such as lactate or glucose (White and Landau 1965), and then absorbed.

In the small intestine of the rat, about 10% of fructose is converted to glucose and about 60% to lactate. The remaining 30% passes into the portal venous blood.

In the guinea-pig, 55%–80% of fructose is converted to glucose, 10% to lactate, and the remaining 10%–35% passes into the portal vein unchanged (Herman 1974).

In man, up to 80%–90% of fructose is absorbed as such from the jejunum (Cook 1969), and there is only minor conversion of fructose to lactate during absorption (Herman 1974). The conversion of some fructose to glucose by the human intestine has also been demonstrated.

The transformation of fructose into glucose and lactic acid in the small intestine occurs via the hexokinase and fructokinase portions of the glycolytic pathway (Fig. 4.4).

In the case of man and the rat, the very low level of jejunal glucose-6-phosphatase (EC 3.1.3.9) caused the reduced conversion of fructose into glucose and the increased absorption of fructose unchanged.

Once fructose reaches the blood-stream, it is rapidly utilized, as demonstrated by several studies in human subjects and animals (Smith et al. 1953).

Administered in a peripheral vein at a dose of 0.5 g/kg body weight, over 30 min, to normal human beings, fructose disappears from the circulation twice as fast as glucose,its half-life being about 18 min (Smith et al. 1953), compared with 43 min for glucose (Conard 1955).

### 4.3.2 Transport

In human beings and laboratory animals, the liver and, to a lesser extent, the kidney and small intestine are the main sites of fructose metabolism. These tissues possess a number of enzymes that are responsible for a specialized metabolic pathway of fructose metabolism.

Fructose does not enter the hepatocytes freely, as evidenced by the steep gradient between its extra- and intracellular concentration. Computation from the kinetics of fructose uptake and metabolism in the perfused rat liver are in accord with a carrier-mediated transport with a $K_m$ of 67 mM and a $V_{max}$ equal to 30 μM/min per gram of tissue (Sestoft and Fleron 1974).

Competition studies indicate that fructose is transported into the liver cell, at least in part, by the same system as glucose and galactose (Craik and Elliott 1980; Hooper and Short 1977). In the kidney, fructose is actively reabsorbed by the tubule, and its reabsorption is inhibited by glucose.

The utilization of fructose by peripheral tissues seems to be negligible: glucose, which is always present in the blood, prevents both the uptake and phosphorylation of fructose by hexokinase (HK) (EC 2.7.1.1) (Hue 1974). However, adipose tissue could be an exception (Froesch and Ginsberg 1962).

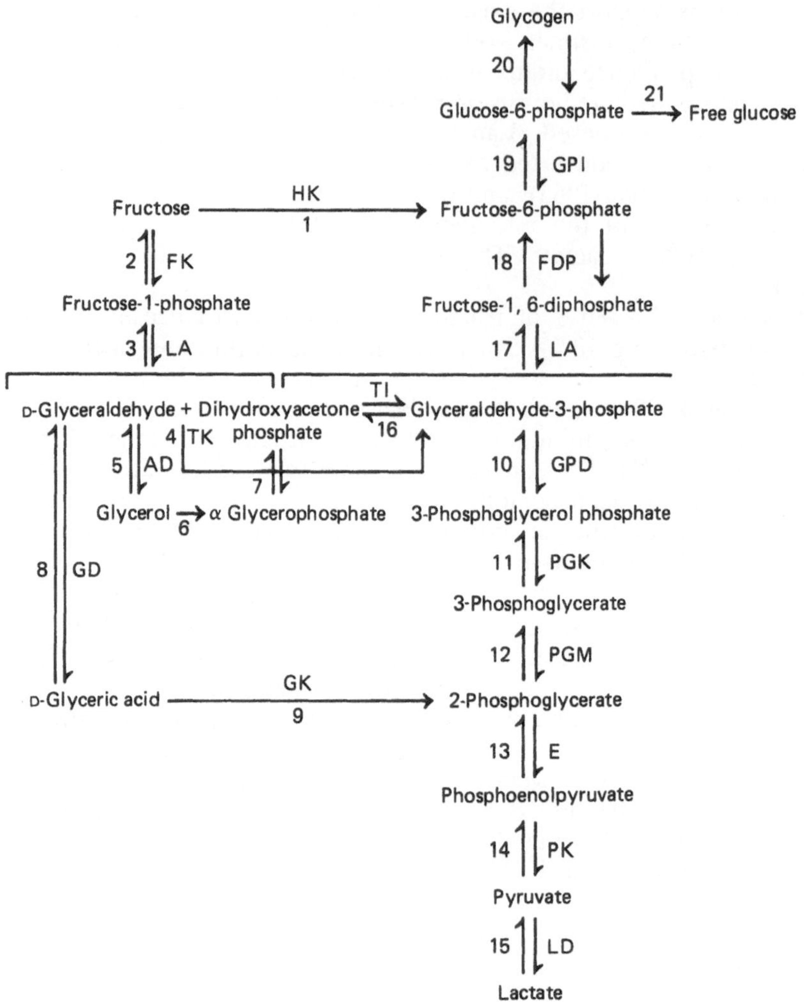


**Fig. 4.4.** The metabolic fate of fructose.

In adipose tissue, fructose is transported by at least two different carriers. A specific fructose transporter requires a fructose concentration of about 25 mM for half-maximal saturation and is insensitive to glucose and insulin (Craik and Elliott 1980). Fructose can also enter the adipocyte by the glucose carrier, which has a 5- to 10-fold lower $K_m$ than its specific carrier. The $V_{max}$ is approximately 50% lower and is roughly doubled by insulin. However, participation of the latter carrier in the transport of fructose inside the adipocytes under normal physiological conditions is unlikely, because of the competition exerted by glucose (Halperin and Cheema-Dhadli 1982; Schoenle et al. 1979).

### 4.3.3 Metabolic Pathway of Fructose in the Liver

#### *4.3.3.1 Phosphorylation of Fructose*

Liver hexokinase (HK) (EC 2.7.1.1) can catalyse the phosphorylation of glucose, fructose, and mannose, but the affinity of the enzyme for glucose is 20 times that for fructose (Sols and Crane 1954). Therefore, glucose prevents the phosphorylation of fructose into fructose-6-phosphate (Reaction 1, Fig. 4.4). Furthermore, glucokinase (EC 2.7.1.2), which is also present in the liver, is inactive with respect to fructose (Vinuela et al. 1963).

Fructose is phosphorylated into fructose-1-phosphate by fructokinase (FK) (EC 2.7.1.4) (Reaction 2, Fig. 4.4) (Hers 1952; Cori et al. 1951; Leuthardt and Testa 1951). The phosphoryl donor is the $Mg^{2+}$-ATP complex. The $K_m$ for fructose is about 0.5 mM (Adelman et al. 1967; Parks et al. 1957; Hers 1952); however, the values of the affinity constant for $Mg^{2+}$-ATP, obtained by different authors, vary between 0.2 and 2 mM (Sanchez et al. 1971; Adelman et al. 1967; Parks et al. 1957).

The $V_{max}$ in crude extract of rat and human liver reaches 2–3 μM/min per gram of tissue at 22–25 °C or about 10 μM/min per gram at 37 °C (Sanchez et al. 1971; Heinz et al. 1968; Adelman et al. 1967).

An extremely rare, harmless inborn error of metabolism in human beings, known as "essential fructosuria", is caused by the absence of fructokinase (Schapira et al. 1961–62). It is characterized by the presence of fructose in the urine after the ingestion of foods containing fructose. "Essential fructosuria" is inherited as an autosomal trait (Lasker 1941).

#### *4.3.3.2 Splitting of Fructose-1-Phosphate (Reaction 3, Fig. 4.4)*

Fructose-1-phosphate is split to D-glyceraldehyde and dihydroxyacetone phosphate by liver aldolase or aldolase B (LA) (EC 4.1.2.13) (Hers 1952; Leuthardt and Testa 1951). The same enzyme catalyzes the splitting of the glycolytic-glucogenetic intermediate, fructose-1,6-diphosphate, into D-glyceraldehyde-3-phosphate and dihydroxyacetone phosphate. The enzyme has the same $V_{max}$ for both fructose esters (Steinmann and Gitzelmann 1981; Hers 1952), reaching 2.3 μM/min per gram of tissue at 25 °C in the rat (Woods et al. 1970) and human liver (Steinmann and Gitzelmann 1981; Heinz et al. 1968). However, its $K_m$ for fructose-1-phosphate is several orders of magnitude higher than that for fructose-1,6-diphosphate, respective values of approximately 1 mM and 4–12 μM having been reported in human liver (Koster et al. 1975).

In addition, aldolase B (EC 4.1.2.13) catalyses the condensation of the trioses to fructose-1,6-diphosphate. An inborn deficiency in the ability of aldolase B to split fructose-1-phosphate is found in human beings with hereditary fructose intolerance (Gitzelmann et al. 1983). These subjects are asymptomatic as long as they do not eat any food containing fructose, but suffer severe hypoglycaemia and vomiting shortly after the ingestion of the ketohexose. The disorder is inherited as an autosomal recessive trait, and its frequency has been estimated at approximately 1:30 000 (Cornblath and Schwartz 1976).

#### 4.3.3.3 *Metabolism of* D-*Glyceraldehyde*

There are three mechanisms by which D-glyceraldehyde, which is formed in the liver on the splitting of fructose-1-phosphate, can be converted into a triose phosphate:

1. In the most important reaction, D-glyceraldehyde is directly phosphorylated into D-glyceraldehyde-3-phosphate by triokinase (TK) (EC 2.7.1.28), which is located in the cytoplasm (Hers and Kusaka 1953) (Reaction 4, Fig. 4.4).
   The preferential phosphoryl donor is adenosine triphosphate (ATP), but studies have shown that inosine triphosphate (ITP) can be utilized at 14% and guanosine triphosphate (GTP) at 10% of the rate of ATP utilization (Frandsen and Grunnet 1971).
   The maximal activity of rat liver triokinase (EC 2.7.1.28) is approximately 1.5 μM/min per gram at 37°C (Hers 1962), while the $K_m$ values are 0.01 mM for D-glyceraldehyde (Sillero et al. 1969) and 0.77 mM for $Mg^{2+}$-ATP (Frandsen and Grunnet 1971).
2. D-Glyceraldehyde is reduced to glycerol by NADH and alcohol dehydrogenase (AD) (EC 1.1.1.1) or by NADPH and aldose reductase (EC 1.1.1.21) (Reaction 5, Fig. 4.4), followed by phosphorylation to α-glycerophosphate and subsequent oxidation to dihydroxyacetone phosphate (Reactions 6, 7, Fig. 4.4) (Wolf and Leuthardt 1953).
3. D-Glyceraldehyde is oxidized to glyceric acid by glyceraldehyde dehydrogenase (GD) (EC 1.2.1.3) (Lamprecht and Heinz 1958), and glyceric acid is then phosphorylated to 2-phosphoglyceric acid by the enzyme D-glycerate kinase (GK) (EC 2.7.1.31) (Reactions 8, 9, Fig. 4.4) (Ichiara and Greenberg 1959).

#### 4.3.3.4 *Fate of Glyceraldehyde-3-Phosphate*

Glyceraldehyde-3-phosphate is an important intermediate of the glycolytic-gluconeogenic pathway.

In the second stage of glycolysis (Reaction 10, Fig. 4.4), glyceraldehyde-3-phosphate is oxidized to 3-phosphoglyceroyl phosphate; the enzyme catalysing this reaction is the glyceraldehyde phosphate dehydrogenase (GPD) (EC 1.2.1.9).

Phosphoglycerate kinase (PGK) (EC 2.7.2.3) catalyses the transfer of the acyl phosphate group from 3-phosphoglyceroyl phosphate to ADP, with formation of 3-phosphoglycerate, which is converted to 2-phosphoglycerate by the enzyme phosphoglyceromutase (PGM) (EC 2.7.5.3).

The dehydration of 2-phosphoglycerate to phosphoenol pyruvate is catalysed by enolase (E) (EC 4.2.1.11), and the transfer of the phosphate group from phosphoenol pyruvate to ADP, yielding free pyruvate, is catalysed by pyruvate kinase (PK) (EC 2.7.1.40).

Finally, pyruvate is reduced to lactate by lactate dehydrogenase (LD) (EC 1.1.1.17).

In the gluconeogenic pathway (Reactions 17–19, Fig. 4.4), glyceraldehyde-3-phosphate can be converted into dihydroxyacetone phosphate by the enzyme triosephosphate isomerase (TI) (EC 5.3.1.1) (Reaction 16, Fig. 4.4).

The condensation of glyceraldehyde-3-phosphate with dihydroxyacetone phosphate produces fructose-1,6-diphosphate. This reaction is catalysed by liver aldolase (LA) (EC 4.1.2.13).

Fructose (hexose) diphosphatase (FDP) (EC 3.1.3.11) removes the 1-phosphate group from fructose-1,6-diphosphate, yielding fructose-6-phosphate.

In the last step of gluconeogenesis, fructose-6-phosphate is reversibly converted into glucose-6-phosphate by glucose phosphate isomerase (GPI) (EC 5.3.1.9).

A biosynthetic pathway starting from glucose-6-phosphate leads to formation of the glycogen, a storage polymer (Reaction 20, Fig. 4.4); furthermore, in some tissues (particularly the liver, kidney, and intestinal epithelium), glucose-6-phosphate may be dephosphorylated to form free glucose (Reaction 21, Fig. 4.4).

### 4.3.4 Metabolism of Fructose in Other Tissues

The kidney and small intestine possess the same metabolic pathways of fructose metabolism that have been described for the liver. The three enzymes, fructokinase (EC 2.7.1.4), aldolase B (EC 4.1.2.13), and triokinase (EC 2.7.1.28) convert fructose to intermediates of the glycolytic-gluconeogenic pathway.

In the brain (Slein et al. 1950), erythrocytes, and leukocytes, fructose is phosphorylated by hexokinase (EC 2.7.1.1) to fructose-6-phosphate (Reaction 1, Fig. 4.4). These tissues metabolize fructose almost as fast as glucose, but only in the absence of glucose. In the presence of a physiological concentration of glucose (approximately 5mM), the utilization of 5 mM fructose is about 90% inhibited (Froesch and Ginsberg 1962).

Muscle may participate to some extent in the metabolism of fructose, but muscle uses little fructose compared with glucose. Fructose oxidation to $CO_2$ and incorporation into glycogen by the rat diaphragm in vitro proceeds at a rate of one-eighth to one-quarter that of glucose (Renold and Thorn 1955).

The utilization of fructose by adipose tissue may be of more importance. Fructose is phosphorylated by hexokinase (EC 2.7.1.1) to fructose-6-phosphate, as previously described. No inhibition of glucose occurs at this stage, because of the virtual absence of free glucose (which is bound to its carrier) in the adipocyte (Froesch and Ginsberg 1962).

### 4.3.5 End-Products of Fructose Metabolism

Studies of the individual organs that possess the specialized pathway of fructose metabolism show that, in accord with its entrance in the glycolytic-glucogenetic pathway, fructose is converted mainly to glucose and lactic acid.

In isolated rat liver perfused with fructose, about 50%–75% of the ketose is converted to glucose and 20%–25% to lactate plus pyruvate (Sestoft and Fleron 1974; Woods et al. 1970; Exton and Park 1967).

Quantitatively, glycogen constitutes the third most important end-product of fructose metabolism. In rat liver perfused with 20 mM fructose for 1 hour, 8% of the ketose was recovered as glycogen (Renold and Thorn 1955).

Studies with radioactive fructose have also shown a conversion of the ketose into triglycerides, which has been estimated to represent 1%–3% of its overall metabolism (Nikkila 1969; Zakim et al. 1969; Bar-on and Stein 1968).

Further minor end-products of fructose metabolism include ketone bodies, glycerol, sorbitol, and $CO_2$.

### 4.3.6 Accumulation of Fructose Metabolites

Within minutes of oral, and especially parenteral, administration of fructose, a rapid accumulation of up to 10 μmol/g fructose-1-phosphate (normally not a detectable metabolite) can be seen in the liver (Burch et al. 1969), kidney (Burch et al. 1980), and small intestine (Kjerulf-Jenson 1942) of laboratory animals and normal human beings. The accumulation of fructose-1-phosphate demonstrates that the velocity of its formation is much faster than the rate of its further metabolism.

The metabolism of fructose also brings about the 2- to 10-fold elevations of the concentration of most glycolytic-gluconeogenic intermediates (Woods et al. 1970; Burch et al. 1969). The values reached are, nevertheless, much smaller than those for fructose-1-phosphate.

A review on fructose intolerance has recently been written by Schapira (1986).

## 4.4 References


Adelman RC, Ballard FJ, Weinhouse S (1967) Purification and properties of rat liver fructokinase. J Biol Chem 242: 3360–3365

Allen RJL, Leahy JS (1966) Some effects of dietary dextrose, fructose, liquid glucose and sucrose in the adult small rat. Br J Nutr 20: 339–347

Bar-on H, Stein Y (1968) Effect of glucose and fructose administration on lipid metabolism in the rat. J Nutr 94: 95–105

Benedict FG, Carpenter RM (1918) Food ingestion and energy transformation with special reference to the stimulating effects of nutrients. Carnegie Research Institute, Washington DC, pp 47–250 (No 261)

Blair DB, Tuba J (1963) Rat intestinal sucrase. I. Intestinal distribution and reaction kinetics. Can J Biochem Physiol 41: 905–916

Blair DB, Yakimets W, Tuba J (1963) Rat intestinal sucrase. II. The effects of rat age and sex and of diet on sucrase activity. Can J Biochem Physiol 41: 917–929

Brook M, Noel P (1969) Influence of dietary liquid glucose, sucrose or fructose on body fat formation. Nature (Lond) 222: 562–563

Burch HB, Max P Jr, Chyu K, Lowry OH (1969) Metabolic intermediates in liver of rats given large amounts of fructose or dihydroxyacetone. Biochem Biophys Res Commun 34: 619–626

Burch HB, Choi S, Dence CN, Alvey TR, Coley BR, Lowry OH (1980) Metabolic effects of large fructose loads in different parts of the rat nephron. J Biol Chem 255: 8239–8244

Conrad V (1955) Mesure de l'assimilation du glucose: bases théoretiques et application cliniques. Acta Gastroenterol Belg 18: 655–705

Cook GC (1969) Absorption products of D(–)fructose in man. Clin Sci 37: 675–687

Cook GC (1971) Absorption and metabolism of D(–)fructose in man. Am J Clin Nutr 24: 1302–1307

Cori CF (1925) The fate of sugar in the animal body. I. The rate of absorption of hexoses and pentoses from the intestinal tract. J Biol Chem 66: 691–715

Cori GT, Ochoa S, Stein MW, Cori CF (1951) The metabolism of fructose in liver. Isolation of fructose-1-phosphate and inorganic pyrophosphate. Biochim Biophys Acta 7: 304–317

Cornblath M, Schwartz R (1976) Disorders of carbohydrate metabolism in infancy, 2nd edn. Saunders, Philadelphia, Pennsylvania, p 322

Craik JD, Elliott KRF (1980) Transport of D-glucose and D-galactose into isolated rat hepatocytes. Biochem J 192: 373–375

Crane RK (1960) Intestinal absorption of sugars. Physiol Rev 40: 789–825

Crossley JN, Macdonald I (1970) The influence in male baboons of a high sucrose diet on the portal and arterial levels of glucose and fructose following a sucrose meal. Nutr Metab 12: 171–178

Dahlqvist A (1974) Physiology of carbohydrate digestion and absorption. In: Sipple HL, McNutt KW (eds) Sugars in nutrition. Academic Press, New York, p 189

Dahlqvist A, Borgström B (1961) Digestion and absorption of disaccharides in man. Biochem J 81: 411–418

Dahlqvist A, Lindberg T (1966) Development of the intestinal disaccharidase and alkaline phosphatase activities in the human fetus. Clin Sci 30: 517–528

Deren JJ, Bloitman SA, Zamcheck N (1967) Effect of diet upon intestinal disaccharidases and disaccharide absorption. J Clin Invest 46: 186–195

Drewnowski A, Brunzell JD, Sande K, Iverius PH, Greenwood MR (1985) Sweet tooth reconsidered: taste responsiveness in human obesity. Physiol Behav 35: 617–622

Es AJH van, Groot L de, Vogt JE (1986) Energy balances of eight volunteers fed on diets supplemented with either lactitol or saccharose. Br J Nutr 56: 545–554

Exton JH, Park CR (1967) Control of gluconeogenesis in liver. I. General features of gluconeogenesis in the perfused livers of rats. J Biol Chem 242: 2622–2636

Fehilly AM, Phillips KM, Sweetman PM (1984) A weighed dietary survey of men in Caerphilly, South Wales. Hum Nutr Appl Nutr 38A: 270–276

Finer N (1985) Sugar substitutes in the treatment of obesity and diabetes mellitus. Clin Nutr 4: 207–214

Frandsen EK, Grunnet N (1971) Kinetic properties of triokinase from rat liver. Eur J Biochem 23: 588–592

Fridhandler L, Quastel JH (1955) Absorption of sugars from isolated surviving intestine. Arch Biochem Biophys 56: 412–423

Froesch ER, Ginsberg JL (1962) Fructose metabolism of adipose tissue. I. Comparison of fructose and glucose metabolism in epididymal adipose tissue of normal rats. J Biol Chem 237: 3317–3324

Garn SM, Solomon MA, Cole PE (1980) Sugar-food intake of obese and lean adolescents. Ecol Food Nutr 9: 219–222

Gitzelmann R, Steinmann B, Van den Berghe G (1983) Essential fructosuria, hereditary fructose intolerance, and fructose 1,6-diphosphatase deficiency. In: Stanbury JB, Wyngaarden JB, Fredrickson DS, Goldstein JL, Brown MS (eds) The metabolic basis of inherited disease, 5th edn. McGraw-Hill, New York, pp 118–140

Grinker J (1978) Obesity and sweet taste. Am J Clin Nutr 31: 1078–1087

Guyton AC (1981) Transport through the cell membrane. Textbook of medical physiology. Saunders, Philadelphia, Pennsylvania, pp 840–841

Halperin ML, Cheema-Dhadli S (1982) Comparison of glucose and fructose transport into adipocytes of the rat. Biochem J 202: 717–721

Heinz F, Lamprecht W, Kirsch J (1968) Enzymes of fructose metabolism in human liver. J Clin Invest 47: 1826–1832

Herman RH (1974) Hydrolysis and absorption of carbohydrates and adaptive responses of the jejunum. In: Sipple HL, McNutt KW (eds) Sugars in nutrition. Academic Press, New York, pp 145–172

Hers HG (1952) La fructokinase du fois. Biochim Biophys Acta 8: 416–423

Hers HG (1962) Triokinase. In: Colowich SP, Kaplan NO (eds) Methods in enzymology. Academic Press, New York, vol 5, p 362–364

Hers HG, Kusaka T (1953) Le metabolisme du fructose-1-phosphate dans le foie. Biochim Biophys Act 11: 427–437

Hooper RH, Short AH (1977) The hepatocellular uptake of glucose, galactose, and fructose in conscious sheep. J Physiol (Lond) 264: 523–539

Hue L (1974) The metabolism and toxic effects of fructose. In: Sipple HL, McNutt KW (eds) Sugars in nutrition. Academic Press, New York, pp 357–371

Ichiara A, Greenberg DM (1959) Studies on the purification and properties of D-glyceric acid kinase. J Biol Chem 225: 949–958

Kjerulf-Jensen K (1942) The hexosemonophosphoric acids formed within the intestinal mucosa

during absorption of fructose, glucose, and galactose. Acta Physiol Scand 4: 225–248
Koster JF, Slee RG, Fernandez J (1975) On the biochemical basis of hereditary fructose intolerance. Biochem Biophys Res Commun 64: 289–294
Lamprecht W, Heinz F (1958) Isolierung von Glycerin-aldehydehydrogenase aus Rattenleber. Zur Biochemie des Fructosestoffwechsels. Z Naturforsch 13b: 464–465
Lasker M (1941) Essential fructosuria. Hum Biol 13: 51–63
Lehninger AL (1976) Biochemistry. The molecular basis of cell structures and function, 2nd edn. Worth Publishers, New York, pp 793–794
Leuthardt F, Testa E (1951) Die phosphorylierung der fructose in der leber. II. Mitteiling. Helv Chim Acta 34: 931–938
Macdonald I (1981) Sucrose. In: Howard AN (ed) Nutritional problems in modern society. John Libbey, London, pp 60–67
Macdonald I (1984) Differences in dietary induced thermogenesis following the ingestion of various sugars. Ann Nutr Metab 28: 226–230
Macdonald I (1986) Dietary carbohydrates and energy balance. Prog Biochem Pharmacol 21: 181–191
Macdonald I, Grenby TH, Fisher MA, Williams CA (1981) Differences between sucrose and glucose diets in their effects on the rate of body weight change in rats. J Nutr 111: 1543–1547
Nikkila EA (1969) Control of plasma and liver triglyceride kinetics by carbohydrate metabolism and insulin. Adv Lipid Res 7: 63–134
Nordstrom C, Dahlqvist A, Josefsson L (1968) Quantitative determination of enzymes in different parts of the villi and crypts of rat small intestine. Comparison of alkaline phosphatase, disaccharidases, and dipeptidases. J Histochem Cytochem 15: 713–721
Parks RE, Ben-Gershom E, Lardy HA (1957) Liver fructokinase. J Biol Chem 227: 231–242
Porikos KP, Booth G, Van Itallie TB (1977) Effect of covert nutritive dilution on the spontaneous food intake. Am J Clin Nutr 30: 1638–1644
Rechcigl M Jr (ed) (1978) CRC handbook series in nutrition and food. Section E: nutritional disorders. I. Effect of nutrient excesses and toxicities in animals and man. CRC Press, Boca Raton, Florida
Reddy BS, Pleasants JR, Wostmann BS (1968) Effect of dietary carbohydrates on intestinal disaccharidases in germ free and conventional rats. J Nutr 95: 413–419
Reiser S, Michaelis OE, Putney J, Hallfrisch J (1975) Effect of sucrose feeding on the intestinal transport of sugars in two strains of rats. J Nutr 105: 894–905
Reiser S, Hallfrisch J (1977) Insulin sensitivity and adipose tissue weight of rats fed starch or sucrose diets ad libitum or in meals. J Nutr 107: 147–155
Reiser S, Hallfrisch J, Putney J, Lev F (1976) Enhancement of intestinal sugar transport by rats fed sucrose as compared to starch. Nutr Metab 20: 461–470
Renold AE, Thorn GW (1955) Clinical usefulness of fructose. Am J Med 19: 163–168
Richardson JF (1972) The sugar intake of business men and its inverse relationship with relative weight. Br J Nutr 27: 449–460
Rodin J (1977) Implications of responsiveness to sweet taste for obesity. In: Weiffenbach J (ed) Taste and development – the genesis of sweet preference (DHEW Publ No 77–1068)
Rosensweig NS, Herman RH (1968) Control of jejunal sucrase and maltase activity by dietary sucrose or fructose in man: a model for the study of enzyme regulation in man. J Clin Invest 47: 2253–2262
Rosensweig NS, Herman RH (1970) Dose response of jejunal sucrase and maltase activities to isocaloric high and low carbohydrate diets in man. Am J Clin Nutr 23: 1373–1377
Rubner M (1931) History of the development of energy utilization in vertebrates. Sber Preuss Adad 17: 313
Sanchez JJ, Gonzales NS, Pontis HG (1971) Fructokinase from rat liver. I. Purification and properties. Biochim Biophys Acta 227: 67–78
Schapira F (1986) Hereditary fructose intolerance (L'intolérance héréditaire au fructose). Cah Nutr Diet 21: 72–75
Schapira F, Schapira G, Dreyfus JC (1961–62) La lesion enzymatique de la fructosurie benigne. Enzym Biol Clin 1: 170–175
Schoenle E, Zapf J, Foresch ER (1979) Transport and metabolism of fructose in fat cells of normal and hypophysectomized rats. Am J Physiol 237: E325–E330
Sestoft L, Fleron P (1974) Determination of the kinetic constants of fructose transport and phosphorylation in the perfused rat liver. Biochim Biophys Acta 345: 27–38
Sharief N, Macdonald I (1982) Different effects of various carbohydrates on the metabolic rate in rats. Ann Nutr Metab 26: 66–72

Sillero MAG, Sillero A, Sols A (1969) Enzymes involved in fructose metabolism in liver and the glyceraldehyde metabolic crossroads. Eur J Biochem 10: 345–350
Slein MW, Cori GT, Cori CF (1950) Comparative study of hexokinase from yeast and animal tissue. J Biol Chem 186: 763–780
Smith LH Jr, Ettinger RH, Seligson D (1953) A comparison of the metabolism of fructose and glucose in hepatic disease and diabetes mellitus. J Clin Invest 32: 273–282
Sols A, Crane RK (1954) Substrate specificity of brain hexokinase. J Biol Chem 210: 581–595
Steinmann B, Gitzelmann (1981) The diagnosis of hereditary fructose intolerance. Helv Paediatr Acta 36: 297–300
Van den Berghe G (1986) Fructose: metabolism and short-term effects on carbohydrate and purine metabolic pathways. Prog Biochem Pharmacol 21: 1–32
Vinuela E, Salas M, Sols A (1963) Glucokinase and hexokinase in liver in relation to glycogen synthesis. J Biol Chem 238: 1175–1177
White LW, Landau BR (1965) Sugar transport and fructose metabolism in human intestine in vitro. J Clin Invest 44: 1200–1213
Wilson TH (1962) Intestinal absorption. Saunders, Philadelphia, Pennsylvania, pp 69–72
Wilson TH, Vincent TN (1955) Absorption of sugars in vitro by the intestine of the golden hamster. J Biol Chem 216: 851–866
Wolf HP, Leuthardt F (1953) Ueber die Glycerin-dehydrogenase der Leber. Hev Chim Acta 36: 1463–1467
Wood SD, Reid JT (1975) The influences of dietary fat metabolism and body fat composition in meat feeding and nibbling rats. Br J Nutr 34: 15–24
Woods HF, Eggleston LV, Krebs HA (1970) The cause of hepatic accumulation of fructose-1-phosphate on fructose loading. Biochem J 119: 501–510
Zakim D, Herman RH, Gordon WC Jr (1969) The conversion of glucose and fructose to fatty acids in human liver. Biochem Med 2: 427–437

Chapter 5

# Biochemical Aspects

## 5.1 Sucrose

### 5.1.1 Effects on Plasma Constituents

#### 5.1.1.1 Plasma Glucose

It has been reported that the plasma glucose reponses to the oral administration of various carbohydrates are less with equal amounts of sucrose compared with glucose (Ferlito et al. 1978) in the fasted and the diabetic rat. In fasting man, the glucose response to sucrose was similar to that to glucose and did not appear to be dose related (Macdonald et al. 1978). Compared with fructose and sorbitol, the peak increment in plasma glucose was greatest after sucrose meals, in both normal and diabetic subjects (Akgun and Ertel 1980). In normal subjects and in diabetics given fructose- or sucrose-containing cakes and ice cream, the serum glucose was lower after fructose administration (Crapo et al. 1982). When sucrose is compared with glucose, so that both carbohydrates contain the same amount of glucose, the plasma glucose curves are similar (Crapo et al. 1976). It has been suggested that sucrose intake at USA levels, by carbohydrate-sensitive persons, can produce undesirable changes in parameters associated with glucose tolerance (Reiser et al. 1981b). The modifications that differences in amount and kind of digested carbohydrate can produce in the glucose response have been reviewed by Reaven (1979). The "glycaemic index' (the relative rise in plasma glucose compared with a standard glucose challenge) of sucrose is 59% (Jenkins et al. 1981).

On a long-term basis, the effect of giving 70% of energy as sucrose to female Wistar rats for 6–10 weeks is an increase in the serum glucose and insulin levels after an oral glucose load (Vrána et al. 1974). In rabbits, a 39.5% sucrose-as-energy diet for 12 weeks did not alter the glucose tolerance test (Heidecker et al. 1985), nor did a 34.5% w/w sucrose diet given to female mini-pigs, genetically selected, for 4–5 weeks (Phillips et al. 1982). In baboons, a 75% sucrose-as-energy diet resulted in improved glucose tolerance (Coltart and Crossley 1970). In man, an 80% sucrose-as-energy diet, consumed for up to 10 weeks, led to an improvement in the glucose tolerance test in normal subjects (Anderson et al. 1973). After 4 weeks on a 42% energy-intake sucrose diet, the fasting plasma glucose of young women was raised, as it was after a comparable glucose diet (Kelsay et al. 1974b).

In 35- to 45-year-old men and women sensitive to carbohydrates, 30% of energy as sucrose or wheat starch in a 43% carbohydrate diet for 6 weeks raised fasting serum glucose and insulin levels (Reiser et al. 1979). In normal men and women, 50–107 g of fructose or sucrose per day, for 7–14 days, did not affect fasting glucose or insulin levels in the serum (Bossetti et al. 1984).

### *5.1.1.2 Plasma Lipids*

The ingestion of 68% sucrose by rats for 30 days resulted in an increase in the fasting plasma triglyceride fraction, especially compared with other carbohydrates (Sheorain et al. 1980; Bruckdorfer et al. 1974b); this is associated with increased production of very low density lipoprotein (VLDL) particles (Kazumi et al. 1985), an effect that is antagonized by dietary carnitine (Richter 1985). There are several studies on rats that report an increase in serum triglyceride concentration after high levels of sucrose in the diet (Hostmark et al. 1982; Kelly et al. 1980; Shiff et al. 1971; Vijayagopalan and Kurup 1970). An overall increase in plasma lipids has been found in rats on high sucrose diets containing pectin (Urbanowicz et al. 1985), with the reverse occurring when starch replaced sucrose. A similar response is seen in young men on high sucrose diets, when the raised serum triglyceride level is partly offset by a high fibre content in the diet (Albrink and Ullrich 1986).

In the pig, plasma cholesterol, triglyceride, and phospholipid levels do not seem to be affected by dietary sucrose (Bruckdorfer and Yudkin 1975), whereas the baboon responds in a manner similar to the rat (Srinivasan et al. 1979; Macdonald and Keyser 1977).

In rats weaned on various dietary carbohydrates, sucrose is associated with higher plasma triglyceride levels than starch or glucose, and the raised plasma triglyceride concentration normally found in pregnancy is accentuated by sucrose (Bourne et al. 1975). In diabetic rats on a 60% sucrose diet, the plasma VLDL increased five-fold, though the high-density lipoprotein (HDL) also increased (Bar-on et al. 1976).

The plasma cholesterol level in rats (Anderson 1969) and monkeys (Srinivasan et al. 1979) is less influenced by dietary sucrose than is the triglyceride response.

After an acute load of carbohydrate in fasting man, no alterations in plasma cholesterol or triglyceride levels are seen (Ferlito et al. 1978), or may even fall (Macdonald et al. 1978), a fall that is not insulin dependent. The effects of dietary carbohydrate on plasma cholesterol levels are much less than those following dietary fat (McGandy et al. 1966).

In long-term feeding experiments in man, the replacement of sucrose by glucose polymers (partial starch hydrolysate) led to a fall in plasma cholesterol and phospholipid levels (Bagley et al. 1976) as did a replacement by leguminous seeds, wheat, and potatoes (Grande et al. 1965). In human beings on a high sucrose diet, there was a significant increase in fasting plasma triglyceride levels, not seen on a high starch diet (Macdonald 1964), and the primary responsibility for the elevated plasma triglyceride is an inadequate removal of triglyceride from the plasma (Bolzano et al. 1972). In patients with non-insulin-

dependent diabetes mellitus, a high sucrose diet caused an elevation of VLDL triglyceride and total plasma cholesterol, but no change in fasting plasma glucose or insulin (Coulston et al. 1985).

For review articles on this subject, see Macdonald (1986, 1973).

#### *5.1.1.3 Plasma Insulin*

When sucrose is ingested on a weight-for-weight basis with other carbohydrates, such as fructose, glucose, and sorbitol (at 0.25, 0.5, 0.75, and 1 g/kg body weight), the plasma insulin response is approximately half that after glucose ingestion (Macdonald et al. 1978); however, when sucrose is ingested in amounts that contain the same quantity of glucose (e.g., 50 g glucose versus 100 g sucrose), the plasma insulin response is 20% more with sucrose (Crapo et al. 1976), perhaps because some of the fructose in sucrose has been converted to glucose. When comparing the plasma insulin response to sucrose, fructose, and sorbitol in non-insulin-dependent diabetics, it was found that the peak increment was greatest after sucrose and lowest after sorbitol (Akgun and Ertel 1980). On the other hand, in normal persons and non-insulin diabetics, the fasting levels of plasma insulin were not affected by the amount of sucrose in the diet (Coulston et al. 1985). It has been reported that insulin demand and secretion rate are not influenced by the type of carbohydrate given, and that intestinal uptake of carbohydrate is the rate-limiting step rather than hydrolysis (Lutjens et al. 1975). In a brief report, it was stated that dietary sucrose given to rats results in a loss of sensitivity to insulin by adipose tissue (Bruckdorfer et al. 1974a). Reaven (1979) gives a review on how differences in amount and kind of ingested carbohydrate can modify the plasma glucose response.

#### *5.1.1.4 Plasma Lactate, Pyruvate, and Urate Levels*

After the acute ingestion of various amounts of sucrose, plasma lactate, pyruvate, and urate levels increase (Macdonald et al. 1978), with the lactate : pyruvate ratio in blood falling after a breakfast without sugar (Kelsay et al. 1974a). The salivary lactate level correlates well with plasma lactate concentration. In two children with glucose-6-phosphate deficiency, plasma lactate increased during sucrose or lactose feeding and decreased with maltose feeding (Fernandes 1974).

### **5.1.2 Effects on Lipogenesis**

#### *5.1.2.1 Hepatic Lipogenesis*

There are conflicting views concerning the effects of sucrose ingestion on hepatic lipogenesis. Some reports state that, in rats, it is increased by all sugars with an equal effect from sucrose and maltose (Michaelis et al. 1975; Naismith and Rana 1974), and that it is the disaccharide configuration ("disaccharide effect") that causes an increased induction of various liver enzymes (Michaelis et al. 1975). On the other hand, it has been found that sucrose induced greater liver lipid deposition in rats than did dextrin (Marshall and Womack 1954; Litwack et al. 1952), and that fructose was similar to sucrose in this respect (Harper et al. 1953). Compared with glucose, a diet of 73% sucrose, fed to female rats for 2–3 weeks, caused increased liver triglyceride and stimulated fatty acid synthesis (Holt et al. 1979). Studies on rabbits showed that sucrose gave rise to a fatty liver (Macdonald 1962) compared with other dietary carbohydrates. In rats on a 58% sucrose, low-fat diet, the lipogenic enzymes increased dramatically, more so in young animals than in adults (Boll et al. 1982); it has been suggested that this induction may be mediated by lactate in the portal blood rather than by fructose (Cha and Randall 1982). Dietary sucrose given to the rat seemed to increase glucose-6-phosphate dehydrogenase and malic enzyme activity in the liver to a greater extent than glucose or fructose (Michaelis and Szepesi 1973); this is not due to contaminants in the sucrose (Michaelis and Szepesi 1974). Using isolated hepatic mitochondria, 65% sucrose-as-energy diets increased incorporation of tritium into fatty acid (Bouillon and Berdanier 1983).

Hepatocytes from sucrose-fed rats doubled the synthesis and secretion of triglyceride compared with controls; they also showed a two- to three-fold increase in apolipoprotein synthesis (Boogaerts et al. 1984). A sucrose-rich diet also decreases total triglyceride lipase activity in the rat (Basilico et al. 1983), as well as increasing fatty acid synthetase activity (Bourne et al. 1975). It has been suggested that the main difference in the effect of fructose on blood triglyceride levels is not in the stimulation of lipogenesis, but in the greater conversion of fructose to lipids in the liver (Bar-on and Stein 1968). After a 48-hour fast, re-feeding of sucrose had a greater lipogenic effect than starch (Baltzell and Berdanier 1985). In the cholesterol-fed rat, sucrose in the diet was associated with lower plasma cholesterol levels than starch in the diet (Anderson 1969).

Sucrose fed with beef tallow or safflower oil to rats resulted in greater hepatic lipid infiltration than starch in the diet (Carroll and Williams 1982).

In a study on man, it was shown that sucrose enhances triglyceride synthesis in the liver (Cahlin et al. 1973).

The formation of glycogen in the liver from intravenous fructose is a direct uptake of fructose and not via the liver converting it to glucose (Bergstrom and Hultman 1967). Shafrir (1985) gives a review on the effects of sucrose and fructose on carbohydrate and lipid metabolism and its resulting consequences.

### *5.1.2.2 Non-Hepatic Lipogenesis*

After feeding weanling rats diets containing starch, sucrose, maltose, fructose, or glucose for 50 days, it was claimed that lipogenesis in adipose tissue was depressed by sucrose and fructose (Naismith and Rana 1974). A similar finding was reported in pregnant and lactating rats, when depressed fat synthesis was produced by high fructose, but not by sucrose, diets (Bourne et al. 1975). The brown adipose tissue was increased in rats overfed with sucrose, but only in the lean, and not obese, animals (Holt et al. 1983).

In 35 patients operated on for gallstones, it was shown that a diet containing 800 kcal of sucrose, given for 2 weeks to normolipaemics, increased the lipoprotein lipase activity in adipose tissue (Cahlin et al. 1973).

### 5.1.3 Effects on Cell Components

An overall effect on cell components has been observed in rat hepatocytes following the ingestion of large amounts of fructose as a 3% solution (Gottschall and McMillan 1985). It is believed that this was due to an intracellular build-up of fructose-1-P, thereby disturbing osmotic equilibrium. In the intestinal epithelium of the rat, sucrose feeding increases the RNA polymerase activity in both villus and crypt cells (Raul and Von der Decken 1985). Reducing sugars in cigarettes induces alkali labile sites in DNA (Morita et al. 1985), whereas adding sugars, including sucrose, reduced the mutagenicity of cigarette smoke condensate (Sato et al. 1979). Dietary sucrose did not alter the adenine nucleotide or phosphate content of liver and blood, but, in diabetic animals, blood phosphate levels rose (Kang 1979).

Two reviews relevant to this subject include one on the morphological changes of organs after sucrose or fructose feeding (Poulsom 1986) and another on the effect of sucrose and fructose on carbohydrate and lipid metabolism (Shafrir 1985).

### 5.1.4 Effects on Mineral Metabolism

The effects of sugars and amino acids on the increased excretion of calcium and magnesium in the urine is accepted, but there is no evidence to support the view that sucrose is different in this respect. However, in rats on a magnesium-deficient diet, sucrose, instead of starch, caused a more marked increase in plasma and liver lipids, lactate, and glycerophosphate (Rayssiguier et al. 1981). Recently, fructose or a fructose-containing carbohydrate, such as sucrose, markedly reduced the bioavailability of dietary copper (Reiser et al. 1985), and thereby increased the severity of copper deficiency in rats (Reiser et al. 1983).

### 5.1.5 Effects of Sex Difference

The sex difference in the metabolic response to sucrose or fructose is well documented (Macdonald 1976), and the insulin and glycogen responses of mice to diets high in sucrose and various fats vary depending on the sex of the animal (Bonnevie-Nielsen 1980).

In women using oral contraceptives and on a 43% sucrose-as-energy diet for 3 weeks, it was found that the glucose and insulin responses to a sucrose meal were significantly greater than on a high starch diet (Behall et al. 1980a). This would support the view that the hyperlipidemia in pregnancy is due to enhanced synthesis of endogenous lipid in the liver (Ferlito et al. 1979). However, those using oral contraceptives and on high sucrose or starch diets showed no difference in the serum trigyceride and cholesterol responses compared with those not using oral contraceptives (Behall et al. 1980b).

### 5.1.6 Effects of Exercise

The increased respiratory quotient after sucrose ingestion during prolonged aerobic exercise is the result of increased glycolysis resulting from an increased supply of pyruvate (Benade et al. 1973). The rate of resynthesis of muscle glycogen after exercise appeared to be the same for glucose, fructose, or sucrose (Costill 1985). The diurnal changes in liver and skeletal muscle did not affect the glycogen peaks, as these occurred about 8 hours after sucrose ingestion, regardless of the time of feeding (Suzuki et al. 1983).

When fructose or glucose are given intravenously in man, it was shown that, during rest, glucose and fructose produced similar moderate rises of muscle glycogen, but, after work, glycogen resynthesis was greater with glucose (Bergstrom and Hultman 1967).

With prolonged exercise, a dose of 75 g fructose caused less fluctuations in plasma glucose, insulin, and gastrointestinal polypeptide (GIP) than glucose or placebo, but was no more effective in sparing glycogen (Koivisto et al. 1985). The increase in plasma triglyceride seen with fructose in the diet can be prevented when the subject is on an exercise programme (Hostmark and Blom 1985).

## 5.2 Fructose

### 5.2.1 Effects on Plasma Constituents

The serum glucose response to oral fructose is less than that to glucose, as is serum insulin (Macdonald et al. 1978), glycogen, and growth hormone (Bohannon et al. 1980). When $^{14}$C-labelled fructose is given to an adult man, 31% of the infused label appeared as blood glucose (Atwell and Waterhouse 1971). However, the serum lactate increases after fructose ingestion (Bergstrom et al. 1968) and is thus contraindicated in acidosis. In a 2-year study on man, no consistent difference was found in serum glucose, insulin, triglyceride, urate, lactate, or pyrimate levels after the prolonged consumption of fructose or sucrose (Huttunen et al. 1976).

### 5.2.2 Effects on Lipogenesis

When fructose replaces glucose in the diet of rats, there is an increase in fatty acid synthetase activity in the liver, but a decrease in depot fat (Bourne et al. 1975); this is compatible with the reported enhanced rate of fatty acid synthesis as a cause of hypertriglyceridaemia in normal man (Zakim 1972). After 4 days on a 69% fructose-as-energy diet, the liver lipid of rats increased (Sugawa-Katayama and Morita 1975). The liver lipid response to dietary fructose in rats is related to the sex of the animal; in female animals, the response is similar to various dietary carbohydrates; in males, a higher content of liver lipid was reported after a starch diet, but the highest serum lipid concentration followed a fructose diet, both diets containing no fat (Takemoto 1975).

This apparent difference in responses of liver lipid content could be due to species and strain differences, as it has been shown that a diet high in fructose raised the serum triglyceride level in rats, but not in ducks or pigs, and fatty acid synthesis was increased in rat, but not chicken, liver (Waterman et al.

1975). In rabbits, a fructose diet for 12 weeks increased cholelithiasis (Heidecker et al. 1985). In diabetic rats on a 25% fructose or glucose diet, fructose inhibited hepatic cholesterol synthesis, with extra-hepatic synthesis not being affected (Feingold and Moser 1985).

In húman liver, the ketohexokinase activity is four times greater than glucokinase, implying that fructose is metabolized faster than glucose (Heinz et al. 1968); when $^{14}$C-labelled fructose or $^{14}$C-labelled glucose is given to fasting man, the serum triglyceride has significantly greater radioactivity after fructose ingestion in the ensuing few hours (Macdonald 1968). Impaired insulin binding and insulin sensitivity is seen in subjects who consumed 100 kcal of fructose per day for one week, but not when equivalent amounts of glucose were taken (Beck-Nielsen et al. 1980).

The nutritional significance of fructose is reviewed by Wang and Van Eys (1981), and fructose metabolism in adipose tissue is reviewed by Froesch (1972).

### 5.2.3 Effects on Adenine Nucleotides and Muscle Glycogen

When livers from normal and diabetic rats are perfused with a solution containing fructose, the level of cyclic adenosine monophosphate (AMP) increases, there is activation of glycogen phosphorylase, increase in protein kinase activity and fructose-1-P, and a decrease in adenosine triphosphate (ATP) (Johnson and Miller 1982). The effect of fructose on pyruvate kinase seems to be directly on the liver (Noguchi et al. 1985). Fructose given intravenously to rats also depletes liver ATP and inorganic phosphate (Maenpaa et al. 1968).

In healthy man, intravenous fructose (1.5 g/kg body weight per hour) increases serum uric acid levels (Heukenkamp and Zollner 1972) and decreases ATP and adenosine diphosphate (ADP) in the liver to half its initial value, with AMP unchanged. These changes are not seen with intravenous glucose (Bode et al. 1971). Fructose-1-P increases in the liver four-fold and glycerol phosphate increases three-fold after intravenous fructose and sorbitol (Bode et al. 1973). All these findings suggest that, in clinical medicine, intravenous fructose may have disadvantages compared with glucose.

After a 4-hour infusion of fructose in normal man, the increase in muscle glycogen was the same as after a glucose infusion, but, in glycogen-depleted muscle, glucose gave twice as much muscle glycogen as fructose. Liver glycogen was higher after fructose than glucose infusion. Fructose is therefore taken up by muscle, but is less effective than glucose in promoting glycogen deposition in glycogen-depleted muscle (Bergstrom et al. 1972).

### 5.2.4 Fructose and Lactic Acidosis

In man, in the post-operative period, the serum lactate concentration rises more after fructose than glucose infusions (Hessov 1974). In two children with metabolic acidosis, the serum pH fell with the bicarbonate level during fructose infusion (Anderson et al. 1969). After intravenous fructose, the raised lactic acid can be partly reduced by the simultaneous infusion of amino acids

(Bergstrom et al. 1972). The ingestion of fructose leads to significant increases in serum lactate levels in healthy man (Macdonald et al. 1978).

In animals, the hyperlipidaemia and lactic acidosis seen after fructose feedings are not seen in Zucker rats, though it did produce hyperinsulinaemia in non-diabetic obese animals (Koh et al. 1985). It has also been reported that, in the intestinal mucosa of the guinea-pig, the conversion of fructose to glucose occurs by the same mechanism as that in the liver involving fructose-1-P and triose-P as intermediates (Ginsburg and Hers 1960); hence, fructose does not elevate serum triglycerides in the guinea-pig (Bar-on and Stein 1968).

### 5.2.5 Effects on the Kidney

All the enzymes involved in fructose metabolism, except for glycerate kinase, have been found in the cortex and medulla of human kidney, though low in the medulla compared with the cortex (Heinz et al. 1975).

Rats on a diet containing up to 68% fructose for 6 months showed a significant increase in kidney weight, and there was histological evidence of renal damage after fructose and sucrose diets, but not after a starch diet (Boot-Handford and Heath 1981).

## 5.3 References


Akgun S, Ertel NH (1980) A comparison of carbohydrate metabolism after sucrose, sorbitol, and fructose meals in normal and diabetic subjects. Diabetes Care 3: 582–585

Albrink MJ, Ullrich IH (1986) Interaction of dietary sucrose and fibre on serum lipids in healthy young men fed high carbohydrate diets. Ansen J Clin Nutr 43: 419–428

Anderson G, Brohult J, Sterner G (1969) Case report: increasing metabolic acidosis following fructose infusion in two children. Acta Paediatr Scand 58: 301–304

Anderson JW, Herman RH, Zakim D (1973) Effect of high glucose and high sucrose diets on glucose tolerance of normal men. Am J Clin Nutr 26: 600–607

Anderson TA (1969) Effect of carbohydrate source on serum and hepatic cholesterol levels in the cholesterol fed rat. Proc Soc Exp Biol Med 130: 884–887

Atwell ME, Waterhouse C (1971) Glucose production from fructose. Diabetes 20: 193–199

Bagley R, Ford M, Green LF (1976) Changes in plasma lipids after substituting glucose-syrup solids for table sucrose. Proc Nutr Soc 35: 68A–69A

Baltzell JK, Berdanier CD (1985) Effect of the interaction of dietary carbohydrate and fat on the response of rats to starvation-refeeding. J Nutr 115: 104–110

Bar-on H, Roheim PS, Eder HA (1976) Hyperlipoproteinemia in Streptozoticin-treated rats. Diabetes 25: 509–515

Bar-on H, Stein Y (1968) Effect of glucose and fructose administration on lipid metabolism in the rat. J Nutr 94: 95–105

Basilico MZ, Gutman R, Yommi MR, Francone O, Lombardo YB (1983) Post-heparin plasma hepatic triglyceride lipase and monoglyceride hydrolase activities in hyperlipemia induced by a sucrose rich diet. Biomed Pharmacol 37: 36–41

Beck-Nielsen H, Pedersen O, Lindskov HO (1980) Impaired cellular insulin binding and insulin sensitivity induced by high-fructose feeding in normal subjects. Am J Clin Nutr 33: 273–278

Behall KM, Moser PB, Kelsay JL, Prather ES (1980a) The effect of kind of carbohydrate in the diet and use of oral contraceptives on metabolism of young women. II. Serum glucose, insulin and glycogen. Am J Clin Nutr 33: 1041–1048

Behall KM, Moser PB, Kelsay JL, Prather ES (1980b) The effect of kind of carbohydrate in the diet and use of oral contraceptives on metabolism of young women. III. Serum lipid levels. Am J Clin Nutr 33: 825–831

Benade AJS, Jansen CR, Rogers GG, Wyndham CH, Strydom NB (1973) The significance of an increased RQ after sucrose ingestion during prolonged aerobic exercise. Pflugers Arch 342: 199–206

Bergstrom J, Hultman E (1967) Synthesis of muscle glycogen in man after glucose and fructose infusion. Acta Med Scand 182: 93–107

Bergstrom J, Hultman E, Roch-Norlund AE (1968) Lactic acid accumulation in connection with fructose infusion. Acta Med Scand 184: 359–364

Bergstrom J, Furst P, Gallyas F et al. (1972) Aspects of fructose metabolism in normal man. Acta Med Scand (Suppl) 542: 57–64

Bode CH, Schumacher H, Goebell H, Zelder O, Pelzel H (1971) Fructose induced depletion of liver adenine nucleotides in man. Horm Metab Res 3: 289–290

Bode CH, Zelder O, Rumpelt HJ, Wittkamp U (1973) Depletion of liver adenosine phosphates and metabolic effects of intravenous infusion of fructose or sorbitol in man and in the rat. Eur J Clin Invest 3: 436–441

Bohannon NV, Karam JH, Forsham PH (1980) Endocrine responses to sugar ingestion in man: advantages of fructose over sucrose and glucose. J Am Diet Assoc 76: 555–560

Boll M, Bruckner E, Berndt J (1982) Age-related differences in dietary regulation of lipogenic enzymes in rat liver. Hoppe-Seyler's Z Physiol Chem 363: 103–106

Bolzano K, Sanhofer F, Sailer S, Braunsteiner H (1972) The effect of oral administration of sucrose on the turnover rate of plasma free fatty acids on the esterification rate of plasma free fatty acids to plasma triglycerides in normal subjects, patients with primary endogenous hypertriglyceridemia and patients with well controlled diabetes mellitus. Horm Metab Res 4: 439–446

Bonnevie-Nielsen V (1980) Experimental diets affect pancreatic insulin and glycogen differently in male and female mice. Metabolism 29: 386–391

Boogaerts JR, Malone-McNeal M, Archambault-Scheynayder J, Davis RA (1984) Dietary carbohydrate induces lipogenesis and very low density protein synthesis. Am J Physiol 246: E77–E83

Boot-Hanford R, Health H (1981) The effect of dietary fructose and diabetes on the rat kidney. Br J Exp Pathol 62: 398–406

Bossetti BM, Kocher LM, Moranz JF, Falko JM (1984) The effects of physiologic amounts of simple sugars on lipoprotein, glucose and insulin levels in normal subjects. Diabetes Care 7: 309–312

Bouillon DJ, Berdanier CD (1983) Effect of maternal carbohydrate intake on mitochondrial activity and on lipogenesis by the young and mature progeny. J Nutr 113: 2205–2216

Bourne AR, Richardson DP, Bruckdorfer KR, Yukdin J (1975) Some effects of different dietary carbohydrates on pregnancy and lactation in rats. Nutr Metab 19: 73–90

Bruckdorfer KR, Yudkin J (1975) A comparison of dietary starch and dietary sucrose in the pig. Nutr Metab 19: 225–232

Bruckdorfer KR, Kang SS, Yudkin J (1974a) Insulin sensitivity of adipose tissue of rats fed with various carbohydrates. Proc Nutr Soc 33: 4A–5A

Bruckdorfer KR, Kang SS, Khan IH, Bourne AR, Yudkin J (1974b) Diurnal changes in the concentrations of plasma lipids, sugars, insulin and corticosterone in rat fed diets containing various carbohydrates. Horm Metab Res 6: 99–106

Cahlin E, Jonsson J, Persson B et al. (1973) Sucrose feeding in man. Effects on substrate incorporation into hepatic triglycerides and phosphoglycerides in vitro and on removal of intravenous fat in patients with hyperlipoproteinaemia. Scand J Clin Lab Invest 32: 21–33

Carroll C, Williams L (1982) Choline deficiency in rats as influenced by dietary energy sources. Nutr Rep Intern 25: 773–782

Cha CJ, Randall HT (1982) Effects of substitution of glucose-oligosaccharides by sucrose in a defined formula diet or interstitial disaccharides, hepatic lipogenic enzymes and carbohydrate metabolism in young rats. Metabolism 31: 57–66

Coltart TM, Crossley JN (1970) Influence of dietary sucrose on glucose and fructose tolerance and triglyceride synthesis in the baboon. Clin Sci 38: 427–437

Costill DL (1985) Carbohydrate nutrition before, during and after exercise. Fed Proc 44: 364–368

Coulston AM, Hollenbeck CB, Donner C, Williams R, Chiou YM, Reaven GM (1985) Metabolic effects of added dietary sucrose in individuals with non insulin dependent diabetes mellitus. Metabolism 34: 962–966

Crapo P, Reaven G, Olefsky J (1976) Plasma glucose and insulin responses to orally administered single and complex carbohydrates. Diabetes 25: 741–747

Crapo P, Scarlett JA, Kolterman OG (1982) Comparison of the metabolic responses to fructose and sucrose sweetened foods. Am J Clin Nutr 36: 256–261

Feingold KR, Moser AH (1985) Effect of glucose or fructose feeding on cholesterol synthesis in diabetic animals. Am J Physiol 249. Gastrointest Liver Physiol 12: 6634–6641

Ferlito S, Maugeri D, Lo Furno R, Calafato M (1978) Effetti di un carico orale di glucosio, di fruttosio e di saccarosio sui livelli glicemici, insulinemici e lipidemici in soggetti normali. Arch Sci Med 135: 447–460

Ferlito S, Nolfo G, Caruso L, Calafato M, Lo Furno R (1979) Comportamento della glicemia, della insulinemia e del quadro lipidemico dopo carico orale di glucosio, saccarosio e fruttosio in donne gravide normali. Min Gin 31: 805–814

Fernandes J (1974) The effect of disaccharides on the hyperlactacidaemia of glucose-6-phosphate deficient children. Acta Paediatr Scand 63: 695–698

Froesch ER (1972) Fructose metabolism in adipose tissue. Acta Med Scand (Suppl) 542: 37–42

Ginsburg V, Hers HG (1960) On the conversion of fructose to glucose by guinea pig intestine. Biochim Biophys Acta 38: 427–434

Gottschall EG, McMillan DB (1985) Structural changes in rat hepatocytes following ingestion of sugar solutions. Acta Anat (Basel) 123: 178–188

Grande F, Anderson JT, Keys A (1965) Effect of carbohydrates of leguminous seeds, wheat, and potatoes on serum cholesterol concentration in man. J Nutr 86: 313–317

Harper AE, Monson WJ, Arata DA, Benton DA, Elvehjem CA (1953) Influence of various carbohydrates on utilization of low protein rations by white rat, comparison of several proteins and carbohydrates. J Nutr 51: 523–537

Heidecker HA, Borgman RF, Maurice DV (1985) Influence of dietary carbohydrates upon the metabolism of lipids and minerals in rabbits. Nutr Res 5: 983–992

Heinz F, Lamprecht W, Kirsch J (1968) Enzymes of fructose metabolism in human liver. J Clin Invest 47: 1826–1832

Heinz F, Schlegel F, Krause PH (1975) Enzymes of fructose metabolism in human small intestine mucosa. Enzyme 19: 93–101

Hessov I (1974) Effects of fructose and glucose infusions on blood acid-base equilibrium in the post-operative period. Acta Chir Scand 140: 347–351

Heukenkamp PU, Zollner N (1972) The comparative metabolism of carbohydrates administered intravenously. Nutr Metab 14 (Suppl): 58–73

Holt PR, Dominguez AA, Kwartier J (1979) Effect of sucrose feeding upon intestinal and hepatic lipid synthesis. Am J Clin Nutr 32: 1792–1798

Holt S, York DA, Fitzsimons TR (1983) The effects of corticosterone cold exposure and overfeeding with sucrose on brown adipose tissue of obese Zucker rats (fa/fa). Biochem J 214: 215–223

Hostmark AT, Blom PCS (1985) Previous exercise nullifies the plasma triacylglycerol response to repeated fructose ingestion in young men. Acta Physiol Scand 125: 553–554

Hostmark AT, Spydevold O, Lystad E, Eilertsen E (1982) Plasma lipoproteins in rats fed starch, sucrose, glucose or fructose. Nutr Rep Int 25: 161–167

Huttunen JK, Makinen KK, Scheinin A (1976) Turku sugar studies. XI. Effects of sucrose, fructose and xylitol diets on glucose lipid and urate metabolism. Acta Odontol Scand 33(Suppl 70): 239–245 reprinted 34: 345–351

Jenkins DJA, Wolever TMS, Taylor RH et al. (1981) Glycemic index of foods: a physiological basis for carbohydrate exchange. Am J Clin Nutr 34: 362–366

Johnson PR, Miller TB (1982) Adverse effects of fructose in perfused livers of diabetic rats. Metabolism 31: 12–125

Kang SS (1979) The effects of dietary sucrose and streptozotocin diabetes on blood and liver constituents. Nutr Metab 23: 327–334

Kazumi T, Vranic M, Steiner G (1985) Changes in very low density lipoprotein particle size and production in response to sucrose feeding and hyperinsulinaemia. Endocrinology 117: 1145–1150

Kelly TJ, Holt PR, Ai-Lien WU (1980) Effect of sucrose on intestinal very low density lipoprotein production. Am J Clin Nutr 33: 1033–1040

Kelsay JL, Behall KM, Clark WM (1974a) Glucose, fructose, lactate and pyruvate in blood and lactate and pyruvate in parotid saliva in response to sugars with and without other foods. Am J Clin Nutr 27: 819–825

Kelsay JL, Behall KM, Holden JM, Prather ES (1974b) Diets high in glucose or sucrose in young women. Am J Clin Nutr 27: 926–936

Koh ET, Mueller J, Osilesi O, Knehans A, Reiser R (1985) Effects of fructose feeding on lipid parameters on obese and lean, diabetic and non diabetic Zucker rats. J Nutr 115: 1274–1284

Koivisto VA, Harkonen M, Karonen SL et al. (1985) Glycogen depletion during prolonged

exercise. Influence of glucose, fructose or placebo. J Appl Physiol 58: 731–737
Litwack G, Hankes LV, Elvehjem CA (1952) Effect of factors other than choline on liver fat deposition. Proc Soc Exp Biol Med 1: 441–445
Lutjens A, Verleur H, Plooij M (1975) Glucose and insulin levels on loading with different carbohydrates. Clin Chim Acta 62: 239–243
Macdonald I (1962) Some influence of dietary carbohydrate on liver and depot lipids. J Physiol (Lond) 162: 334–344
Macdonald I (1964) The influence of dietary carbohydrates on the lipid pattern in serum and in adipose tissue. Clin Sci 27: 23–30
Macdonald I (1968) Ingested glucose and fructose in serum lipids in healthy men and after myocardial infarction. Am J Clin Nutr 21: 1366–1373
Macdonald I (1973) Effects of dietary carbohydrates on serum lipids. Prog Biochem Pharmacol 8: 216–241
Macdonald I (1976) Sex differences in the metabolic response to dietary carbohydrates. In: Berdanier CD (ed) Carbohydrate metabolism. Hemisphere, Washington, pp 211–222
Macdonald I (1986) Dietary carbohydrate and energy balance. Prog Biochem Pharmacol 21: 181–191
Macdonald I, Keyser A (1977) Some effects in baboons of chronic ingestion of glycerol with sucrose or glucose. Am J Clin Nutr 30: 1661–1669
Macdonald I, Keyser A, Pacy D (1978) Some effects in man of varying the load of glucose, sucrose, fructose or sorbitol on various metabolites in blood. Am J Clin Nutr 31: 1305–1311
Maenpaa PG, Raivio KO, Keromaki MP (1968) Liver adenine nucleotides: fructose induced depletion and its effects on protein synthesis. Science 161: 1253–1254
Marshall MW, Womack MJ (1954) Influence of carbohydrates, nitrogen source and prior state of nutrition balance and liver composition in adult rat. J Nutr 52: 51–64
McGandy RB, Hegsted DM, Mars ML, Stare FJ (1966) Dietary carbohydrate and serum cholesterol levels in man. Am J Clin Nutr 18: 237–242
Michaelis OE, Szepesi B (1973) Effects of various sugars on hepatic glucose-6-phosphate dehydrogenase, malic enzyme and total liver lipid of the rat. J Nutr 103: 697–705
Michaelis OE, Szepesi B (1974) The mechanism of a specific metabolic effect of sucrose in the rat. J Nutr 104: 1597–1609
Michaelis OE, Nace CS, Szepesi B (1975) Demonstration of a specific metabolic effect of dietary disaccharides in the rat. J Nutr 105: 1186–1191
Morita J, Veda K, Nanjo S, Komano T (1985) Sequence specific damage of DNA induced by reducing sugars. Nucleic Acids Res 13: 449–458
Naismith DJ, Rana IA (1974) Sucrose and hyperlipidaemia. Nutr Metab 16: 285–294
Noguchi T, Inove H, Tanaka T (1985) Transcriptional and post-transcriptional regulation of L-type pyruvate kinase in diabetic rat liver by insulin and dietary fructose. J Biol Chem 26: 14393–14397
Phillips RW, Westmoreland N, Panepinto L, Case GL (1982) Dieteary effects on metabolism of Yucatan miniature swine selected for low and high glucose utilization. J Nutr 112: 104–111
Poulsom R (1986) Morphological changes of organs after sucrose or fructose feeding. Prog Biochem Pharmacol 21: 104–134
Raul F, Von der Decken A (1985) Changes in chromatin structure and transcription activity by starvation and dietary sucrose in mature and immature intestinal epithelial cells of the rat. Cell Mol Biol 31: 299–304
Rayssiguier Y, Gueux E, Weiser D (1981) Effect of magnesium deficiency on lipid metabolism in rats fed a high carbohydrate diet. J Nutr 111: 1876-1883
Reaven GM (1979) Effects of differences in amount and kind of dietary carbohydrate on plasma glucose and insulin responses in man. Am J. Clin Nutr 32: 2568-2578
Reiser S, Handler HB, Gardner LB, Hallfrisch JG, Michaelis OE, Prather ES (1979) Isocaloric exchange of dietary starch and sucrose in humans. II. Effects fasting blood insulin, glucose and glucagon and on insulin and glucose response to a sucrose load. Am J Clin Nutr 32: 2206–2216
Reiser S, Bickard MC, Hallfrisch J, Michaelis OE, Prather ES (1981a) Blood lipids and their distribution in lipoproteins in hyperinsulinemic subjects fed three different levels of sucrose. J Nutr 11: 1045-1057
Reiser S, Bohn E, Hallfrisch J, Michaelis OE, Keeney M, Prather ES (1981b) Serum insulin and glucose in hyperinsulinaemic subjects fed three different levels of sucrose. Am J Clin Nutr 34: 2348–2358

Reiser S, Ferretti FJ, Fields M, Smith JC (1983) Role of dietary fructose in the enhancement of mortality and biochemical changes associated with copper deficiency in rats. Am J Clin Nutr 38: 214-222

Reiser S, Smith JC, Mertz W et al. (1985) Indices of copper status in humans consuming a typical American diet containing either fructose or starch. Am J Clin Nutr 42: 242-251

Richter V (1985) Sucrose induced hyperlipoproteinaemia and carnitine. Nahrung 29: 626–627

Sato S, Ohka T, Nagao M, Tsuji K, Kosuee T (1979) Reduction in mutagenicity of cigarette smoke condensate by added sugars. Mutat Res 60: 155–161

Shafrir E (1985) Effect of sucrose and fructose on carbohydrate and lipid metabolism and the resulting consequences. In: Beitner R (ed) Regulation of carbohydrate metabolism. CRC Press, Boca Raton, Florida, vol 2, pp 95–140

Sheorain VS, Mattock MB, Subrahmanyam D (1980) Mechanism of carbohydrate induced hypertriglyceridemia: plasma lipid metabolism in mice. Metabolism 29: 924–929

Shiff TS, Roheim PS, Eder HA (1971) Effects of high sucrose diets and 4-aminopyrazolopyrimidine on serum lipids and lipoproteins in the rat. J Lipid Res 12: 596-603

Srinivasan SR, Clevidence BA, Pargaonkar PS, Radhakrishenamurthy B, Berenson GS (1979) Varied effects of dietary sucrose and cholesterol on serum lipids, lipoproteins and apolipoproteins in rhesus monkeys. Atherosclerosis 33: 301–314

Sugawa-Katayama Y, Morita N (1975) Effects of a high fructose diet on lipogenic enzyme activities in some organs of rats fed *ad libitum*. J Nutr 105: 1377–1383

Suzuki M, Ide K, Saituh S (1983) Diurnal changes in liver and skeletal muscle of rats in relation to the feeding time of sucrose. J Nutr Sci Vitaminol (Tokyo) 29: 545–552

Takemoto T-I (1975) Sex difference in the response to different kinds of dietary carbohydrates in rats. Tohoku J Exp Med 115: 213–222

Urbanowicz M, Chalcarz W, Jelone B, Czanocinska J (1985) Plasma lipid and protein responses in rats fed diets containing pectin and varying levels of sucrose and starch. Nutr Pfts Lutem 32: 649–658

Vijayagopalan P, Kurup PA (1970) Effect of dietary starches on the serum, aorta, and hepatic lipid levels in cholesterol-fed rats. Atherosclerosis 11: 252–264

Vrána A, Slabochova Z, Fabry P, Kazdová L (1974) Influence of diet with a high starch or sucrose content on glucose tolerance, serum insulin level and insulin sensitivity in rats. Physiol Bohemoslov 23: 305–310

Wang YM, Van Eys J (1981) Nutritional significance of fructose and sugar alcohols. Ann Rev Nutr 1: 437–475

Waterman RA, Romsos DR, Tsai AC, Miller ER, Leveille GA 1975) Effects of dietary carbohydrate source on growth, plasma metabolites and lipogenesis in rats, pigs and chicks. Proc Soc Exp Biol Med 150: 220–225

Zakim D (1972) The effect of fructose on hepatic synthesis of fatty acids. Acta Med Scand (Suppl) 542: 205–214

Chapter 6

# Health Aspects

## 6.1 Sucrose, Glucose, Fructose

### 6.1.1 Diabetes

In some studies, an impaired glucose tolerance test was reported after rats were given high levels of dietary sucrose (Kanarek and Orthen-Gambill 1982; Hallfrisch et al. 1979a; Laube et al. 1978; Romsos and Levielle 1974). Maltose also had the same effect as sucrose compared with a starch-rich diet (Romsos and Levielle 1974). However, other studies have reported an unchanged or slightly improved glucose tolerance (Vallerand et al. 1986; Portha et al. 1982; Lin and Anderson 1977; Eaton and Kipnis 1969). Some in vitro studies suggest that basal and/or insulin-stimulated glucose uptake and metabolism are increased in isolated adipocytes of rats fed a high sucrose diet (Oka et al. 1980; Olefsky and Saekow 1978).

In a recent study, a chronic (4 weeks) high sucrose diet (63% of energy) fed to normal rats did not lead to a deterioration of glucose tolerance (Kergoat et al. 1987); furthermore, this study provided direct evidence that insulin-mediated glucose uptake by peripheral tissues is enhanced, while the liver is resistant to insulin action due to a diminished ability of insulin to suppress the hepatic glucose output.

In animals with diabetes or induced diabetes, it has been found that a diet with 60% of the energy as sucrose, glucose, or fructose leads to a rapid induction of diabetes in diabetic-prone mice (Leiter et al. 1983), and sucrose, compared with starch, augments the symptoms of streptozotocin-induced diabetes (Hallfrisch et al. 1979b). The impairment of carbohydrate and lipid metabolism by a high glucose diet was much smaller than that induced by fructose or sucrose (Cohen 1978).

It should be noted that Cohen et al. (1972) were able to induce diabetes in a highly inbred group of rats fed sucrose, but the animals were especially selected for this purpose (Nuttall and Gannon 1981). In addition, sucrose represented a high proportion of the total energy intake (70%). Furthermore, the synthetic sucrose-rich diet used was copper deficient, and it has been shown that the absence of copper reduced the plasma insulin response to an oral glucose load and glucose incorporation into adipose tissue lipids and diaphragm glycogen (Michaelis et al. 1975; Michaelis and Szepesi 1974).

It has been suggested that fructose has an inhibitory effect on insulin secretion and that this is mediated by signals from the gastrointestinal tract (Hara and Saito 1981). It was reported many years ago that the impaired ability of fructose-fed rats to utilize glucose may be due to reduced glucokinase activity resulting from enzyme induction in the diet (Hill et al. 1954). It has been suggested that even low sucrose levels in a high fat diet can produce higher insulin levels than starch in the diet – a "relative insulin resistance' (Hallfrisch et al. 1981).

Specific organ lesions have been reported in diabetic animals, with retinal changes predominating. Compared with dietary starch, sucrose did not alter the concentration of glucose, alcohol, sorbitol, or fructose in the normal rat retina, but they were markedly elevated in the diabetic state (Heath and Hamlett 1976). In sucrose-induced retinopathy, it was found that fructose was a dietary microangiopathic agent, and the lesions that developed were similar to those in diabetic rats (Boot-Hanford and Heath 1980).

Since retinal and glomerular changes occurred at normal blood-glucose levels in sucrose-, but not in starch-fed, control rats, fructose-initiated intracellular protein–hexose interaction may be the speculative mechanism of non-hyperglycaemic microangiopathy (Shafrir, 1985).

A significant increase in retinal lactate was found in rats given starch or sucrose diets, and the lactate : pyruvate ratio was increased in the sucrose-fed animals. In diabetic animals, the lactate was not further elevated after a sucrose diet (Heath et al. 1975). The conduction velocity of motor nerves was significantly impaired by a fructose (via sorbitol accumulation) diet given to diabetic rats (Hotta et al. 1985).

In rats bred for impaired glucose tolerance, a high sucrose diet gave rise to diabetes and renal arteriosclerosis, whereas starch-fed siblings developed none of these changes (Cohen 1972). Similarly, in streptozotocin-induced diabetic rats, 68% sucrose in the diet caused a greater deterioration in the diabetic state, compared with controls, and the kidney lesions were more severe in the animals on sucrose (Gonnermann et al. 1982). Other pathological changes seen in sucrose-fed diabetic rats include testicular degeneration (Rosenmann et al. 1974) and an increase in haemoglobin A (HbA) (Powell et al. 1982).

In patients with maturity onset diabetes, Type II (NIDDM), sucrose produces a steeper blood glucose increase and greater insulin requirement than fructose and sorbitol (Kaspar and Irsigler 1980), but significantly lower (40%–50%) than the response observed with glucose or with some starch foods (Jenkins et al. 1981; Crapo et al. 1976). A carbohydrate enriched meal (adding 90 g of CHO including 50 g of sucrose to the regular evening meal), taken at 1900 hours, causes a slight but significant rise in fasting plasma glucose level in Type I (IDDM) diabetes, but not in NIDDM (Belkhadir et al. 1985). In both normal and hyperinsulinemic men, up to 15% of dietary energy intake as fructose increases the glucose and insulin response after a sucrose load (Hallfrisch et al. 1983). A more moderate amount of fructose (30 g), given daily over 6 weeks to eight NIDDM subjects failed to show any deleterious effects on body weight, HbA, fasting plasma glucose and insulin, plasma total and HDL, cholesterol, plasma phospholipids, and uric acid levels (Grigoresco et al. 1985).

After sucrose hyperalimentation for 7 days, both insulin and c-peptide levels increased, with deterioration in glucose tolerance, but with no evidence of any decrease in hepatic insulin extraction (Olsen et al. 1983).

On the other hand, a survey of 10 000 men provided no evidence that sugar energy is related to the incidence of diabetes (Kahn et al. 1971). Many years ago, Himsworth (1935) compiled a considerable amount of evidence in support of a negative association between carbohydrate ingestion and risks of diabetes. Similar conclusions have been drawn by epidemiologists in the field of diabetes (Keen et al. 1979; West 1978).

Studies carried out more than 10 years ago indicated that fructose and sorbitol have no advantage over sucrose in respect to effects on blood glucose, in patients with well-regulated adult onset diabetes (Lenner 1976). Later studies on healthy volunteers indicated that plasma glucose is more stable, and lower levels of insulin are required to maintain euglycaemia when sucrose, rather than corn syrup, is the source of dietary sugar (Thompson et al. 1978). More recently, other studies have supported the view that the inclusion of moderate amounts of sugar in the diets of patients with diabetes may carry little risk of creating adverse metabolic effects (Steel et al. 1983). Furthermore, in single-meal studies, partial substitution of sucrose for other dietary carbohydrates did not significantly exaggerate the glycemic response in patients with either insulin-dependent or non-insulin-dependent diabetes (Slama et al. 1984; Bantle et al. 1983). In pump-treated diabetics, inclusion of a moderate amount of sucrose in the diet did not impair metabolic control (Chantelau et al. 1985). In NIDDM subjects, sucrose or honey at breakfast have no greater acute hyperglycaemic effect than bread does (Bornet et al. 1985), and no deleterious effects of simple versus complex sugars have been noted in other studies (Hollenbeck et al. 1985). Furthermore, a more recent study concluded that ingestion of moderate amounts of sucrose by insulin and non-insulin-dependent diabetics did not produce any detectable deterioration of metabolic control (Petersen et al. 1986).

In studies on patients with non-insulin diabetes, given diets containing either 1% or 16% of calories as sucrose for 15 days, Coulston et al. (1985) found that higher sucrose intake led to higher glucose and triglyceride concentrations throughout the day, as well as higher levels of fasting total and VLDL triglyceride and cholesterol concentrations. A study from the same group, in which patients with NIDDM were given diets containing proportionate increases in sucrose and total carbohydrate content, demonstrated similar effects of the higher carbohydrate intake on post-prandial glucose, insulin, and triglyceride concentrations, glucosuria, as well as higher levels of total and VLDL triglyceride and cholesterol, and lower levels of HDL (Coulston and Swislocki 1985).

Other reviews related to dietary carbohydrate and diabetes have been published by Taskinen (1986), Nuttall (1983), Walker (1977), and Bierman and Nelson (1975).

### 6.1.2 Obesity

An important question is whether sucrose plays a unique role in increasing body weight, rather than acting as a source of dietary energy like any other food. Rats given a 32% sugar solution showed increased voluntary activity, but only the females became obese (Hirsch et al. 1982). When granulated sucrose was added to a standard diet, rats gained significantly more weight, expressed per kilocalorie consumed, than controls. Rats given sucrose or glucose in the drinking-water had more brown adipose tissue than those animals receiving granulated sucrose or fructose in their drinking water (Kanarek and Orthen-Gambill 1982). In another study, normal rats, aged 6 weeks, given a high sucrose diet (62%) for 3 weeks, did not show increased body weight gain or mean daily food intake, compared with a chow-fed rat group (Kergoat et al. 1987).

In 1982, a hypothesis was formulated that sucrose effects first hyperglycaemia and then reactive hypoglycaemia, and that this elicits hunger and hence obesity (Geiselman and Novin 1982). This gave rise to considerable comment and a reply from the original author to the comments in 1985 (Geiselman 1985). Another possible explanation for associating carbohydrate foods with obesity was the suggestion that, due to a diminished level of serotonin (5HT) in obese human beings (both diabetic and non-diabetic), there was an increased craving for carbohydrate-rich foods (Ashley et al. 1985).

In 28 volunteers on a diet containing normal amounts of refined carbohydrate with very little complex carbohydrate, there was increased energy intake, decreased fibre intake, and decreased intake of nearly all vitamins and minerals (Heaton et al. 1983). The increase in the metabolic rate after sucrose ingestion was greater than after glucose, maltose, or lactose ingestion, and this difference was much less marked in obese subjects (Sharief and Macdonald 1982). There are reviews on obesity and diabetes (Renold et al. 1978) and dietary carbohydrates, mainly disaccharides and monosaccharides, and energy (Macdonald 1986; Shafrir 1985).

### 6.1.3 Oral and Dental Effects

There is considerable literature on this subject, both in experimental animals and man, as well as in vitro studies. The simple experiment of subjecting animals (usually rats) to diets with varying proportions and types of dietary carbohydrate has revealed that, at concentrations up to 10% sucrose or glucose in the diet, both carbohydrates are equally cariogenic (Huxley 1971); at 33%, sucrose produced more caries than glucose or fructose, which were equally cariogenic (Grenby and Hutchinson 1969). In another study, 24% of energy as sucrose was the upper threshold level for caries (Green and Hartles 1970). A marked increase in fissure and smooth-surface caries has been reported with diets containing 20%–40% sucrose, no further caries occurring at concentrations higher than 40% sucrose in the diet. This study noted an increased eating frequency with increasing sucrose concentrations (Hefti and Schmid 1979). When 56% of the dietary energy is supplied either by sucrose or a mixture of glucose and fructose (invert sugar), given in equal amounts, the caries score was 1.5 times higher in the sucrose group. No proof of adaptation of oral flora to metabolize invert sugar in a more cariogenic way was found (Birkhed et al. 1981). There is a tendency for fewer fissure caries lesions to be found in rats given maltose compared with those given sucrose, though the difference is not sigtnificant (Skinner et al. 1982a). Nevertheless, in rats, it was found that starch is more cariogenic than bread, and sucrose even more so, but no difference in cariogenicity was found between raw and cooked starch (Grenby 1967). Confirmation that cooked wheat starch is cariogenic, but less so than sucrose, is reported in a study that also showed that 36 meals alternating cooked wheat starch and powdered sucrose were as cariogenic as the same number of meals of sucrose. This study also reported that equal parts of wheat starch and sucrose in the animals' diet were as cariogenic as sucrose alone (Firestone et al. 1982). Studies on the cariogenicity of dry breakfast cereals in rats showed that pre-sweetened cereals were more conducive to smooth-surface caries than non-pre-sweetened cereals, but no such difference was seen in occlusal caries (McDonald and Stookey 1977).

Using extracted teeth, it was found that the cariogenic potential of various breakfast cereals did not depend on the sugar content, the amount of plaque formed, or the amount of sugar retained by the teeth, but on the buffering capacity of the cereal (Katz et al. 1975).

In a study of human foods given to rats, the results did not support the view that sucrose present naturally in foods is different in cariogenicity from the same sugar when it is refined and added to foods (Stephan 1966). A report suggests that repair of the damage produced by fermentable sugars can occur if artificial sweeteners are consumed between meals, perhaps by increasing salivary flow (Leach and Green 1981). The basal composition of the food can affect the cariogenicity of sucrose. Thus, a diet of natural ingredients containing 15% sucrose and higher, increased the incidence of caries in rats, in a linear fashion, whereas a diet with purified ingredients containing sucrose above 15% did not (Huxley 1977).

The results of studies on plaque suggest that dietary carbohydrate forms an extracellular polysaccharide that is difficult to dislodge, thereby entrapping organisms (Jong et al. 1985). The adherence of streptococci to hydroxyapatite spheroids was greatly increased by sucrose, but glucose did not have a similar effect (Adshead et al. 1983). It has been proposed that sucrose-binding of bacteria to enamel is due to the rigid 1,3-linked glucan produced by *Streptococcus mutans* (Rolla et al. 1985).

Studies on the oral and dental effects of sucrose and other carbohydrates in man are more difficult to carry out, especially in relation to caries. In studies using dental students who consumed fructose, glucose, sucrose, or xylitol for 4 days, it was found that the least amount of plaque was seen with xylitol, intermediate with fructose and glucose, and highest with sucrose (Scheinin and Mäkinen 1971). Again using dental students, a positive correlation was reported between caries experience and plaque found in 3 days on a normal diet (Grenby 1974). The same report noted a significant correlation between plaque formation and the daily consumption of sugar in tea and coffee. Demineralization was found in eight volunteers rinsing nine times a day with 50% sucrose solution, and this was not reduced by a calcium glycerophosphate rinse (Edgar et al. 1978). In a study of 4000 adolescents, those with a high measure of decayed, missing, and filled teeth reported a 10-fold increase in intake of sugar-containing foods compared with the lowest intakes and that the obese adolescents did not have a greater intake of sugar-containing foods (Garn et al. 1980).

Though there is an ecological effect of sucrose on oral microflora, there is a high inter-subject and site variation, so it would be unwise to use microbiological criteria in oral dietary assessments (Minah et al. 1985). Using plaque pH as a criterion, it has been recommended that beverages should have a sucrose content no greater than 2% (Maiwald et al. 1984). In patients with denture stomatitis, a sucrose mouth rinse aggravated the condition (Olsen and Birkeland 1976). It has also been noted that the polysaccharide matrix of plaque formed in the presence of maltose is different from that of sucrose (Skinner et al. 1982b).

Factors that modify the oral response to diet include not only sucrose but the consistency of the food, its time of consumption, and the conditions under which it is eaten (Glass and Fleisch 1974). The sugar substitution of a single food item will not reduce caries significantly; the whole spectrum of foods needs to be modified (Newbrun 1978). Chewing gum significantly reduced plaque formation during a 5-day period, with no difference between sugared or sugar-free gum (Addy et al. 1982).

In school children, there seemed to be no difference in dental caries experience whether they ate ready-to-eat cereals or other dietary regimes (Rowe et al. 1974). In younger children, prolonged suckling of comforter feeding bottles containing milk, fruit juices, and syrup sweetened by addition of sugar was an extremely significant aetiological factor in rampant caries of deciduous teeth (Winter et al. 1966). Sugar-based medicines continuously administered to children may cause dental caries and gingivitis (Roberts and Roberts 1979).

The volume of saliva in the mouth before swallowing may be an important determinant of caries susceptibility (Lagerlof and Dawes 1985a). The time between sucrose ingestion and peak salivary flow rate was independent of sucrose concentration, but the peak flow did decline to a level that was linearly related to the sucrose concentration (Lagerlof and Dawes 1985b). The clearance of glucose and sucrose from the saliva was rapid (0–6 minutes after a test solution), and the rate of clearance was dependent on rate of salivary flow (Sreenby et al. 1985).

There are numerous reviews on sucrose and oral diseases; these include:

Oral effects of sugars and sweeteners (Kleinberg 1985).

The use of sugar in foods (Koivistoinen and Hyvonen 1985).

Factors associated with the acceptance of sugar and sugar substitutes by the public (Mackay, 1985).

Sucrose in the dynamics of the carious process (Newbrun, 1982a).

Sugar and dental caries: a review of the evidence (Rugg-Gunn and Edgar 1984).

Sucrose and dental caries: a review of human studies (Newbrun, 1982b).

Sugar restriction and substitution for prevention of caries (Newbrun and Frostell 1978).

Diet and dental caries (Andlaw 1977).

The role of dietary carbohydrates in plaque formation and oral disease (Brown 1975).

Sugar and dental decay (Finn and Glass 1975).

Sugar and the formation of dental plaque (Makinen 1974).

Dietary sugars and sugar substitutes (Scheinin 1974).

### 6.1.4 Atherosclerosis and Coronary Heart Disease

The relationship between dietary sucrose intake and atherosclerosis and coronary heart disease (CHD) has been suggested because of the ready association between these two parameters (Yudkin 1978, 1972). Some evidence to support this association was suggested by the finding that those who develop CHD consumed more sucrose than a control group, but the difference was not statistically significant (Paul et al. 1968), and increases in serum triglyceride levels were found in CHD patients receiving simple carbohydrates compared with controls (Palumbo et al. 1977). It has been shown that patients suffering from coronary artery disease do not consume more sucrose than control subjects apparently unaffected (Nuttall and Gannon 1981; Grande 1974; Groen et al. 1966). In a group of hyperlipidaemic patients, sucrose did not seem to have any effect; it was suggested, however, that his may depend on the type of hyperlipidaemia, the relative proportions of sucrose and polysaccharide, and the kind and amounts of fat in the diet (Little et al. 1970). In persons who are carbohydrate sensitive, it has been suggested that sucrose, at levels of intake common in the USA, could lead to a blood profile consistent with CHD (Reisser et al. 1981). After a small cardiac infarct, sucrose is an effective anti-lipolytic agent (Tansey et al. 1979). In a study on 11 non-insulin-dependent diabetics, it was found that 50% of energy as carbohydrate, with 16% of this as sucrose, leads to changes in plasma lipids and post-prandial glucose that have been identified as risk factors for CHD (Coulston et al. 1985).

On the other hand, a study in the United Kingdom revealed that CHD patients took slightly, but not significantly, more sucrose than controls, and it was noted that more CHD patients were smokers and tended to take more sucrose in hot drinks than non-smokers (Burns-Cox et al. 1969). No evidence was found of a higher sucrose intake in 334 men with CHD, but women with angina had a slightly, but not significantly, higher intake of sucrose (Elwood et al. 1970). In 100 male patients with CHD, it was not possible to confirm an unusually high sucrose intake (Finegan et al. 1968). A longitudinal study of CHD did not reveal any correlation with any dietary component except coffee (Paul et al. 1963) and, more recently, a survey failed to reveal a relationship between dietary sucrose and the incidence of CHD (Garcia-Palmieri et al. 1980).

There have been several reviews of the role of dietary carbohydrate, particularly sucrose, in the aetiology of CHD, including Grande (1975), Keys (1975), and Opie (1975). A review of dietary carbohydrate and atherosclerosis was published by Story (1982) and, more recently, an account of the effects of dietary carbohydrates in metabolic disturbances in man was published (Taskinen 1986).

### 6.1.5 Renal Effects

The majority of investigations on this subject have been carried out on rats. Long-term feeding resulted in nephrocalcinosis, mostly in the cortico-medullary region, and basophilic deposits in the tubules with increased excretion of *N*-acetyl-β-glucosaminidase in the urine (Kang et al. 1979). Another study reported more nephrocalcinosis in rats given sucrose, compared with animals given starch, glucose, or fructose (Oxley et al. 1979). In streptozotocin-diabetic rats, sucrose feeding induced glomerular basement membrane changes similar to those found in diabetes in man (Taylor et al. 1980).

When sucrose is given subcutaneously for up to 96 hours, changes are seen in the cells of the proximal convoluted tubules (PCT) (Perazzo and Monserrat 1978); after 5 days of hypertonic sucrose, an intra-arterial injection of sucrose was given, and it was found that the controls had greater renal damage than those that were given sucrose (Fiorica and Iampietro 1966).

### 6.1.6 Blood Pressure

Though no studies of the effects of sucrose-containing diets on blood pressure in rats have been reported, three studies have measured the catecholamine output. In one, it was found that dietary sucrose induced sympathetic hyperactivity (Cottet-Emard et al. 1980); another found increased norepinephrine turnover in cardiac muscle when the drinking-water contained sucrose (Landsberg and Young, 1982); the third reported no change in resting plasma catecholamine levels after 3 days of sucrose supplementation (Weick et al. 1983). When 77% of caloric intake is given to rhesus monkeys as sucrose, an increase in systolic blood pressure was observed (Clevidence et al. 1981).

In man, there also seems to be some uncertainty. The sodium retention caused by a high carbohydrate intake is counter-regulated by a reduction in plasma aldosterone, and a high carbohydrate diet did not cause an increase in blood pressure in normal man (Affarah et al. 1986). Sucrose and glucose were reported to transiently elevate blood pressure (Hodges and Rebello 1983), and carbohydrate-sensitive individuals receiving a diet containing various levels of sucrose for 6 weeks were found to have a raised diastolic pressure, when 33% of the dietary energy was present as sucrose (Israel et al. 1983).

The effects of dietary carbohydrates on blood pressure have been reviewed (Karanja and McCarron 1986), as well as the effects of sucrose ingestion on blood pressure (Preuss and Fournier 1982).

### 6.1.7 Wound Healing

There are reports that sucrose applied directly to wounds and damaged skin areas aids the healing process, especially in chronic ulceration (Gordon et al. 1985; Trouillet et al. 1985; Hersage and Montenegro 1984). One report suggests that the ingestion by man of glucose, fructose, sucrose, honey, or orange juice significantly decreases the capacity of neutrophils to engulf bacteria, whereas starch did not have this effect (Sanchez et al. 1973). Acute renal failure has been reported after the use of granulated sugar in deep infected wounds (Debure et al. 1987).

### 6.1.8 Gastrointestinal Function

The increase in serum insulin levels reported after a period of increased sucrose intake may be mediated by an effect on gastric inhibitory peptide by a sucrose load (Reiser et al. 1980). A review of sugar-induced diarrhoea in children (Gracey and Burke 1973) and one on the effect of the type of dietary carbohydrate on small intestinal functions (Reiser and Lewis 1986) have been published as has a full account of sugar transport and fructose metabolism in human intestine (White and Landau 1965).

Sucrose, together with electrolytes, has been shown to be useful in rehydrating infants with acute diarrhoea (Chatterjee et al. 1977); more recently, glucose polymers of DE 10, 15, and 24 have also been shown to be suitable for infants recovering from acute diarrhoea (Lebenthal et al. 1983).

Ingestion of fructose and sucrose increases the rate of removal of intravenous alcohol, compared with glucose (Soterakis and Iber 1975). It has been postulated that a high intake of refined carbohydrate may be partly responsible for the increased incidence of Crohn's disease (Martini and Brandes 1976).

The scientific evidence that supports the hypothesis that sucrose increases gallstone formation is limited to a study on guinea-pigs in which ingestion of sucrose and invert sugar were compared. Sucrose led to lower levels of glucuronic acid in the faeces with more in the urine and more bile conjugation of injected phenophthalein (Ahrens et al. 1985). Rabbits receiving 39.5% sucrose, glucose, or fructose as energy in the diet had an increase in cholelithiasis compared with starch-fed animals (Heidecker et al. 1985).

### 6.1.9 Behaviour and Nervous System Changes

It has been reported that, compared with a diet rich in starch, one containing 68% of energy as sucrose affected the retinal vascular system in rats, particularly in diabetic animals (Papachristodoulou et al. 1976). Though having no effect on activity levels, a bolus dose of sucrose, glucose, or fructose by stomach tube raised the brainstem and cerebral cortex serotonin concentration, and glucose and fructose decreased dopamine concentration in the corpus striatum (Crane and Lachance 1983). This study confirmed some earlier findings indicating that diet can influence brain serotonin levels in rats (Colmenares et al. 1975). It has been proposed that the serotonin release in the brain comprises part of a behavioural feed-back loop reducing the tendency to consume more carbohydrate (Wurtman and Wurtman 1979). It has been found that the greater the carbohydrate content of the diet the greater the nocturnal activity of the rat, with no correlation between the amount of energy consumed and activity patterns (Chiel and Wurtman 1981). In rats, limited access to a sucrose solution may lead to over-eating and major changes in circadian feeding behaviour (Hirsch and Walsh 1982). It has been proposed that noradrenaline in the paraventricular nucleus region may play a specific and unique role in the control of carbohydrate ingestion (Leibowitz et al. 1985).

An early claim that sugar had adverse consequences on behaviour was published by Seham and Seham (1929), who claimed that there was a connection between hyperactivity and sugar consumption. In a more recent study, 21 children, recruited for their reported behavioural reactivity to sugar, were given doses of 1.5 g/kg body weight of sucrose, glucose, and a saccharine placebo. None of the sugars produced any signs of hyperactivity (Behar et al. 1984). Ferguson et al. (1986) also failed to detect any adverse effects of sucrose on normal pre-school children and children thought to be sugar-reactive by their parents. Yet another study undermines the hypothesis that sucrose plays a major role in the behaviour of hyperactive bodys aged 7–12 years (Wolraich et al. 1985). However, a study by Prinz et al. (1980) suggests that sucrose consumption adversely affects the behaviour of hyperactive children.

At a more practical level, it was found that the addition of 14.85% sucrose improved the acceptability of a UNICEF formula for children (Grewal et al. 1973).

It has been proposed that sucrose acts by somehow causing the release of endorphins, thus explaining sweet snacking at times of stress, binge eaters preferring sweets during a binge (Fullerton et al. 1985). This would be consistent with the findings that the anti-opiate Nalaxone significantly reduces food intake with no perception of hunger (Cohen et al. 1985).

In a cross-over study with 40 young men, it was reported that auditory reaction time was significantly reduced following a high carbohydrate meal compared with a high protein meal (Lieberman et al. 1986).

In 1977, it was reported that the ingestion of sucrose stimulated sympathetic nervous system activity (Young and Landsberg 1977). However, more recently, it has been shown that several carbohydrates have the same effect when sympathetic activity was measured in brown adipose tissue (Walgren et al. 1987).

Reviews in this area include: the psychological correlates of sugar consumption (Moskowitz 1978), dietary influences on the synthesis of neurotransmitters in the brain (Growdon and Wurtman 1979), and diet and juvenile delinquency (Gray and Gray 1983). Reviews on claims of antisocial behaviour from consumption of sugar (Harper and Gans 1986) and on carbohydrate intakes and children's behaviour (Kruesi 1986) have also been published.

### 6.1.10 Sucrose Intolerance

Apart from the very rare condition where there is a congenital absence of sucrase in the intestine, there are two reports relating to sucrose intolerance in dwellers in Arctic regions. In 1973, it was reported that a 50-g oral sucrose load given to six Eskimos, who have previously complained of diarrhoea and abdominal cramps after eating sucrose-containing foods, caused a flat blood glucose response in five of them. None was intolerant to milk, and the results were not due to monosaccharide malabsorption (Bell et al. 1973). A more recent report found a high incidence of sucrose intolerance among children in Greenland (Gudman-Hoyer 1985).

## 6.2 General Reviews on Health Aspects of Sucrose

There are many reviews on the health aspects of sucrose, including:

Metabolic responses to dietary carbohydrates: interaction of dietary and hereditary factors (Cohen 1986).

Evaluation of health aspects of sugars contained in carbohydrate sweeteners (Glinsmann et al. 1986).

Effects of dietary sucrose or fructose on carbohydrate and lipid metabolism (Vrána and Kazdová 1986).

Metabolic effects of high sucrose or fructose intake (Vrána and Fabry 1983).

The nutritional significance of sucrose consumption 1970–1980 (Lee 1981).

Sucrose and disease (Nuttall and Gannon 1981).

Carbohydrates, sucrose and human disease (Bierman 1979a).

Carbohydrate and sucrose intake in the causation of atherosclerotic heart disease, diabetes mellitus and dental caries (Bierman 1979b).

Role of sugar in modern nutrition (Stare 1975).

## 6.3 References


Addy M, Perriam E, Sterry A (1982) Effects of sugared and sugar-free chewing gum on the accumulation of plaque and debris on the teeth. J Clin Periodontol 9: 346–354

Adshead VM, Parke JM, Chambers PJ, Davies RM, Cole JA (1983) An in vitro study of the role of sucrose and interactions between oral bacteria in possible mechanisms of dental plaque formation. Arch Oral Biol 28: 723–727

Affarah HB, Hall WD, Heymsfield SB, Kutner M, Wells JO, Tuttle EP (1986) High carbohydrate diet: antinatriuretic and blood pressure response in normal man. Am J Clin Nutr 44: 341–348

Ahrens RA, Garland SL, Kigutha HN, Russek E (1985) The disaccharide effect of sucrose feeding on glucuronide excretion and bile concentration of injected phenolphthalein in guinea-pigs. J Nutr 115: 288–291

Andlaw RJ (1977) Diet and dental caries – a review. J Hum Nutr 31: 45–52

Ashley DVM, Fleury MO, Golay A, Maeder E, Leathwood PD (1985) Evidence for diminished brain 5-hydroxy tryptamine biosynthesis in obese diabetic and non diabetic humans. Am J Clin Nutr 42: 1240–1245

Bantle JP, Taine DC, Castle GW, Thomas JW, Hoogwerf BJ, Goetz FC (1983) Postprandial glucose and insulin response to meals containing different carbohydrates in normal and diabetic subjects. N Engl J Med 309: 7–12

Behar D, Rapoport SL, Adams AS, Berg CJ, Cornblath M (1984) Sugar challenge testing with children considered behaviourally "sugar reactive". Nutr Behav 1: 277–288

Belkhadir J, Rosset T, Elgrably F et al. (1985) Effect of an extra intake of carbohydrate at dinner on morning after fasting plasma glucose levels in types I and II diabetes. Br Med J 291: 1608

Bell RR, Draper HH, Bergan JG (1973) Sucrose, lactose and glucose intolerance in northern Alaskan Eskimos. Am J Clin Nutr 26: 1185–1190

Bierman EL (1979a) Carbohydrates, sucrose and human disease. Am J Clin Nutr 32: 2712–2722

Bierman EL (1979b) Carbohydrate and sucrose intake in the causation of atherosclerotic heart disease, diabetes mellitus, and dental caries. Am J Clin Nutr 32: 2644–2647

Bierman EL, Nelson R (1975) Carbohydrates, diabetes, and blood lipids. World Rev Nutr Diet 22: 280–287

Birkhed D, Topitsoglou V, Edwardson S, Frostell G (1981) Cariogenicity of invert sugar in long term rat experiments. Caries Res 15: 302–307

Boot-Hanford R, Heath H (1980) Identification of fructose as the retinopathic agent associated with the ingestion of sucrose-rich diets in the rat. Metabolism 29: 1247–1252

Bornet F, Haardt MJ, Costagliola D, Blayo A, Slama G (1985) Sucrose or honey at breakfast have no additional acute hyperglycaemic effect over an hyperglucidic amount of bread in Type 2 diabetic patients. Diabetologia 28: 213–217

Brown AT (1975) The role of dietary carbohydrates in plaque formation and oral disease. Nutr Rev 33: 353–361

Burns-Cox CJ, Doll R, Ball KP (1969) Sugar intake and myocardial infarction. Br Heart J 31: 485–490

Chantelau EA, Gosseringer G, Sonnenberg GE, Berger M (1985) Moderate intake of sucrose does not impair metabolic control in pump-treated diabetic out-patients. Diabetologia 28: 204–207

Chatterjee A, Jalan KN, Agarwal SK et al. (1977) Evaluation of a sucrose electrolyte solution for oral rehydration in acute infantile diarrhoea. Lancet I: 1333–1335
Chiel HJ, Wurtman RJ (1981) Short term variations in diet composition change the pattern of spontaneous motor activity in rats. Science 263: 676–678
Clevidence BA, Srinivasan SR, Webber LS, Radhakrishnamurthy B, Dalferes E, Berenson GS (1981) Serum lipoprotein and blood pressure levels in rhesus monkeys fed sucrose diets. Biochem Med 25: 186–197
Cohen AM (1972) Effect of sucrose feeding on glucose tolerance. Acta Med Scand (Suppl) 542: 173–179
Cohen AM (1978) Genetically determined response to different ingested carbohydrates in the production of diabetes. Horm Metab Res 10: 86–92
Cohen AM (1986) Metabolic responses to dietary carbohydrates: interactions of dietary and hereditary factors. Prog Biochem Pharmacol 21: 74–103
Cohen AM, Teitelbaum A, Saliternick R (1972) Genetics and diet as factors in development of diabetes mellitus. Metabolism 21: 235–240
Cohen MR, Cohen RM, Pickar D, Murphy KL (1985) Nalaxone reduces food intake in humans. Psychosom Med 47: 132–138
Colmenares JL, Wurtman RJ, Fernstrom JD (1975) Effects of ingestion of a carbohydrate fat meal on the levels and synthesis of 5-hydroxyindoles in various regions of the rat central nervous system. J Neurochem 25: 825–829
Cottet-Emard JM, Peyrin L, Bonnod J (1980) Dietary induced changes in catecholamine metabolites in rat urine. J Neural Transm 48: 189–201
Coulston AM, Swislocki ALM (1985) Metabolic effects of high carbohydrate, moderate sucrose diets in patients with non-insulin-dependent diabetes mellitus (NIDDM). Diabetes 34 (Suppl): 34A
Coulston AM, Hollenbeck CB, Donner C, Williams R, Chiou YM, Reaven GM (1985) Metabolic effects of added dietary sucrose in individuals with non-insulin-dependent diabetes mellitus. Metabolism 34: 962–966
Crane SC, Lachance PA (1983) Effects of sucrose, glucose and fructose on spontaneous activity and brain monoamines in rat pups. Nutr Rep Intern 28: 991–997
Crapo PA, Reaven G, Olefsky JM (1976) Plasma glucose and insulin responses to orally administered simple and complex carbohydrates. Diabetes 25: 741–747
Debure A, Gachot B, Lacour B, Kreis H (1987) Acute renal failure after use of granulated sugar in deep infected wounds. Lancet I: 1034–1035
Eaton RP, Kipnis DM (1969) Effects of high carbohydrate diets on lipid and carbohydrate metabolism in the rat. Am J Physiol 217: 1160–1165
Edgar WM, Geddes DAM, Jenkins ON, Rugg-Gun AJ, Howell R (1978) Effects of calcium glycerophosphate and sodium fluoride on the induction in vivo of caries-like changes in human dental enamel. Arch Oral Biol 23: 655–661
Elwood PC, Moore S, Waters WE, Sweetnam P (1970) Sucrose consumption and ischaemic heart disease in the community. Lancet I: 1014–1016
Ferguson HB, Stoddard C, Simeon JG (1986) Double blind challenge studies of behavioural and cognitive effects of sucrose-aspartame ingestion in normal children. Nutr Rev 44: 144–150
Finegan A, Hickey N, Maurer B, Mulcahy R (1968) Diet and coronary heart disease: dietary analysis on 100 male patients. Am J Clin Nutr 21: 143–148
Finn SB, Glass RB (1975) Sugar and dental decay. World Rev Nutr Diet 22: 304–326
Fiorica V, Iampietro PF (1966) Tolerance to sucrose in adapted and non-adapted rats. Proc Soc Exp Biol Med 122: 647–652
Firestone AR, Schmid R, Muhlemann HR (1982) Cariogenic effects of cooked wheat starch alone or with sucrose and frequency-controlled feedings in rats. Arch Oral Biol 27: 759–763
Fullerton DT, Getto CJ, Swift WJ, Carlson IH (1985) Sugar opioids and binge eating. Brain Res Bull 14: 673–680
Garcia-Palmieri MR, Sorlie P, Tillotson J, Costas R, Cordero E, Rodriquez M (1980) Relationship of dietary intake to subsequent coronary heart disease incidence: the Puerto Rico Heart Health Program. Am J Clin Nutr 33: 1818–1827
Garn SM, Solomon MA, Schaefer A (1980) Internal validation of sugar-food intakes in obese adolescents. Am J Clin Nutr 33: 1890–1891
Geiselman PJ (1985) Appetite, hunger and obesity as a function of dietary sugar intake. Appetite 6: 64–79
Geiselman PJ, Novin D (1982) The role of carbohydrates in appetite, hunger and obesity. J Intake Res 3: 203–223

Glass RL, Fleisch S (1974) Diet and dental caries: dental caries incidence and the consumption of ready-to-eat cereals. J Am Diet Assoc 88: 807–813

Glinsmann WH, Brausquin H, Park YK (1986) Evaluation of health aspects of sugars contained in carbohydrate sweeteners. Food and Drug Administration, Washington DC, USA

Gonnermann B, Schaffer-Spiegel R, Laube H, Schatz H (1982) The effect of a saccharose-rich diet on carbohydrate and lipid metabolism of streptozotocin-diabetic rats and genetically determined 'diabetic' mice (gg-diab). Int J Obes 6: 41–48

Gordon H, Middleton K, Seal D, Sullens K (1985) Sugar and wound healing. Lancet II: 663–664

Gracey M, Burke V (1973) Sugar induced diarrhoea in children. Arch Dis Child 48: 331–336

Grande F (1974) Sugars in cardiovascular disease. In: Sipple HL, McNutt KW (ed) Sugars in nutrition. Academic Press, New York, pp 401–437

Grande R (1975) Sugar and cardiovascular disease. World Rev Nutr Diet 22: 248–269

Gray GE, Gray LK (1983) Diet and juvenile delinquency. Nutr Today May-June: 14–22

Green RM, Hartles RL (1970) The effects of diets containing varying percentages of sucrose and maize starch on caries in the albino rat. Caries Res 4: 188–192

Grenby TH (1967) Investigations in experimental animals on the cariogenicity of diets containing sucrose and/or starch. Caries Res 1: 208–221

Grenby TH (1974) Dental plaque formation in relation to type of carbohydrate. Proc Nutr Soc 33: 24A–25A

Grenby TH, Hutchinson JB (1969) The effects of diets containing sucrose, glucose or fructose on experimental dental caries in two strains of rats. Arch Oral Biol 14: 373–380

Grewal T, Gopaldas T, Haltenberger P, Ramakrishnan I, Ramachandran G (1973) Influence of sugar and flavour on the acceptability of instant CSM: trials on young children from an urban orphanage. J Food Sci Technol 10: 149–152

Grigoresco C, Halfon JJP, Bros M et al. (1985) Effet metabolique de la prise quotidienne de 30 g fructose pendant 2 mois. Diabetes Metab 11: 327–328

Groen JJ, Balogh M, Yaron E, Cohen AM (1966) Effect of interchanging bread and sucrose as main source of carbohydrate in a low fat diet on the serum cholesterol levels of healthy volunteer subjects. Am J Clin Nutr 19: 46–58

Growdon JH, Wurtman RJ (1979) Dietary influences on the synthesis of neurotransmitters in the brain. Nutr Rev 37: 129–136

Gudman-Hoyer E (1985) Sucrose malabsorption in children. A report of thirty-one Greenlanders. J Pediatr Gastroenterol Nutr 4: 873–877

Hallfrisch J, Lazar R, Jorgenson C, Reiser S (1979a) Insulin and glucose responses in rats fed sucrose or starch. Am J Clin Nutr 32: 787–793

Hallfrisch J, Lazar R, Reiser S (1979b) Effect of feeding sucrose or starch to rats made diabetic with streptozotocin. J Nutr 109: 1909–1915

Hallfrisch J, Cohen L, Reiser S (1981) Effects of feeding rats sucrose in a high fat diet. J Nutr 111: 531–536

Hallfrisch J, Ellwood KC, Michaelis OE, Reiser S, O'Dorisio TM, Prather EM (1983) Effects of dietary fructose on plasma glucose and hormone responses in normal and hyperinsulinaemic men. J Nutr 133: 1819–1826

Hara E, Saito M (1981) Impaired insulin secretion after oral sucrose and fructose in rats. Endocrinology 109: 966–970

Harper AE, Gans DA (1986) Claims of antisocial behaviour from consumption of sugar. Food Tech 40: 142–149

Heath H, Hamlett VC (1976) The sorbitol pathway. Effect of streptozotocin induced diabetes and the feeding of a sucrose-rich diet on glucose, sorbitol and fructose in the retina, blood and liver of rats. Diabetologia 12: 43–46

Heath H, Kang SS, Phillipou D (1975) Glucose, glucose-6-phosphate, lactate and pyruvate content of the retina, blood and liver of streptozotocin-diabetic rats fed sucrose or starch-rich diets. Diabetologia 11: 57–62

Heaton KW, Emmett PM, Henry CL, Thornton JR, Manhire A, Hartog M (1983) Not just fibre. The nutritional consequences of refined carbohydrate foods. Hum Nutr Clin Nutr 37C: 31–35

Hefti A, Schmid R (1979) Effect on caries incidence in rats of increasing dietary sucrose levels. Caries Res 13: 298–300

Heidecker HA, Borgman RF, Maurice DV (1985) Influence of dietary carbohydrates upon the metabolism of lipids and minerals in rabbits. Nutr Res 5: 983–992

Hersage L, Montenegro JR (1984) Treatment of suppurating wounds by application of sucrose. Gaz Med 91: 59–62

Hill R, Baker N, Chaikoff IL (1954) Altered metabolic patterns induced in the normal rat by feeding an adequate diet containing fructose as the sole carbohydrate. J Biol Chem 209: 705–716

Himsworth HP (1935) Diet and the incidence of diabetes mellitus. Clin Sci 2: 117–148

Hirsch E, Walsh M (1982) Effect of limited access to sucrose on over-eating and patterns of eating. Physiol Behav 29: 129–134

Hirsch E, Ball E, Godkin L (1982) Sex differences in the effects of voluntary activity on sucrose induced obesity. Physiol Behav 29: 253–262

Hodges RE, Rebello T (1983) Carbohydrates and blood pressure. Ann Intern Med 98: 838–841

Hollenbeck CB, Coulston AM, Donner C, Williams RA, Reaven GM (1985) The effects of variations in percent of naturally occurring complex and simple carbohydrates on plasma glucose and insulin response in individuals with non-insulin-dependent diabetes mellitus. Diabetes 34: 151–155

Hotta N, Kakuta H, Fukasawa H et al. (1985) Effects of a fructose rich diet and the aldose reductase inhibitor ONO-2235 on the development of diabetic neuropathy in streptozotocin treated rats. Diabetologia 28: 176–180

Huxley HG (1971) The cariogenicity of various percentages of dietary sucrose and glucose in experimental animals. NZ Dent J 67: 85–98

Huxley HG (1977) The cariogenicity of dietary sucrose at various levels in two strains of rat under unrestricted and controlled frequency feeding conditions. Caries Res 11: 237–242

Israel KD, Michaelis OE, Reiser S, Keeney M (1983) Serum uric acid, inorganic phosphorus, and glutamic oxaloacetic transaminase and blood pressure in carbohydrate sensitive adults consuming three different levels of sucrose. Ann Nutr Metab 27: 425–435

Jenkins DJA, Wolever TMS, Taylor RH et al. (1981) Glycemic index of foods: a physiological basis for carbohydrate exchange. Am J Clin Nutr 34: 362–366

Jong MA, Kieboom CWA, Likassen JAM, Hoeven JS (1985) Effects of dietary carbohydrates on the numbers of *Streptococcus mutans* and *Actinomyces viscosus* in dental plaque of mono-infected gnotobiotic rats. J Dent Res 64: 1134–1137

Kahn HA, Herman JB, Medalie JH, Neufeld HN, Riss E, Goldburt U (1971) Factors related to diabetes incidence: a multivariate analysis of two years' observation on 10 000 men. The Israel Ischaemic Heart Disease Study. J Chronic Dis 23: 617–629

Kanarek RB, Orthen-Gambill N (1982) Differential effects of sucrose, fructose and glucose on carbohydrate-induced obesity in rats. J Nutr 112: 1546–1554

Kang SS, Price RG, Yudkin J, Worcester NA, Bruckdorfer KR (1979) The influence of dietary carbohydrate and fat on kidney calification and the urinary excretion of *N*-acetyl-β-glucosaminidase (EC 3.2.1.30). Br J Nutr 41: 65–71

Karanja N, McCarron D (1986) Effects of dietary carbohydrates on blood pressure. Prog Biochem Pharmacol 21: 248–265

Kaspar L, Irsigler K (1980) Vergleich des Blutglukoseanstiegs und des Insulinbedarfs nach oralen Gaben von Saccharose Fruktose Sorbit oder einer Mischung Fructose-Sorbit. Wien Klin Wochenschr 92: 683–687

Katz S, Olson BL, Park KC (1975) Factors related to the cariogenic potential of breakfast cereals. Pharmacol Ther Dent 2: 109–131

Keen H, Thomas BJ, Jarrett RJ, Fuller JH (1979) Nutrient intake, adiposity and diabetes in man. Br Med J i: 655–658

Kergoat M, Bailbe D, Portha B (1987) Effect of high sucrose diet on insulin secretion and insulin action: a study in the normal rat. Diabetologia 30: 252–258

Keys A (1975) Coronary heart disease. The global picture. Atherosclerosis 22: 149–192

Kleinberg I (1985) Oral effects of sugar and sweeteners. Int Dent J 35: 180–189

Koivistoinen P, Hyvonen L (1985) The use of sugar in foods. Int Dent J 35: 175–179

Kruesi MJP (1986) Carbohydrate intake and childrens' behaviour. Food Technol 40: 150–151

Lagerlof F, Dawes C (1985a) The effect of swwallowing frequency on oral sugar clearance and pH changes by *Streptococcus mitior* in vivo after sucrose ingestion. J Dent Res 64: 1229–1232

Lagerlof F, Dawes C (1985b) Effect of sucrose as a gustatory stimulus on the flow rates of parotid and whole saliva. Caries Res 19: 206–211

Landsberg L, Young JB (1982) Effects of nutritional status on autonomic nervous system function. Am J Clin Nutr 32: 1234–1240

Laube H, Wojcikowski C, Schatz H, Pfeifer EF (1978) The effect of a high maltose and sucrose feeding on glucose tolerance. Horm Metab Res 10: 192–195

Leach SA, Green RM (1981) Reversal of fissure caries in the albino rat by sweetening agents. Caries Res 15: 508–511

Lebenthal E, Heitlinger L, Lee PC et al. (1983) Corn syrup sugars: in vitro and in vivo digestibility and clinical tolerance in acute diarrhoea of infancy. J Paediatr 103: 29–34

Lee VA (1981) The nutritional significance of sucrose consumption 1970–1980. CRC Crit Rev Food Sci Nutr 47: 1–47
Leibowitz SF, Weiss GF, Yee F, Tretter JB (1985) Noradrenergic innervation of the paraventricular nucleus: specific role in control of carbohydrate ingestion. Brain Res Bull 14: 561–567
Leiter EH, Coleman DL, Ingram DK, Reynolds MA (1983) Influence of dietary carbohydrate on the induction of diabetes in C57BL/KsJ-db/db diabetes mice. J Nutr 113: 184–195
Lenner RA (1976) Specially designed sweeteners and food for diabetics: a real need? Am J Clin Nutr 29: 726–733
Lieberman HR, Spring BJ, Garfield G (1986) The behavioural effects of food constituents: strategies used in studies of amino acids, proteins, carbohydrates and caffeine. Nutr Rev 44(Suppl): 61–70
Lin W, Anderson JW (1977) Effects of high sucrose or starch bran diets on glucose and lipid metabolism on normal and diabetic rats. Nutr 107: 584–595
Little JA, Birchwood BL, Simmons DA et al. (1970) Interrelationship between the kinds of dietary carbohydrate and fat in hyperlipoproteinaemic patients. Atherosclerosis 11: 173–181
Macdonald I (1986) Dietary carbohydrate and energy balance. Prog Biochem Pharmacol 21: 181–191
Mackay DAM (1985) Factors associated with the acceptance of sugar and sugar substitutes by the public. Int Dent J 25: 201–209
Maiwald HJ, Greiling B, Callmeier S, Westphal S, Taeufel A (1984) The cariogenicity of refreshing beverages. Biol Abstr 79: 184
Makinen KK (1974) Sugars and the formation of dental plaque. In: Sipple H, McNutt KW (ed) Sugars in nutrition. Academic Press, New York, pp 645–687
Martini GA, Brandes JW (1976) Increased consumption of refined carbohydrates in patients with Crohn's disease. Klin Wochenschr 54: 367–371
McDonald JL, Stookey GK (1977) Animal study concerning the cariogenicity of dry breakfast cereals. J Dent Res 56: 1001–1006
Michaelis OE, Szepesi B (1974) The mechanism of a specific metabolic effect of sucrose in the rat. J Nutr 104: 1597–1609
Michaelis OE, Nace CS, Szepesi B (1975) Demonstration of a specific metabolic effect of dietary disaccharides in the rat. J Nutr 105: 1186–1191
Minah GE, Solomon ES, Chu K (1985) The association between dietary sucrose consumption and microbial population shifts at six oral sites in man. Arch Oral Biol 30: 397–401
Moskowitz HR (1978) Psychological correlates of sugar consumption. In: Guggenheim B (ed) Health and sugar substitutes. Proceedings of the ERGOB conference on sugar substitutes, Geneva, 30 Oct–1 Nov 1978. Karger, Basel, pp 10–16
Newbrun E (1978) Criteria indicative of cariogenicity or non cariogenicity of foods and beverages. In: Guggenheim B (ed) Health and sugar substitutes. Proceedings of the ERGOB conference on sugar substitutes, Geneva, 30 Oct–1 Nov 1978. Karger, Basel, pp 253–258
Newbrun E (1982a) Sucrose in the dynamics of the caries process. Int Dent 32: 13–23
Newbrun E (1982b) Sugar and dental caries. A review of human studies. Science 217: 418–428
Newbrun E, Frostell G (1978) Sugar restriction and substitution for caries prevention. Caries Res 12(Suppl 1): 65–73
Nuttall RQ (1983) Diet and the diabetic patient. Diabetic Care 6: 197–207
Nuttall RQ, Gannon MC (1981) Sucrose and disease. Diabetic Care 4: 305–310
Oka Y, Akanuma Y, Kasuga M, Kosaka K (1980) Effects of a high glucose diet on insulin binding and on insulin action in rat adipocytes. Diabetologia 19: 468–474
Olefsky JM, Saekow M (1978) The effects of dietary carbohydrate content on insulin binding and glucose metabolism by isolated rat adipocytes. Endocrinology 103: 2252–2263
Olsen I, Birkeland JM (1976) Initiation and aggravation of denture stomatitis by sucrose rinses. Scand J Dent Res 84: 94–97
Olsen ME, Faber OK, Binder C (1983) Hepatic extraction of insulin after carbohydrate hyperalimenation. Acta Endocrinol (Copenh) 102: 416–419
Opie LH (1975) Dietary sucrose in relation to the development of ischaemic heart disease. Am Heart J 89: 674–675
Oxley JA, Bruckdorfer KR, Edwards G, Yudkin J (1979) The effect of dietary carbohydrates and other diet constituents on nephrocalcinosis in the rat. Proc Nutr Soc 38: 85A
Palumbo PJ, Briones ER, Nelson RA, Kottke BA (1977) Sucrose sensitivity of patients with coronary artery disease. Am J Clin Nutr 30: 394–401
Papachristodoulou D, Heath H, Kang SS (1976) The development of retinopathy in sucrose fed and streptozotocin diabetic rats. Diabetologia 12: 367–374
Paul O, Lepper MH, Phelan WH et al. (1963) A longitudinal study of coronary heart disease.

Circulation 28: 20–31
Paul O, MacMillan A, McKean H, Park H (1968) Sucrose intake and coronary heart disease. Lancet II: 1049–1051
Perazzo JC, Monserrat AJ (1978) Effects of repeated injections of sucrose on the kidney: role of the route of administration. IRCS medical science. Cell and membrane biology. Kidneys and urinary system. Metabolism and nutrition. Pathology 6: 41
Petersen DB, Lambert J, Gerring S et al. (1986) Sucrose in the diet of diabetic patients: just another carbohydrate? Diabetologia 29: 216–220
Portha B, Giroix MH, Picon L (1982) Effect of diet on glucose tolerance and insulin response in chemically diabetic rats. Metabolism 21: 1194–1199
Powell HC, Ivor LP, Costello ML, Wolf PL (1982) Elevated haemoglobin A in streptozocin diabetic rats and in rats on sucrose and galactose enriched diets. Clin Biochem 15: 133–137
Preuss HG, Fournier RD (1982) Effects of sucrose ingestion on blood pressure. Life Sci 30: 879–886
Prinz RJ, Roberts WA, Hantman E (1980) Dietary correlates of hyperactive behaviour in children. J Consult Clin Psychol 48: 760–769
Reiser S, Lewis CG (1986) Effect of the type of dietary carbohydrate on small intestine functions. Prog Biochem Pharmacol 21: 135–159
Reiser S, Michaelis OE, Cataland S, O'Dorisio TM (1980) Effect of an isocaloric exchange of dietary starch and sucrose in humans on the gastric inhibitory polypeptide response to a sucrose load. Am J Clin Nutr 33: 1907–1911
Reiser S, Bickard MC, Hallfrisch J, Michaelis OE, Prather ES (1981) Blood lipids and their distribution in lipoproteins in hyperinsulinemic subjects fed three different levels of sucrose. J Nutr 11: 1045–1057
Renold AE, Berger M, Jeanrenaud B (1978) Obesity and diabetes. In: Guggenheim B (ed) Health and sugar substitutes. Proceedings of the ERGOB conference on sugar substitutes, Geneva, 30 Oct–1 Nov 1978. Karger, Basel, pp 17–21
Roberts IF, Roberts GJ (1979) Relation between medicines sweetened with sucrose and dental disease. Br Med J ii: 14–16
Rolla G, Scheie AA, Ciardi JE (1985) Role of sucrose in plaque formation. Scand J Dent Res 93: 105–111
Romsos DR, Leveille GA (1974) Effect of meal frequency and diet composition on glucose tolerance in the rat. J Nutr 104: 1503–1512
Rosenmann E, Palti Z, Teitelbaum A, Cohen AM (1974) Testicular degeneration in genetically selected sucrose fed diabetic rats. Metabolism 23: 343–348
Rowe HG, Anderson RH, Wanninger LA (1974) Effects of ready to eat breakfast cereals on dental caries experience in adolescent children. A three year study. J Dent Res 53: 33–36
Rugg-Gunn AJ, Edgar WM (1984) Sugar and dental caries. A review of the evidence. Community Dent Health 1: 85–92
Sanchez A, Reeser JL, Lau HS et al. (1973) Role of sugars in human neutrophilic phagocytosis. Am J Clin Nutr 26: 1180–1184
Scheinin A (1974) Dietary sugars and sugar substitutes. Int Dent J 23: 427–431
Scheinin A, Mäkinen KK (1971) The effect of various sugars on the formation and chemical composition of dental plaque. Int Dent J 21: 302–321
Seham M, Seham G (1929) The relation between malnutrition and nervousness. Am J Dis Child 37: 1–38
Shafrir E (1985) Effect of sucrose and fructose on carbohydrate and lipid metabolism and the resulting consequences. Regulation of carbohydrate metabolism, vol 2. CRC Press, Boca Raton, Florida, pp 95–140
Sharief NN, Macdonald I (1982) Differences in dietary induced thermogenesis with various carbohydrates in normal and overweight men. Am J Clin Nutr 35: 267–272
Skinner A, Connolly P, Naylor MN (1982a) Influence of the replacement of dietary sucrose by maltose in solid and in solution on rat caries. Caries Res 16: 443–452
Skinner A, Connolly P, Naylor MN (1982b) The influence of the replacement of dietary sucrose by maltose on the formation and biochemistry of human dental plaque. Arch Oral Biol 27: 603–608
Slama G, Haardt MJ, Jean-Joseph P et al. (1984) Sucrose taken during mixed meal has no additional hyperglycaemic action over isocaloric amounts of starch in well controlled diabetes. Lancet II: 122–124
Soterakis J, Iber FL (1975) Increased rate of alcohol removal from blood with oral fructose and sucrose. Am J Clin Nutr 28: 254–257

Sreenby LM, Chatterjee R, Kleinberg I (1985) Clearance of glucose and sucrose from the saliva of human subjects. Arch Oral Biol 30: 269–274

Stare FJ (1975) Role of sugar in modern nutrition. World Rev Nutr Diet 22: 239–247

Steel JM, Mitchell D, Prescott RL (1983) Comparison of the glycaemic effect of fructose, sucrose and starch containing mid-morning snacks in insulin dependent diabetics. Hum Nutr Appl Nutr 37A: 3–8

Stephan RM (1966) Effects of different types of human foods on dental health in experimental animals. J Dent Res 45: 1551–1561

Story JA (1982) Dietary carbohydrate and atherosclerosis. Fed Proc 41: 2797–2800

Tansey MJB, Opie LH, Kennelly BM (1979) The effects of oral sucrose and of estimated infarct size on plasma free fatty acids, plasma glucose and serum insulin in the early stages of acute myocardial infarction. Eur J Clin Invest 9: 81–88

Taskinen MR (1986) Effects of dietary carbohydrates in metabolic disturbances in man. Prog Biochem Pharmacol 21: 160–180

Taylor SA, Price RG, Kang SS, Yudkin J (1980) Modification of the glomerular basement membrane in sucrose-fed and streptozotocindiabetic rats. Diabetes 19: 364–372

Thompson RG, Hayford JT, Danney MM (1978) Glucose and insulin responses to diet. Effect of variations in source and amount of carbohydrate. Diabetes 27: 1010–1026

Trouillet JL, Chastre J, Fagon JY, Pierre J, Domart Y, Gilbert C (1985) Use of granulated sugar in treatment of open mediastinitis after cardiac surgery. Lancet II: 180–184

Vallerand AL, Lupien J, Bukowiecki LJ (1986) Synergistic improvement of glucose tolerance by sucrose feeding and exercise training. Am J Physiol 250: E607–E614

Vrána A, Fabry P (1983) Metabolic effects of high sucrose or fructose intake. World Rev Nutr Diet 42: 56–101

Vrána A, Kazdová L (1986) Effects of dietary sucrose or fructose on carbohydrate and lipid metabolism. Prog Biochem Pharmacol 21: 59–73

Walgren MC, Young JB, Kaufnan LN, Llandsberg L (1987) The effects of various carbohydrates on sympathetic activity in heart and interscapular brown adipose tissue of the rat. Metabolism 36: 585–594

Walker ARP (1977) Sugar intake and diabetes mellitus. S Afr Med J 49: 842–851

Weick BG, Ritter S, McCarty R (1983) Plasma catecholamines in fasted and sucrose supplemented rats. Physiol Behav 30: 247–252

West KM (1978) Epidemiology of diabetes and its vascular lesions. Elsevier, New York

White LW, Landau BR (1965) Sugar transport and fructose metabolism in human intestine in vitro. J. Clin Invest 44: 1200–1213

Winter GB, Hamilton MC, James PMC (1966) Role of the comforter as an aetiological factor in rampant caries of the deciduous condition. Arch Dis Child 41: 207–212

Wolraich M, Milich R, Stumbo P, Schultz F (1985) Effects of sucrose ingestion on the behaviour of hyperactive boys. J Paediatr 106: 675–682

Wurtman JJ, Wurtman RJ (1979) Drugs that enhance central serotoninergic transmission diminish elective carbohydrate consumption in rats. Life Sci 24: 895–904

Young JB, Landsberg L (1977) Stimulation of the sympathetic nervous system during sucrose feeding. Nature (Lond) 269: 615–661

Yudkin J (1972) Sucrose and cardiovascular disease. Proc Nutr Soc 31: 331–337

Yudkin J (1978) Dietary factors in arteriosclerosis: sucrose. Lipids 13: 370–372

Chapter 7

# Safety Aspects

## 7.1 General Observations

In the scientific literature, not many studies on the safety of sucrose per se can be found, presumably because the long history of its use justified the assumption of its safety, coupled with the observation that its forms of biological availability, namely glucose and fructose, are acknowledged and typical components of many living organisms, including the human body (Mann 1964).

Since sucrose has often been used as a control in many toxicological studies (for example, those involving other sweeteners), adequate data are available to permit an evaluation of its safety. In addition, it has been utilized as a component of the experimental animal diet in many nutritional and biochemical studies.

As a food and food ingredient, sucrose is ingested; therefore, feeding studies are relevant and are discussed in this context. Nevertheless, exposure to sucrose by other routes is also dealt with in this chapter.

Apparent discrepancies were frequently noted in studies conducted with experimental animals, and these were often attributed to shortfalls inherent to the conditions of the experimental design or to the purity of the material used. On the other hand, when extrapolating animal data to human beings, certain modulating factors should be considered, including route of administration, duration of exposure, and species, strain, sex, and age differences, with regard to sensitivity to sucrose.

For example, BHE – a strain developed by crossing albino (Yale strain origin) and black-and-white hooded rats (Pennsylvania State College origin) – and Holtzman strains of rats show similar responses to sucrose feeding; these responses differed from the responses obtained with Wistar rats. Sucrose diets fed to the first two strains produced larger livers, a lower percentage of liver protein, increased liver fat and cholesterol, a greater degree of kidney damage, and higher serum cholesterol, while the Wistar strain was not affected. When fed stock diets, all three strains grew at about the same rate; but on sucrose diets, the BHE and Holtzman strains ate more and weighed more than Wistar rats (Marshall and Hildebrand 1963). Similarly, sucrose had some effect on the growth rate of male rats, but the effect disappeared after 6 months

(Dalderup and Visser 1969). Lakshmanan et al. (1981) observed that food utilization for weight gain was less efficient for young Wistar rats than for young BHE rats, as a response to a high-protein diet containing sucrose or cornstarch. Sex and age, as well as dietary components, may affect responses to sucrose. Compared with starch, sucrose caused a fall in liver fat when diets were fed from weaning, and an increase was observed when diets were fed from 3 months of age (Al Nagdy et al. 1966). More extensive effects are found in mature rats than in young ones (Chevalier et al. 1972). On the other hand, dietary sucrose early in life has been reported to potentiate effects later in life, in some animals species (Moser and Berdanier 1974). The mechanism of these effects is not well defined, and several hypotheses have been presented (Naismith and Khan 1971, 1970; Bender and Thadani 1970; Nikkila 1969)

Many studies (e.g., Takemoto 1975) have described the differences in response to sucrose and other carbohydrates by male and female rats; for example, depression in growth was greater in male than in female rats (Al Nagdy et al. 1970).

Dietary factors, such as the presence of fat and protein in the diet, have been identified as modulating the biological effects of sucrose in animals (Al Nagdy et al. 1966; Carroll 1963; Harper et al. 1953) and human beings (Macdonald 1967, 1966; McGandy et al. 1966). Contradictory results may be the outcome of apparently small differences in experimental procedures (Bender and Damji 1972) and different responses of species to the same dietary manipulation (St Clair et al. 1971; Schultz and Grande 1968; Taylor et al. 1967; Grant and Fahrenbach 1959).

The duration of the experimental feeding also plays a role, since adaptation can occur with time, so that the effect found in short-term experiments are decreased or disappear when the experimental period is lengthened (Nestel et al. 1970; Dalderup and Visser 1969; Rifkind et al. 1966; Lees and Frederickson 1965; Macdonald and Roberts 1965; Fillios et al. 1958).

These examples illustrate the need to consider physiological and other factors in interpreting data to assess safety.

Age-associated pathology and its impact on toxicity testing has been examined by Grice and Burek (1984). The use of laboratory animals in nutrition research and toxicological testing has been reviewed and discussed by Waddel and Desai (1981).

## 7.2 Acute Exposure

In his extensive review on acute toxicity data and lethal doses, Spector (1956) reported no lethal dose of sucrose by any route of administration. No acute intoxications (accidental or otherwise) in human beings have been reported in the literature.

### 7.2.1 Oral

Boyd et al. (1965) determined that the oral $LD_{50}$ of sucrose for Wistar albino rats is 35.4 ± 7 g/kg body weight. Adult males weighing 300–500 g were given doses ranging from 5 to 80 g/kg in aqueous solutions, by intragastric cannula, in a volume of 60 ml/kg. All animals given 5–24 g/kg survived; those given 60–80g/kg died within 5 hours. Female rats treated similarly showed an $LD_{50}$ of 29.7 ± 3.7 g/kg, which, though lower than that for males, was not significantly so. Death from the higher dose was apparently due to marked capillary congestion of the brain and meninges and, possibly, from a fulminating gastroenteritis. Death from lower doses was marked by renal lesions and vacuolar degeneration of the convoluted tubules. Survivors showed polydipsia, polyuria, alkalinuria, and some hyperthermia, between 3 and 6 days following treatment.

### 7.2.2 Parenteral

Some of the early acute toxicity studies by the parenteral route were reported by Boyd et al. (1965) and include, for example, studies on the excretion of sucrose following dosages of the order of 2–10 g/kg body weight, given intravenously to dogs, and no lethal effects were observed, with most of the sucrose eliminated in the urine (Kuriyama 1917).

Subcutaneous and intraperitoneal injection to mice of a 50% sucrose solution at a dosage of 30–40 g/kg body weight produced a hypothermic "osmonarkose" and death in about 6 hours (Hausmann 1925).

Death of rabbits was observed when sucrose was administered intravenously for 6 hours at dosages exceeding 50 g/kg body weight (Helmholz and Bollman 1939).

### 7.2.3 Factors Affecting Toxicity

Constantopoulos and Boyd (1968a) studied the influence of a number of factors on the toxicity of sucrose in Wistar albino rats obtained from the Canadian Breeding Laboratory, St Constant, Quebec. The $LD_{50}$ for doses given in aqueous solution by intragastric cannula to fasting animals was not significantly different from that of sucrose given immediately after feeding. A single dose of 38 g/kg (the $LD_{50}$) given in water in volumes of 35–60 ml/kg resulted in death rates proportional to the volume administered. The mortality rate for a dose of 37.5 g/kg body weight in 50 ml/kg solution was higher for female rats than for males, and the pre-mortem symptoms were also more pronounced in the female. The mortality rate for older, heavier rats was 40% versus 9.8% for younger, lighter animals. Successive dosing with sublethal doses of sucrose builds a tolerance to otherwise lethal doses. Dosing for 5 days a week versus 7 days a week, for 5 weeks, did not produce any significant differences in mortality rates. Seasonal differences in dosing (winter or spring) were also not significant.

### 7.2.4 Comparative Toxicity

#### *7.2.4.1 Sugar and Drugs*

Studies were designed by Constantopoulos and Boyd (1968b) to rank the $LD_{50}$ doses of a number of drugs and to compare them with the $LD_{50}$ of sucrose.

The $LD_{50}$ (100 days) for sucrose was first determined by giving several hundred young male albino rats from the Canadian Breeding Laboratory, St Constant, Quebec, doses varying from 20–45 g/kg body weight daily, by intragastric cannula, for up to 100 days or until 60% of the animals had died. The maximum dose of sucrose with no deaths up to 100 days was about 20 g/kg body weight per day; the lowest dose resulting in 100% deaths within 100 days was about 35 g/kg body weight per day. The $LD_{50}$ (100 days), plus or minus one standard error, was determined to be 28.5 ± 1.3 g/kg body weight per day. The authors have proposed an expression for ranking $LD_{50}$ doses of various drugs by a 100-day $LD_{50}$ index, which gives a ratio of $LD_{50}$ (100 days) over $LD_{50}$ (1 day) or acute toxic dose. A tabular presentation for a number of drugs and sucrose index values are given in the paper.

#### *7.2.4.2 Glucose, Fructose, and Sucrose*

Harper and Worden (1964) reported, in abstract form, comparative toxicity studies with glucose, fructose, and sucrose comprising about 80% of the diet in rats over a period of 26 weeks. In the glucose-fed animals, body weight gain was significantly lower than in controls; fructose increased heart, liver, and kidney weight, liver water, and protein. Sucrose produced less marked effects, and these effects were presumably due to its fructose component. The relative toxicity of the three carbohydrates is: fructose is the most toxic and glucose the least.

## 7.3 Prolonged Administration

### 7.3.1 General Observations

Though no long-term toxicity studies specifically designed to assess the chronic effects of sucrose ingestion have been conducted on laboratory animals, it should be realized that this substance has been a component of the diet of experimental animals in a large number of animal experiments, and it has served as a comparative control in the safety testing of many compounds.

### 7.3.2 Sucrose as a Component of Animal Diets

Sucrose-based diets have been, and are currently, used in many short- and long-term studies designed to provide information on cumulative effects, latency period for development of toxicity, reversibility of toxic effects, and dose–response relationships for xenobiotics, including food additives, pesticide residues, drugs, and environmental contaminants. Sucrose has been, and is currently, utilized as a component of the diet of laboratory animals employed in nutritional and biomedical research, diagnosis, and teaching.

Experimental animal diets are given to both control and test groups. The design of the experiment is such that enough power is built into it to permit the clear differentiation between the effects resulting from test substances and those resulting from experimental animal diets. No significant toxicological effects attributable to sucrose as a component of experimental animal diets have been reported in the literature. Palatability of the diet will influence its acceptability and may play a role in food intake and hence body weight.

The amount of sucrose in experimental animal diets varies substantially. For example, in an 80-week carcinogenicity bioassay with aflatoxin B-1, Wogan and Newberne (1967) administered a diet composed of 7% sucrose; the diet was readily accepted by rats and other rodent species, and supported optimal growth. Friedman et al. (1972) fed groups of experimental animals semi-synthetic diets containing from 65%–77% sucrose for 75 weeks with no adverse effects.

### 7.3.3 Sucrose as a Comparative Control

In studies designed primarily to test the safety of substances such as mannitol, sorbitol, xylitol, hydrogenated glucose syrups, and other sweeteners, sucrose was often supplied as a dietary admixture at levels of 20%, 30%, and 40% to a variety of experimental animals, including mice, rats, rabbits, and dogs, to serve as a comparative control in short- and long-term toxicity studies. No significant adverse effects attributable to sucrose as a comparative control have been reported. The following studies are summarized to serve as an illustration of this point.

Groups of rats (strain and sex not specified) were fed 35% sucrose plus 5% mannitol or 40% sucrose (control group) over a period of 3 months; no toxic effects were reported for the control group. However, the growth curves showed that mannitol was nutritionally inferior to sucrose in supporting growth (Ellis and Krantz 1941). These findings supported earlier observations (Ariyama and Takabasi 1931).

In a 15-day experiment on rats (strain unspecified), sucrose was replaced by 10% or 20% xylitol, respectively. The protein of the diet consisted of casein, while the carbohydrate allowance consisted of starch and sucrose. No adverse effects were reported in the control group fed the experimental diet only (Jursons et al. 1974).

Weanling Wistar-derived male rats were divided into groups of 10 each and received diets containing 0, 5, 10, 20, or 30% of either maltitol or sucrose, or 20% hydrogenated glucose syrup (HGS), for 31 weeks. Body weights were reduced at week 4 in the groups on 20 and 30% maltitol and 20% HGS, but were similar between the treated and control groups at week 8. At termination, body weights and selected organ weights of groups fed 5%–30% maltitol were similar to those of rats fed 5%–30% sucrose (Wada 1972).

In a 2-year toxicity study, eight female and eight male beagle dogs received diets containing 0, 2, 5, 10, and 20% xylitol. Sorbitol (20%) and sucrose (20%) were used as comparative controls. No adverse effects were noted in the control group receiving 20% sucrose (Hummler 1976).

Pure-bred beagle dogs (eight male and eight female animals per group) received diets containing 0 or 20% sorbitol or 20% sucrose as the comparative control. Rice starch was included in the diet of the controls so that the diet consisted of 80% normal diet and 20% carbohydrate. The study was terminated at 104 weeks. No effects were observed in the 20% sucrose group (Heywood et al. 1977).

CFLP male and female mice received diets containing 0, 2, 10, or 20% xylitol or 20% sucrose for 102–106 weeks. Mean intake of 20% sucrose, and 2, 10, and 20% xylitol during the experimental period totalled from 11.3 to 17.4 g/kg per day for males and from 16.5 to 19.6 g/kg per day for females. Female mice that received sucrose exhibited an increased number of hepatocellular tumours (Hunter et al. 1978a). A further comparative statistical analysis was conducted, and the significance of these findings was questioned by the US Food and Drug Administration (US/FDA/STF 1986).

Male and female Sprague Dawley (CD strain) rats received diets containing 2, 5, 10, or 20% xylitol, 20% sorbitol, or 20% sucrose (comparative control) in the diet for 78 weeks. No adverse effects were observed in the group receiving 20% sucrose (Hunter et al. 1978b).

Groups of 20 Sprague Dawley rats, evenly divided by sex, received diets containing 1, 15, or 20% hydrogenated glucose syrup (HGS) for 3 months. The diet of the controls was supplemented with 20% sucrose. No adverse effects were observed in the test or control groups (Coquet et al. 1980).

### 7.3.4 Prolonged Administration/Carcinogenicity

Sucrose has been tested for carcinogenicity in laboratory animals by a variety of routes of administration, either directly or as a component of animals diets, or as a comparative control (Sects. 7.3.2, 7.3.3).

Several epidemiological studies have been carried out to investigate the relationship between dietary factors and the risk of developing cancer; some of these studies include retrospective and case-control investigations on carbohydrate intake, including high sucrose consumption.

#### *7.3.4.1 Oral Administration*

In an 80-week bioassay, undertaken to determine the carcinogenic activity of highly purified aflatoxin B-1 in male and female Fischer rats, Wogan and Newberne (1967) utilized laboratory diets containing up to 7% sucrose. Different amounts of aflatoxin B-1 were added to these laboratory diets, and dimethylsulphoxide (DMSO) was used as a positive control. No evidence of carcinogenicity was observed in control animals that received a laboratory diet containing sucrose.

Friedman et al. (1972) fed rats semi-synthetic diets containing from 65% to 77% sucrose in two independently conducted experiments designed to study the carcinogenic potential of sodium and calcium cyclamates. In the first study, male and female weanling Osborne-Mendel rats were used, and the experiment lasted for 101 weeks; in the second study, male weanling Holtzman rats were used, and the study lasted for 75 weeks. No evidence of carcinogenicity was found in animals that received sucrose (Sect. 7.3.2).

Other long-term feeding studies in which sucrose was utilized as a comparative control are reported in Sect. 7.3.3. No evidence of carcinogenicity has been reported in these studies.

#### 7.3.4.2 Parenteral Administration

Nishiyama (1938) first reported the sarcomagenic effects of glucose solutions. Other investigators, especially in Italy and Japan, confirmed these observations and extended them to other mono-, di-, and polysaccharides (maltose, fructose, galactose, lactose, sucrose, and glycogen) in both rats and mice (Arakawa 1956; Amano and Ito 1943; Cappellato 1942; Zarattini 1941). Apparently, this line of research, which occupied the minds of many investigators from the 1930s to the 1960s, had a direct effect on the safety evaluation of food chemicals such as food additives, particularly food colourings.

Because of the distinct scientific and practical importance of the apparent carcinogenic action of sugars by the parenteral route, and the various fundamental issues involved, Hueper (1965) considered it necessary to test again a series of sugars under experimental conditions duplicating, as closely as practical, those employed by previous investigators. He used special precautions in the administration of the sugars so as to preclude the action of any carcinogenic impurities accidentally introduced into the experimental agents and procedures. Subcutaneous injection of 25% solutions of arabinose, dextrose, galactose, lactose, levulose, maltose, sorbose, and sucrose were given into the region of the nape of the neck of 480 rats and 480 mice (groups of 60 animals), twice a week, for up to 2 years. Only two rats given sorbose solutions developed sarcomas, after 21 months, at the site of injection. No tumours were found at this site in additional series of rats and mice injected similarly with water, or traumatized by needle puncture. While these observations do not support the claim that hypertonic sugar solutions parenterally introduced into rats and mice exert a carcinogenic effect, it was suggested that reported carcinogenicity in other studies may have been due to impurities in the sugar used, e.g., polycyclic hydrocarbons or dibenz [a] anthracene chemicals from charcoal used in filtering and decolourizing processes in sugar manufacture.

Further understanding of the mechanism of the development of subcutaneous sarcomas in rats and mice at the site of injection has been gained from the studies of Grasso and Golberg (1966a,b).

The current thinking on this methodological approach is that the route of administration of repeated injections in rats and mice with consequent development of subcutaneous sarcomas at the site of injection is not relevant to the establishment of the safety of foods or food ingredients (WHO 1987). The use of repeated subcutaneous injections is regarded as an inappropriate methodological approach when testing substances that are going to be ingested, because this route of administration is different from the human one. However, in certain situations, the use of this route may be justified, for example, when a compound is poorly absorbed after oral administration or when one wishes to minimize contamination of the laboratory (WHO 1987; IARC/Montesano et al. 1986; WHO/FAO 1970), which is not the case for sucrose.

#### 7.3.4.3 Tumour Promotion

In a long-term feeding study designed to investigate the carcinogenic and tumour-promotion activity of sodium cyclamate, saccharin, and sucrose, groups of 50 female Swiss mice were fed for 18 months on a standard diet containing

10% sucrose, 5% cyclamate, or 5% saccharin. One-hundred control mice were given the standard diet only. No tumours were observed in animals fed these various sweeteners. As part of the same study, comparable groups of mice of the same strain were given a single intragastric instillation of 50 μg benz [a] pyrene in 0.2 ml polyethylene glycol, 7 days before starting the different test diets. No carcinogenic effects or tumour-promoting activity were observed that could have been attributed to the test substances. Sucrose, even in the presence of benz [a] pyrene was not carcinogenic and had no tumour-promoting effects (Roe et al. 1970).

In an in utero exposure, life-time feeding study, two groups of male and female Sprague Dawley rats received diets containing mixtures of sodium cyclamate and sodium saccharin. The treated groups received a mixture of cyclamate and saccharin (ratio 10:1) admixed in the feed at concentrations of 2% and 5%; another group received a diet containing 20% sucrose and served as a comparative control. After 3 months of exposure, 20 female rats were paired with 10 males in each experimental and control group. The pairing was continued until each group numbered 70 offspring consisting of males and females in equal proportions. No tumour or other manifestations of toxicity were observed. Sucrose, even when tested in a long-term in utero exposure feeding experiment did not elicit any adverse effects (Schmaehl and Habs 1984).

Two tumour-promoting studies were concerned with the effect of dietary carbohydrates, including sucrose, in altering the incidence of mammary tumours induced by dimethylbenz [a] anthracene (DMBA) in female rats. These studies yielded conflicting conclusions regarding the effects of simple sugars compared with complex carbohydrates (Klurfeld et al. 1984; Hoehn and Carroll 1979). The significance of the results obtained in these studies and the statistical manipulation of the data were disputed by the US Food and Drug Administration (US/FDA/STF 1986).

### *7.3.4.4 Human Studies*

When fibre is excluded and considered separately from other carbohydrates, there is very little evidence in the literature that carbohydrates, including sucrose, are carcinogenic to human beings (Newberne and Rogers 1986).

Available reports include epidemiological retrospective and case-control studies. These reports attempt to establish a relationship between dietary exposure to carbohydrates and different types of cancers. Only a few are reported here: Seely and Horrobin (1983); Graham et al. (1982); Lubin et al. (1981); Hems (1978); and Armstrong and Doll (1975). Others may be found in the review articles cited at the end of this section.

In spite of the importance attributed to the role of diet in cancer causation, none of the risk factors for which there is currently sufficient evidence for a causal association with cancer in human beings appears to be directly related to dietary exposure, with the possible exception of excessive intake of alcohol (Tomatis 1986).

Recent reviews on the relationship between cancer, nutrition, and diet comprise Hayashi et al. (1986), Reddy and Cohen (1986), Greenwald et al. (1985), Roe (1983), NRC (1982), and Schoental and Connors (1981).

### 7.3.5 Reproductive Toxicity and Teratogenicity

#### *7.3.5.1 Reproduction*

Whitnah and Bogart (1936) produced reproductive effects in female rats (strain and number of animals not specified) maintained on a diet of which the carbohydrate portion was 30% starch and 20% sucrose; two other groups had diets with 30% starch and 20% lactose, or 50% starch, and showed normal reproductive capacity. After being kept on these diets for about 2 years, females on each diet were changed to regular stock diets, and placed with fertile males for 40 days or until pregnant. One female on sucrose produced a live litter but was sterile thereafter. The sucrose-fed animals matured earlier but showed few oestrus cycles; ovarian abnormalities were cytologically evident. The authors speculated that the action of the sucrose was on the pituitary function, but definite conclusions were not possible.

Groups of 31–33 CFT strain-specific, pathogen-free female weanling rats received diets containing either 20% rice starch, 2, 5, 10, or 20% xylitol, or 20% sucrose for comparison. The test diets were administered for 5 weeks before mating. Parameters investigated included food consumption, body weight changes, and mating performance. No adverse effects were observed (for the teratological part of this study, see Sect. 7.3.5.2) (Palmer and Bottomley 1977).

Groups of pregnant rats of Wister strain were fed ad libitum a commercial pelleted diet throughout pregnancy and lactation. Caffeine was supplied in the drinking-water. Animals were maintained on the same diet during several pregnancies, each terminating with spontaneous delivery on day 22 after mating. When caffeine (10 mg/kg per day) was consumed by the maternal animals together with supplementary sucrose (7 g/kg per day), offspring growth reduction did not occur. In previous studies by the authors, reduced growth of offspring was observed following introduction of caffeine into the normal diet of rats during pregnancy and lactation (Dunlop et al. 1981).

Female Sprague Dawley rats were fed ad libitum purified diets in which the carbohydrate component was either maize, starch, muscovado unrefined sugar, or sucrose. They were kept on these diets from weaning through mating and pregnancy to the end of lactation. No difference was observed in their ability to mate, the progress of pregnancies, or size of litters (Eisa and Yudkin 1985).

In a reproductive toxicity study by Jukes (1986), diets containing 64% sucrose supported normal reproduction in rats (no details on strain, number of animals, or other experimental variables were given; results on some reproductive parameters were reported in tabular form). The same table was reported by Friedman et al. (1972), where it is indicated that this unpublished study was conducted at the WARF Institute Inc., Madison, Wisconsin, USA.

A three-generation reproductive study was conducted on Sprague Dawley, CD-strain-specific, pathogen-free rats with 20 males and 20 females per group. Each group received the test material by dietary administration. A control group received 20% rice starch. One group received 20% sorbitol, and another group received 20% sucrose ad libitum. The pups of the first, second, and third generations ($F_{1a}$, $F_{2a}$, $F_{3a}$) were weighed and killed at 4 days and examined for abnormalities. The $F_{1b}$, $F_{2b}$, and $F_{3b}$ pups were weighed, sexed, and the litters culled to eight per dam. Pups were weighed at 8, 12, and 21 days postpartum. No evidence of reproductive abnormalities or terata or other manifestations of toxicity were observed (Palmer et al. 1978).

### *7.3.5.2 Teratogenicity*

1. *Chick embryo.* Ten sugars (D-glucose, L-glucose, D-galactose, D-mannose, fructose, sucrose, maltose, lactose, trehalose, and raffinose) were tested for teratogenic effects on the chick embryo either by solid application in opened eggs or by injection into the yolk during the first day of incubation. All sugars tested were found to be teratogenic, the range of defects observed being similar to those found when embryos at this stage were treated with other substances (Hughes et al. 1974). An international expert group on the "Principles for Evaluating Health Risks to Progeny Associated with Exposure to Chemicals during Pregnancy", under the auspices of the UNEP/ILO/WHO International Programme on Chemical Safety (IPCS), have affirmed that "methods of testing substances on chick embryo in ova have usually failed to supply reliable results, and hence the predictive value of these tests has not been established" (WHO 1984).

2. *Mouse and rat.* Female albino Wistar-derived rats were administered by gavage doses varying from 50 to 250 mg/kg body weight of sodium cyclamate, 5–25 mg/kg body weight of sodium saccharin, and 2–10 g/kg body weight of sucrose (comparative control), from days 6 to 15 of pregnancy. Tap water served as the solvent. The volumes administered were 3 ml/kg for the cyclamate and saccharin and 20 ml/kg for the sucrose group. Control groups received only tap water. Dams were sacrified on the 20th day of pregnancy, and the foetuses were delivered by caesarean section. Parameters examined included number of implantations, live foetuses, resorptions, malformations, and mean weight of live foetuses. No adverse effects were observed in either control or treated animals (Fritz and Hess 1968)..

In another study designed to investigate the teratogenic potential of sucrose, rats and mice were administered by gavage up to 1600 mg/kg body weight per day of sucrose for 10 consecutive days. No effects on nidation, maternal or foetal survival, or post-mortem abnormalities were observed (Food and Drug Laboratories 1973).

Groups of 31–33 CFT strain-specific, pathogen-free female weanling rats received test diets containing either 20% rice starch, 20% sorbitol, or 20% sucrose for comparison. The test diets were administered for 5 weeks before mating. Parameters investigated included food consumption, body weight changes, and mating performance. Litter data included implantations, pre-implantations, post-implantation loss, litter and mean foetal weights, major malformations and minor abnormalities, and skeletal variants. No clinical signs of toxicity, major malformations, or skeletal variations that could be attributed to treatment were noted (Palmer and Bottomley 1977).

3. *Rabbit.* Cyclamate, saccharin, and sucrose were added to the diet of pregnant white New Zealand rabbits in a variety of doses during days 6–18 of gestation (the period of organogenesis). The doses were administered intragastrically by stomach tube. Sucrose was given in doses of 2, 4, or 10 g/kg per day; sodium cyclamate at 50, 100, or 250 mg/kg per day; and saccharin at 5, 10, or 25 mg/kg per day. Some controls received 20 ml water/kg per day,

and others were left untreated. On day 29 of pregnancy, the animals were sacrificed and foetuses removed. Implantations, absorptions, and corpora lutea were noted; full-term foetuses were killed and prepared for study. Anomalies and malformations appeared equally in test animals and controls. None of the three substances showed a teratogenic or embryotoxic effect (Klotzsche 1969).

Groups of 20 yellow-silver females of a closed, randomized outbred rabbit strain, aged 3–4 months and weighing 2.7–3 kg, received test diets ad libitum containing 2, 5, 10, and 20% xylitol or 20% sucrose or 20% sorbitol baked into the food pellets. Males were untreated. Animals received the test diets from day 7 to day 19 of gestation. Parameters investigated were body weights of females, litter data including implantation and post-implantation loss, litter and mean foetal weights, major and minor anomalies, and skeletal variants. At the 20% level of dietary incorporation, sucrose ingestion was approximately 7–8 g/kg body weight per day. No compound-related effects or skeletal or visceral malformations were noted (Hummler 1978).

## 7.4 Special Studies

### 7.4.1 Mutagenicity

Sucrose has been tested for mutagenicity in mammalian cell mutation assays using mouse lymphoma L5178Y cells with three different protocols. The three tests yielded negative results (J. Becking, personal communication 1986).

The *Salmonella typhimurium* assay without metabolic activation was used to examine the anti-mutagenic effects of products of two non-enzymatic browning reactions obtained by heating a lysine–fructose mixture a 121 °C for 1 hour, and by caramelizing sucrose at 180 °C for 1.5 hours. The anti-mutagenic effect was tested by exposing strain TA1235 in suspension to *N*-methyl-*N'*-nitrosoguanidine (MNNG) in the presence of the browning reaction products. In the case of aflatoxin B-1, strain TA98 was used, and the browning reaction products were added to the pre-carcinogen and an S9 mixture. The mutagenic activity of both carcinogens (MNNG and aflatoxin B-1) was significantly suppressed by the browning reaction products (Chan et al. 1982).

The effects of adding sucrose to high- and low-tar cigarettes on the mutagenicity of their smoke condensates were studied using *Salmonella typhimurium* TA100 and TA98, with and without metabolic activation. The sugars tested were glucose, fructose, galactose, sorbitol, sucrose, and lactose. The lowest mutagenicity observed with these sugars/mg smoke condensate, assayed on TA98 with metabolic activation, was 37% (high-tar cigarette) and 22% (low-tar cigarette) of smoke condensate from untreated cigarettes. Addition of sugars increased the total amount of smoke condensates, but the mutagenicity of the local condensates was also decreased by all sugars, the lowest values being 35% (high-tar cigarette) and 36% (low-tar cigarette) of smoke condensates from cigarettes without added sugar. On assay with TA100, with metabolic activation, a decrease in both specific and total mutagenicity of condensates of high-tar cigarettes was observed with all sugars tested, with the exception of galactose and sucrose. Treatment with glucose, fructose, or sorbitol decreased the specific mutagenicity of condensates of low-tar cigarettes, and sucrose, glucose, and fructose also reduced their total mutagenicity. The effects of added sugars were more marked when assayed on TA98 than on TA100, and, of the sugars tested, fructose and sorbitol had the greatest effects. Addition of sugars had no effect on the mutagenicity of cigarette-smoke condensates without metabolic activation (Sato et al. 1979).

The predictive value of the mutagenicity tests mentioned in this section and other mutagenicity tests has been recently reviewed by IARC/Montesano et al. (1986) and Ashby et al. (1985).

### 7.4.2 Nephrocalcinosis

Sixty male Sprague Dawley rats received diets containing 55% sucrose or 55% starch and 20% fat (butter or polyunsaturated margarine) for 13 months. During the course of the experiment, an increase in urinary *N*-acetyl-β-glucosaminidase (NAG) was noted in the group consuming sucrose. Independent of the saturation state of the fat, urinary NAG is an indicator of renal tubular damage. At the end of the experiment, livers and kidneys were weighed and found to be heavier in the sucrose groups than in the starch groups. Microscopic examination of the kidneys of the rats consuming sucrose showed nephrocalcinosis, tubular damage with basophilic material in the lumina, and thickening of the glomerular basement membrane. Calcium, phosphate, and magnesium levels were increased in the kidneys but not in the plasma of the sucrose-fed rats. In another group of rats, receiving a diet containing 70% sucrose and 2% corn oil, the urinary NAG levels were not elevated, and the microscopic changes were less marked. The authors observed that the results obtained in this study raise the possibility that there may be an interaction between high sucrose and fat in high calcium/high phosphate diets, causing increased deposits of calcium in the kidneys with resultant kidney damage (Kang et al. 1979).

Other studies with nephrocalcinosis-susceptible strains of rats yielded different results. Male Wistar of BHE rats (susceptible to nephrosis) received high protein diets (45% w/w casein) supplemented with 335 carbohydrates in the form of corn starch or sucrose, for 400 days. Nephrocalcinosis was negligible in both strains (Lakshamanan et al. 1981).

The Joint FAO/WHO Expert Committee on Food Additives considered the toxicological significance of kidney lesions found in long-term feeding studies on laboratory rats fed on diet containing 5%–25% of various chemically modified starches and included, in its twenty-sixth report (WHO/FAO 1982), some general considerations on recent research on the toxicological significance of the pelvic form of nephrocalcinosis observed in laboratory rats.

Roe (1979) has briefly, but comprehensively, discussed the problem of mineral deposition in the renal pelvis of rats and the types of nephrocalcinosis.

### 7.4.3 Life Span of Animals

Two studies have suggested that substituting sucrose for starch in the diet of rats may influence the life expectancy of rats susceptible to kidney disease.

Male Wistar rats fed a diet in which 15% of the starch was replaced by sucrose showed an average 80-day reduction in life span. Under identical dietary conditions, female Wistar rats showed only a 22-day reduction in life expectancy (Dalderup and Visser 1969).

Male rats of the carbohydrate-sensitive BHE strain, receiving a diet containing 39% sucrose in place of 39% starch, died 151 days earlier. No effect on life span was found when starch was substituted by sucrose in a strain of rats not susceptible to kidney disease (Durand et al. 1968).

Survival data derived from studies listed in Sects. 7.3.2 and 7.3.3 do not support that feeding sucrose to laboratory animals may cause shortening of their life expectancy.

### 7.4.4 Enzymatic Activity

Experimental observations on the effect of dietary sucrose on enzymatic systems of laboratory animals have been reported in a series of articles dealing with the effects of dietary sucrose on the synthesis and degradation of intestinal sucrase, modulation of RNA polymerase activities in the intestine of the adult rat by dietary sucrose, changing of chromatin structure and transcription activity by starvation and dietary sucrose in mature and immature intestinal epithelial cells in the rat, and other similar experimental findings. At present, the significance of these experimental observations on the adaptive processes triggered in the intestine of laboratory animals by dietary sucrose, as well as other dietary constituents (Warnick and Lazarus 1982; Astrom et al. 1981; Von der Decken et al. 1981), is not defined in terms of their extrapolation to man.

Two reviews on the safety of sucrose are available (US/FDA/STF 1986; SCOGS/FASEB 1976).

## 7.5 References

Al Nagdy S, Miller DS, Qureshi RU, Yudkin J (1966) Metabolic differences between starch and sucrose. Nature (Lond) 209: 81

Al Nagdy S, Miller DS, Yudkin J (1970) Changes in body composition and metabolism induced by sucrose in the rat. Nutr Metab 12: 193

Amano S, Ito S (1943) Glykogen und Geschwulstbildung. Gann 37: 200–203

Arakawa S (1956) Experimental production of sarcoma of mice with the subcutaneous high concentrated sugar solutions, especially lactose and mixture of laevulose and glucose. Gann 48: 363–364

Ariyama H, Takabasi K (1931) The relative nutritive value of various carbohydrates and related compounds. Bull Agric Chem Soc Jpn 6: 1–5

Armstrong B, Doll R (1975) Environmental factors and cancer incidence and mortality in different countries, with special reference to dietary practices. Int J Cancer 15: 617–631

Ashby J, De Serres FJ, Draper M et al. (1985) Evaluation of short-term tests. Report of the International Programme on Chemical Safety's Collaborative Study on In Vitro Assays. Elsevier, Amsterdam (Progress in Mutation Research vol 5)

Astrom S, Arrhenius EK, Von der Decken A (1981) Reduced transcription activity of the rat liver chromatin after protein restriction and selective digestion of nuclei with micrococcus nuclease. J Nutr 111: 1258–1264

Bender AE, Damji KB (1972) Some effects of dietary sucrose. World Rev Nutr Diet 15: 104–155

Bender AE, Thadani PV (1970) Some metabolic effects of dietary sucrose. Nutr Metab 12: 22–39

Boyd EM, Godi I, Abel M (1965) Acute oral toxicity of sucrose. Toxicol Appl Pharmacol 7: 609–618

Cappellato M (1942) Sui sarcomi sperimentali da glucosio nel ratto bianco. Tumori 16: 38–54

Carroll C (1963) Influences of dietary carbohydrate. Fat combinations on various functions associated with glycolysis and lipogenesis in rats. I. Effects of substituting sucrose for rice starch with unsaturated and saturated fat. J Nutr 79: 93–100

Chan RIM, Stich HF, Rosin MP, Powrie WD (1982) Antimutagenic activity of browning reaction products. Cancer Lett 15: 27–33

Chevalier M, Wiley JH, Leveille GA (1972) The age-dependent response of serum triglycerides to dietary fructose. Proc Soc Exp Biol Med 139: 220–223

Constantopoulos G, Boyd EM (1968a) Maximal tolerated amounts of sucrose given by daily intragastric administration to albino rats. Food Cosmet Toxicol 6: 712–727

Constantopoulos G, Boyd EM (1968b) Factors affecting sucrose toxicity. Int Z Klin Pharmakol Toxikol 1: 539–544

Coquet B, Rondot G, Mary MC (1980) Thirteen-week oral toxicity study in the rat with Lycasin 80/55 (unpublished report from the Institut français de recherches et essais biologiques, l'Arbresle, France; submitted to the World Health Organization by Roquettes Frères SA, France. In: Joint FAO/WHO Expert Committee on Food Additives (ed) Toxicological evaluation of certain food additives and contaminants. Cambridge University Press, Cambridge, pp 189–190 (WHO Food Additive Series No 20, 1987)

Dalderup LM, Visser W (1969) Influence of extra sucrose daily food on the life-span of Wistar albino rats. Nature (Lond) 222: 1050–1052

Dunlop M, Court JM, Larkins RG (1981) The effects of maternal carbohydrate (sucrose) supplementation on the growth of offspring of pregnancies with habitual consumption. Biol Neonate 40: 196–198

Durand AMA, Fischer M, Adams M (1968) The influence of type of dietary carbohydrate: effect on histological findings in two strains of rats. Arch Pathol 85: 318–324

Eisa OA, Yudkin J (1985) Some nutritional properties of unrefined sugar and its promotion of the survival of new-born rat. Br J Nutr 54: 593–603

Ellis FM, Krantz JC Jr (1941) Sugar alcohols: metabolism and toxicity studies with mannitol and sorbitol in man and animals. J Biol Chem 141: 147–154

Fillios LC, Naito C, Andrus SB, Portwan OW, Martin RS (1958) Variations in cardiovascular sudanophilia with changes in the dietary level of protein. Am J Physiol 194: 275

Food and Drug Laboratories (1973) Teratologic evaluation of FDA 71-76 (pure cane sugar; sucrose) in mice and rats. Food and Drug Research Laboratories Inc, Waverley, New York (Prepared under DHEW Contract No FDA 21-260)

Friedman L, Richardson HL, Richardson ME, Lethco EJ, Wallace WC, Sauro FM (1972) Toxic response of rats to cyclamates in chow and synthetic diets. J Natl Cancer Inst 49: 751–760

Fritz H, Hess R (1968) Prenatal development in the rat following administration of cyclamate, saccharin and sucrose. Experientia 24: 1140–1141

Graham S, Marshall J, Mettlin C et al. (1982) Diet in the epidemiology of breast cancer. Am J Epidemiol 116: 68–75

Grant NC, Fahrenbach MJ (1959) Effect of dietary sucrose and glucose on plasma cholesterol in chicks and rabbits. Proc Soc Exp Biol Med 100: 250–252

Grasso P, Golberg L (1966a) Early changes at the site of repeated subcutaneous injection of food colorings. Food Cosmet Toxicol 4: 269–282

Grasso P, Golberg L (1966b) Subcutaneous sarcoma as an index of carcinogenicity potency. Food Cosmet Toxicol 4: 297–320

Greenwald P, Ershow AG, Novelli WD, Benton CM (1985) Cancer, diet and nutrition. A comprehensive source book, 1st edn. Marquis WHO's WHO, Chicago, Illinois

Grice HC, Burek JD (1984) Age-associated (geriatric) pathology: its impact on long-term toxicity studies. In: Grice HC (ed) Current issues in toxicology/carcinogenicity studies. Springer, Berlin Heidelberg New York, pp 57–107

Harper AE, Monson WT, Benton DA, Elvehjem CA (1953) The influence of protein and certain amino acids particularly threonine on the deposition of fat in the liver of the rat. J Nutr 50: 383

Harper KH, Worden AN (1964) Comparative toxicity studies on glucose, fructose, and sucrose. Toxicol Appl Pharmacol 6: 365 (Abstract No 70)

Hausmann W (1925) Zur Kenntnis der toxischen Wirkung konzentrierter Zuckerlösungen. Wien Klin Wochenschr 38: 332

Hayashi Y, Nagao M, Sugimura T et al. (ed) (1986) Diet, nutrition and cancer. Japan Scientific Societies Press, Tokyo; VNU Science Press BV, Utrecht

Helmholz HF, Bollman JL (1939) The diuretic action of sucrose and other solutions. Proc Staff Meet Mayo Clin 14: 567–569

Hems G (1978) The contribution of diet and childbearing to breast cancer rates. Br J Cancer 37: 974–982

Heywood R, Chesterman H, Allen TR et al. (1977) Sorbitol toxicity studies in the beagle dog (unpublished report from the Huntingdon Research Center; submitted to the World Health Organization by Hoffmann La Roche and Co, Ltd, Basel). In: Joint FAO/WHO Expert Committee on Food Additives (ed) Summary of toxicological data of certain food additives and contaminants. World Health Organization, Geneva, pp 30–31 (WHO Food Additive Series No 13, 1978)

Hoehn SK, Carroll KK (1979) Effects of dietary carbohydrate on the incidence of mammary tumors induced in rats by 7,12-dimethylbenz(a)anthracene. Nutr Cancer 1: 27–30

Hueper WC (1965) Are sugars carcinogens? An experimental study. Cancer Res 25: 440–443

Hughes AF, Freeman RB, Fadem T (1974) The teratogenic effects of sugars on the chick embryo. J Embryol Exp Morphol 32: 661–674

Hummler H (1976) Chronische Verträglichkeitsprüfung mit Xylit an Hunden (unpublished report submitted to the World Health Organization by Hoffmann La Roche and Co, Ltd, Basel). In: Joint FAO/WHO Expert Committee on Food Additives (ed) Summary of toxicological data of certain food additives and contaminants. World Health Organization, Geneva, pp 131–132 (WHO Food Additive Series No 12, 1977)

Hummler H (1978) Reproduction study in rabbits in oral administration of Ro-06-7045-xylitol phase II teratology study (unpublished report submitted to the World Health Organization by Hoffmann La Roche and Co, Ltd, Basel). In: Joint FAO/WHO Expert Committee on Food Additives (ed) Summary of toxicological data of certain food additives and contaminants. World Health Organization, Geneva, pp 32–33 (WHO Food Additive Series No 13, 1978)

Hunter B, Grahan C, Heywood R, Prentice D, Roe R, Noakes D (1978a) Tumorigenicity and carcinogenicity study with xylitol in long-term dietary administration to mice (unpublished report from the Huntingdon Research Center submitted to the World Health Organization by Hoffmann La Roche and Co, Ltd, Basel). In: Joint FAO/WHO Expert Committee on Food Additives (ed) Summary of toxicological data of certain food additives and contaminants. World Health Organization, Geneva, pp 28–29 (WHO Food Additive Series No 13, 1978)

Hunter B, Colley J, Street A, Heywood R, Prentice D, Magnusson G (1978b) Xylitol tumorigenicity and toxicity study in long-term dietary administration to rats (unpublished report submitted to the World Health Organization by Hoffmann La Roche and Co, Ltd, Basel). In: Joint FAO/WHO Expert Committee on Food Additives (ed) Summary of toxicological data of certain food additives and contaminants. World Health Organization, Geneva, pp 29–30 (WHO Food Additive Series No 13, 1978)

IARC/Montesano R, Bartsch H, Vanio H, Wilbourn J, Yamasaki H (1986) Long-term and short-term assays for carcinogens: a critical approach. A joint publication of the International Agency for Research on Cancer (IARC), Lyon, the International Programme on Chemical Safety (UNEP/ILO/WHO), Geneva, and the Commission of the European Communities (CEC), Luxembourg (IARC Scientific Publication No 83)

Jukes TH (1986) Sugar and health. World Rev Nutr Diet 48: 137–194

Jursons KK, Puzaka JJ, Aalmans AR, Vitolin SP (1974) Izv Akad Nauk Latviiskoi SSR 9: 16–22

Kang SS, Price RG, Yudkin J, Worcester NA, Bruckdorfer KR (1979) The influence of dietary carbohydrate and fat on kidney calcification and the urinary excretion of *N*-acetyl-β-glucosaminidase (EC 3.2.1.30). Br J Nutr 41: 65–71

Klotzsche C (1969) Zur Frage der teratogenen und embriotoxischen Wirkung von Cyclamat Saccharin und Saccharose (Teratogenic and embryotoxic effects of cyclamate, saccharin and sucrose). Arzneimittelforschung 19: 925–928

Klurfeld DM, Weber MM, Kritchevsky D (1984) Comparison of dietary carbohydrates for promotion of DMBA induced mammary tumorigenesis in rats. Carcinogenesis 5: 423–425

Kuriyama S (1917) The fate of sucrose parenterally administered. Am J Physiol 43: 343–350

Lakshmanan FL, Howe JC, Schuster EM, Barnes RE (1981) Response of two strains of rats to a high-protein diet containing sucrose or cornstarch. Proc Soc Exp Biol Med 167: 224–232

Lees RS, Frederickson DS (1965) Carbohydrate induction of hyperlipemia in normal man. Clin Res 13: 327

Lubin JM, Burns PE, Blot WJ, Ziegler RG, Lees AW, Fraumeni JF (1981) Dietary factors and breast cancer. Int J Cancer 28: 685–689

Macdonald I (1966) Lipid response of post-menopausal women to dietary carbohydrate. Am J Clin Nutr 18: 86–90

Macdonald I (1967) Interrelationship between the influence of dietary carbohydrates and fats on fasting serum lipids. Am J Clin Nutr 20: 345–351

Macdonald I, Roberts JB (1965) The incorporation of various $^{14}C$ dietary carbohydrates into serum and liver lipids. Metabolism 14: 991–999

Mann I (1964) The biochemistry of semen and of the male reproductive tract. Methuen, London

Marshall MW, Hildebrand HE (1963) Rat strain differences in response to three diets. J Nutr 79: 227–238

McGandy RB, Hegsted DM, Mars ML, Stare FJ (1966) Dietary carbohydrate and serum cholesterol levels in man. Am J Clin Nutr 18: 237–242

Moser PB, Berdanier CD (1974) Effect of early sucrose feeding on the metabolic patterns of mature rats. J Nutr 104: 687–694

Naismith DJ, Khan NA (1970) Differences in the throughput of triglycerides in the plasma of rats fed various carbohydrates. Proc Nutr Soc 29: 64A (Abstract)

Naismith DJ, Khan NA (1971) Lipoprotein lipase activity in the adipose tissue of rats fed sucrose or starch. Proc Nutr Soc 30: 12A (Abstract)

Nestel PJ, Carroll KF, Havenstein N (1970) Plasma triglycerides response to carbohydrates, fats and caloric intake. Metabolism 19: 1–18

Newberne PM, Rogers AE (1986) The role of nutrients in cancer causation. In: Hayashi Y, Nagao M, Sugimura T et al. (ed) Diet, nutrition and cancer. Japan Scientific Press, Tokyo; VNU Science Press BV, Utrecht

Nikkila EA (1969) Control of plasma and liver triglycerides kinetics by carbohydrates metabolism and insulin. Adv Lipid Res 7: 63–134

Nishiyama Y (1938) Experimentelle Erzeugung der Sarkome bei Ratten durch Wiederholte Injektionen von Glukoselosung. Gann 32: 85–99

NRC (1982) Diet, nutrition and cancer. Committee on diet, nutrition and cancer. Assembly of Life Sciences, National Research Council, National Academy Press, Washington DC

Palmer AK, Bottomley AM (1977) Effect of xylitol during a modified teratology study in rats (unpublished report from the Huntingdon Research Center submitted to the World Health Organization by Hoffman La Roche and Co, Ltd, Basel. In: Joint FAO/WHO Expert Committee on Food Additives (ed) Summary of toxicological data of certain food additives and contaminants. World Health Organization, Geneva, p 32 (WHO Food Additives Series No 13, 1978)

Palmer AK, Bottomley AM, Wight DGD, Cherry CP (1978) Effect of xylitol on reproductive functions of multiple generations in the rat (unpublished report from the Huntingdon Research Center submitted to the World Health Organization by Hoffman La Roche and Co, Ltd, Basel. In: Joint FAO/WHO Expert Committee on Food Additives (ed) Summary of toxicological data of certain food additives and contaminants. World Health Organization, Geneva, pp 31–32 (WHO Food Additives Series No 13, 1978)

Reddy BS, Cohen LA (ed) (1986) Diet, nutrition and cancer: a critical evauation. CRC Press, Boca Raton, Florida

Rifkind BM, Lawson DH, Gale M (1966) Effect of short-term sucrose restriction on serum lipid levels. Lancet II: 1379

Roe DA (1983) Diet, nutrition and cancer: from basic research to policy implications. Proceedings of a workshop at Cornell University. AR Liss, New York

Roe FJC (1979) Mineral deposition in the renal pelvis of rats: a brief review. In: Joint FAO/WHO Expert Committee on Food Additives (ed) Twenty-sixth report. World Health Organization, Geneva, p 11 (WHO Technical Report Series No. 683, 1982)

Roe FJC, Levy LS, Carter RL (1970) Feeding studies on sodium cyclamate, saccharin, and sucrose for carcinogenic and tumor promoting activities. Food Cosmet Toxicol 8: 135–145

Sato S, Ohka T, Nagao M, Tsuji K, Kosuge T (1979) Reduction in mutagenicity of cigarette smoke condensate by added sugars. Mutat Res 60: 155–161

Schmaehl D, Habs M (1984) Investigations on the carcinogenicity of the artificial sweeteners sodium cyclamate and sodium saccharin in rats in a two-generation experiment. Arzneim-Forsch Drug Res 34: 604–606

Schoental R, Connors TA (1981) Dietary influences on cancer: traditional and modern. CRC Press, Boca Raton, Florida

Schultz AL, Grande F (1968) Effects of starch and sucrose on the serum lipids of dogs before and after thyroidectomy. J Nutr 94: 71–73

SCOGS/FASEB (1976) Evaluation of the health aspects of sucrose as a food ingredient. Life

Sciences Research Office (LSRO)/Federation of American Societies for Experimental Biology (FASEB), US Food and Drug Administration, Washington DC (prepared for FDA under contract No FDA/223-75-2004)

Seely S, Horrobin DF (1983) Diet and breast cancer: the possible connection with sugar consumption. Med Hypotheses 11: 319–327

Spector WS (1956) Handbook of toxicology. I. Acute toxicities. Saunders, Philadelphia, Pennsylvania

St Clair RW, Bullock BC, Lehner NDM, Clarkson TB, Lofland HB Jr (1971) Long-term effect of dietary sucrose and starch on serum lipids and atherosclerosis in miniature swine. Exp Mol Pathol 15: 21–33

Takemoto TI (1975) Sex differences in the response of different kinds of dietary carbohydrates in rats. Tohoku J Exp Med 115: 213–222

Taylor DD, Conway ES, Schuster EM, Adams M (1967) Influence of dietary carbohydrates on liver content and on serum lipids in relation to age and strain of rats. J Nutr 91: 275–282

Tomatis L (1986) Diet, nutrition and cancer: concluding remarks and future perspectives. In: Hayashi Y, Nagao M, Sugumura T et al. (ed) Diet, nutrition and cancer. Japan Scientific Societies Press, Tokyo; VNU Science Press BV, Utrecht

US/FDA/STF (1986) Evaluation of health aspects of sugars contained in carbohydrate sweeteners. In: Glinsmann WH, Irausquin H, Youngmee KP (ed) Report from FDA's Sugar Task Force (reprinted from: J Nutr 116: S1–S216)

Von der Decken A, Astrom S, Arrhenius EK (1981) Increased transcription activity of rat liver chromatin after protein restriction and limited digestion of nuclei with micrococcus nuclease. J Nutr 111: 1249–1257

Wada F (1972) The nutritional efficiencies of Maltit (maltitol) and maltit syrup (hydrogenated malt-conversion starch syrup). Proc 24th Jap Food Health Assoc 12–13 October

Waddel CA, Desai ID (1981) The use of laboratory animals in nutrition research. World Rev Nutr Diet 36: 206–222

WARF Institute (1974) Sucrose: a literature review. Wisconsin Alumni Research Foundation, Madison, Wisconsin (prepared for the Sugar Association, Washington DC)

Warnick CT, Lazarus HM (1982) Effect of a protein-free diet and fasting on RNA polymerase activity in mice. J Nutr 112: 293–298

Whitnah CH, Bogart R (1936) Reproductive capacity of female rats as affected by kind of carbohydrates in the ration. J Agric Res 53: 527–532

WHO (1984) EHC 30: Principles for evaluating health risks to progeny associated with exposure to chemicals during pregnancy. World Health Organization, Geneva, 17 pp

WHO (1987) EHC 70: Principles for the safety assessment of food additives and contaminants in food. World Health Organization, Geneva, 174 pp

WHO/FAO (1970) Thirteenth report of the Joint FAO/WHO Expert Committee on Food Additives. World Health Organization, Geneva (WHO Technical Report Series No. 445)

WHO/FAO (1977) Summary of toxicological data of certain food additives. World Health Organization, Geneva (WHO Food Additives Series No 12)

WHO/FAO (1978) Summary of toxicological data of certain food additives and contaminants. World Health Organization, Geneva (WHO Food Additives Series No 13)

WHO/FAO (1982) Twenty-sixth report of the Joint FAO/WHO Expert Committee on Food Additives. World Health Organization, Geneva (WHO Technical Report Series No. 683)

WHO/FAO (1987) Toxicological evaluation of certain food additives and contaminants. Cambridge University Press, Cambridge (WHO Food Additives Series No 20)

Wogan GN, Newberne PM (1967) Dose-response characteristics of aflatoxin B-1 carcinogenesis in the rat. Cancer Res 27: 2370–2376

Zarattini A (1941) Sulla produzione sperimentale del sarcoma nei ratti mediante somministrazione parenterale di glucosio. Tumori 14: 77–84

Chapter 8

# Evaluations

In recent years, there have been reports from several authoritative expert groups on simple and complex carbohydrates that have specifically included sucrose. These reports consist of literature reviews, evaluation of available scientific information, and assessments of the nutritional and safety aspects of sucrose. Some also contain recommendations on its nutritional role, including the possible links with health and disease.

The editors have identified five representative scientific bodies. These are:

1. The British Nutrition Foundation Task Force on Sugars and Syrups (BNF/TFSS 1987);
2. The United States Food and Drug Administration Sugars Task Force (US/FDA/STF 1986);
3. The Panel on Simple and Complex Carbohydrates (PSCC, 1986);
4. The Joint FAO/WHO Meeting on Carbohydrates in Human Nutrition (FAO/WHO 1980); and
5. The Select Committee on GRAS Substances of the Federation of American Societies for Experimental Biology (SCOGS/FASEB 1976).

While the process of ascertaining the validity of a particular experimental animal study or epidemiological report is limited only to one scientific work reported in the pertinent literature, the process of evaluation implies a series of steps, which include a comprehensive collection of as many works as possible, an effort to selectively identify the valid ones, the attempt to interpret the scientific findings based on the collective thinking of the selected experts, and, finally, to reach, wherever possible, a consensus in the light of the guiding principles of science. It is because of the complexity and difficulty inherent to any evaluative process that reports from authoritative expert groups are regarded highly by the scientific community.

Excerpts from the reports of the groups identified above are summarized below. The reader may wish to consult the entire report to evaluate the basis for the conclusions drawn by these expert bodies.

## 8.1 Dental Caries

While a major reduction in the number of potentially cariogenic intakes is likely to mean that less sugar is eaten overall, advice aimed at reducing the total amount of sugar eaten in order to reduce the incidence of caries is as unlikely to be as effective as advice aimed at reducing the frequency of eating and drinking. Unfortunately, it is not yet possible to determine with certainty a safe number of cariogenic intakes below which caries are unlikely to occur (BNF/TFSS 1987).

The aetiology of dental caries is multifactorial, with oral microbial flora, host factors, and dietary factors being the three major factors involved in the development of carious lesions. While it is not possible to quantitate the effects of current sugars consumption on the incidence of dental caries, present evidence supports the contention that the current average of 90th-percentile consumption levels of sugars contribute significantly to caries incidence (US/FDA/STF 1986).

Turning to the possible adverse effects of dietary carbohydrates, and simple carbohydrates in particular, the Panel of Simple and Complex Carbohydrates in Human Nutrition had in mind the recent reports on sucrose and other simple sugars, at levels currently consumed in the USA, developed by Life Sciences Research Office of the Federation of American Societies for Experimental Biology. This report could find no evidence, apart from dental caries, for any untoward pathology arising from the current per capita consumption of sucrose and reaffirmed the GRAS status for sucrose. However, there is a relationship between dental caries and simple carbohydrate (notably sucrose) intake. This positive relationship is based not on the amount of simple carbohydrate consumed, but on the frequency of intake, form of sugar consumed, and length of oral sugar clearance time. Physiological events that magnified this relationship include newly erupted teeth, and teeth in the elderly, in late pregnancy, and during lactation. Though there are numerous pathological situations that also exaggerate the cariogenic response to sucrose, such as various drugs, these were not considered as part of the panel's remit. Nor did the panel consider, in any aspect, the sugar alcohols and the polyols. Some preliminary studies suggest that cooked starch may be cariogenic, but more evidence is needed to confirm this view (PSCC 1986).

Although results demonstrate that dental caries remain a major health problem among children in the USA, it is clear that a substantial and encouraging decrease in the prevalence of the disease has taken place during the past decade. It is conceivable that the frequency of eating sugars has decreased, but data on total per capita sugar consumption show no evidence of reduction during the past 10 years (Brunelle and Carlos 1982).

Dental caries, though inherently multifactorial, can be considered, almost unequivocally, as an infectious, diet-dependent disease. In particular, the influence of a number of carbohydrates, especially sucrose, metabolized by the oral microflora, appears to be of crucial importance in all stages of development of this disease (FAO/WHO 1980).

Reasonable evidence exists that sucrose is a contributor to the formation of dental caries when used at levels that are now current and in the manner now practised (SCOGS/FASEB 1976).

## 8.2 Obesity

Any item of the diet that can be readily eaten, even by people who are not hungry, may lead to over-eating and obesity. It is logical to advise over-weight people to reduce their intake of sugar, since this could effect a decrease in energy intake without an associated decrease in important nutrients (BNF/TFSS 1987).

The available data support the conclusion that sugars do not have a unique role in the aetiology of obesity (US/FDA/STF 1986).

Despite the deep-rooted view of the general public that the consumption of starchy foods and sugar favours the development of obesity, no reliable evidence exists that any specific nutrient contributing excess energy is implicated (FAO/WHO 1980).

## 8.3 Cardiovascular Disease

Though obesity is associated with coronary heart disease (CHD), where multivariate techniques of analysis have been used, it has often not remained a statistically significant predictor of CHD. The obesity effect on CHD may be mainly due to the dietary intake of fat, which accounts for a considerably higher proportion of energy intake than sugar (BNF/TRSS 1987).

Current levels of sugars consumption have not been demonstrated to be an adverse risk factor in terms of blood lipid and lipoprotein profiles for normal individuals. No evidence was found to support the contention that current dietary intake of sugars contributes to the development of hypertension (US/FDA/STF 1986).

The relationship of sucrose to CHD has been examined in a number of papers during the past 20 years. A comprehensive review of the topic concludes that a relationship between sucrose intake and the development of CHD is not supported by the available clinical and epidemiological data (FAO/WHO 1980). There is no conclusive evidence that dietary sugars are an independent risk factor for coronary artery disease in the general population (US/FDA/STF 1986).

The experimental evidence associating sucrose with cardiovascular disease is less than clear. However, there is no evidence that ingestion of sucrose in the concentrations that occur in the average diet causes significant elevations in blood cholesterol or other lipids. Furthermore, it would appear that the primary dietary factors involved in cardiovascular disease are the nature and amount of fat in the diet (SCOGS/FASEB 1976).

## 8.4 Diabetes

There is no strong evidence that a specific component of the diet, such as sugar, causes diabetes. Consequently, there are no reasons that support a change in the 1976 Select Committee on GRAS Substances report conclusion that the consumption of sugars is related to diabetes other than as a non-specific source of calories. There is increasing concern with glycaemic regulation because of its possible causal relation to the development of complications in diabetes mellitus; however, the glycaemic properties of diets (as noted in the review of glucose tolerance) vary according to food sources, method of food preparation, meal pattern, the presence of other dietary constituents, and physical activity. The weight of the current scientific evidence supports the view that present levels in sugar consumption patterns are relatively insignificant factors in the management of diabetes mellitus (US/FDA/STF 1986).

There is no conclusive evidence that the consumption of simple sugars is of aetiological significance in diabetes mellitus (FAO/WHO 1980).

Findings linking ingestion of sucrose with diabetes are essentially circumstantial. There is no plausible evidence that sucrose, except as a non-specific source of excessive calories, is related to the disease (SCOGS/FASEB 1976).

## 8.5 Glucose Tolerance

Since insulin stimulates the production of liver enzymes, a person who has had a high carbohydrate intake some days before a test will appear to have a smaller glucose response than someone who has previously eaten little starch or sugars (BNF/TFSS 1987).

There is no persuasive scientific evidence that sugars, as they are currently consumed by the population of the USA, are an independent risk factor associated with the development of impaired glucose tolerance. Though there is evidence to support the existence of a subset of "carbohydrate-sensitive" males that have Type IV hyperlipoproteinaemia and an increased insulin response to oral sucrose loads, the existence of a unique risk to this population from carbohydrate sweeteners has not been established. The concept that dietary mono- and disaccharides are more contributory to glycaemia than starches is not supported by recent investigations, which show that glycaemic responses to carbohydrate-containing foods vary according to food source, method of food preparation, meal pattern, the presence of other dietary constituents, and physical activity (US/FDA/STF 1986).

## 8.6 Cancer

No adequate evidence in support of a specific role of sugars in breast or colorectal cancer appears to be available (BNF/TFSS 1987).

There is no scientifically based evidence that demonstrates that sugars, as currently consumed, are an independent variable in augmenting the risk of carcinogenesis; specific sugars that are the subject of the current GRAS review process have not been demonstrated to be carcinogenic (US/FDA/STF 1986).

## 8.7 Gallstones

Cross-sectional epidemiological studies do not provide an entirely consistent indication of the influence of dietary factors (on the development of gallstones), though obesity appears to be commonly associated with the condition (BNF/TFSS 1987).

There is no convincing scientific evidence that sugars consumption is an independent risk factor in the production of gallstones (US/FDA/STF 1986).

## 8.8 Urinary Tract Stones

Epidemiological surveys have suggested that obesity is commonly a characteristic of those suffering from this condition (urinary tract stones) (BNF/TFSS 1987).

Acute changes in renal excretion of calcium and magnesium occur following the ingestion of parenteral infusion of sugars and other dietary components, such as amino acids. This response appears to be increased in certain individuals susceptible to nephrolithiasis, and the avoidance of consuming high levels of sugars is one dietary factor that may be important in the medical management of these individuals (US/FDA/STF 1986).

## 8.9 Influence on Behaviour

There is no substantive evidence that the consumption of sugars is responsible for behavioural changes in children or in adults, with the exception of the relatively rare hypoglycaemias that are present in the population.

The suggestion that reactive hypoglycaemia is correlated with behavioural or mood changes in a substantial portion of the population is not supported by experimental evidence.

The theoretical possibility that dietary sugars modify behaviour through effects on neurotransmitter metabolism in the central nervous system has not been documented (US/FDA/STF 1986).

## 8.10 Various Other Health Issues

There are individuals with intolerance to specific sugars (mainly lactose) caused by deficiencies of metabolizing enzymes; such individuals should avoid the consumption of these sugars.

A wide variety of reactions, often described as food allergies, but having no associated immunological reaction or defined reaction mechanism, have been alleged to be causally related to the consumption of many foods or food ingredients, including sugars. Scientific verification of such adverse reactions attributable to sugars consumption was not found.

Food allergies defined in terms of immune response to ingested sugars appear to be extremely rare; however, small amounts of allergens derived from the source of the sugar (e.g., corn, cane, beet) may be present in sufficient quantities to produce allergic reactions in sensitive individuals.

There is no firm evidence that sugars, as currently consumed, interfere with the bioavailability of vitamins, minerals, or trace nutrients; furthermore, there is no scientific evidence supporting the notion that dietary imbalances are preferentially caused by increased sugars consumption (US/FDA/STF 1986).

## 8.11 Overall Assessment

Other than the contribution to dental caries, there is no conclusive evidence on sucrose that demonstrates a hazard to the general public when sucrose is consumed at the levels that are now current and in the manner now practised (US/FDA/STF 1986; SCOGS/FASEB 1976).

## 8.12 References

BNF/TFSS (1987) Sugars and syrups. The report of the British Nutrition Foundation's Task Force. British Nutrition Foundation, London

Brunelle JA, Carlos JP (1982) Changes in the prevalence of dental caries in US schoolchildren, 1961–80. In: Glass RL (ed) The first international conference on the declining prevalence of dental caries. J Dent Res 61: 1346–1351

FAO/WHO (1980) Carbohydrates in human nutrition. Food and Agriculture Organization of the United Nations, Rome (Food and Nutrition Paper No 15)

PSCC (1986) Simple and complex carbohydrates. Proceedings of a conference on diet and health: scientific concepts and principles. Am J Clin Nutr 45: 1039–1040

SCOGS/FASEB (1976) Evaluation of the health aspects of sucrose as a food ingredient. Life Sciences Research Office (LSRO)/Federation of American Societies for Experimental Biology (FASEB), United States Food and Drug Administration, Washington DC (Prepared for US FDA under contract No FDA/223-75-2004)

US/FDA/STF (1986) Evaluation of health aspects of sugars contained in carbohydrate sweeteners. In: Glinsmann WH, Irausquin H, Youngmee KP (eds) Report from FDA's Sugar Task Force (Reprinted from: J Nutr 116: S1–S216)

Chapter 9

# Conclusions and Recommendations

Reports of two expert committees on sugars have recently become available. The expert committees are the British Nutrition Foundation's Task Force on Sugars and Syrups (BNF/TFSS 1987) and the United States Food and Drug Administration (FDA) Sugars Task Force (US/FDA/STF 1986).

These reports give readers a broadly based description of sucrose, including how it is produced and why it is eaten. The reports also show how sucrose, apart from providing sweetness and readily absorbed energy to the human body, plays an important role in food manufacture, for example, as a preservative.

These authoritative scientific bodies' technical reports also deal with nutritional and safety aspects of sucrose and their relevance to human health.

After critically examining the extensive relevant literature and scientifically assessing the factual evidence on the subject, these reports come to conclusions and formulate recommendations aimed at providing assistance to the scientific/medical community, food regulatory authorities, and the general consumer public.

The most salient conclusions and recommendations are outlined below.

1. Sucrose and other sugars (e.g., glucose, lactose, and fructose) occur both as natural constituents of foods and as ingredients of prepared foods, whether home-made or manufactured.
2. The sugars found in food or incorporated during food preparation provide food energy in an inexpensive form and contribute to the acceptability of food providing essential nutrients.
3. Sugars are used in food preparation, partly for sweetness and partly for the other physical and chemical properties they possess. In high concentrations, they also act as food preservatives.
4. Diet is one important factor in the development of dental caries. The frequency of intake of sugar and other fermentable carbohydrates is the significant dietary factor. However, the aetiology of dental caries is multifactorial, with oral microbial flora, host factors, and dietary factors being the three major factors involved in the development of carious lesions.
5. Consumption of sugars is not related to diabetes, other than as a non-specific source of calories.

6. There is no conclusive evidence that dietary sugars are an independent risk factor for coronary artery disease.
7. The available data support the conclusion that sugars do not have a unique role in the aetiology of obesity.
8. There is no convincing scientific evidence that sugars consumption is an independent factor in the production of gallstones.
9. There is no firm evidence that sugars interfere with the bioavailability of vitamins, minerals, or trace nutrients, nor is there scientific evidence supporting the notion that dietary imbalances are preferentially caused by increased consumption.
10. There is no scientifically based evidence that demonstrates that sugars are an independent variable in the risk of carcinogenesis. Specific sugars have not been demonstrated to be carcinogenic.
11. There is no substantive evidence that the consumption of sugars is responsible for behavioural changes in children or adults.
12. The suggestion that reactive hypoglycaemia is correlated with behavioural or mood changes is not supported by experimental evidence.
13. The theoretical possibility that dietary sugars modify behaviour through effects on neurotransmitter metabolism in the central nervious system has not been documented.
14. No evidence was found to support the contention that current dietary intake of sugars contributes to the development of hypertension.
15. Other than the contribution to dental caries, there is no conclusive evidence on sugars that demonstrates a hazard to the general public.

## 9.1 References

BNF/TFSS (1987) Sugars and syrups. The report of the British Nutrition Foundation's Task Force. British Nutrition Foundation, London

US/FDA/STF (1986) Evaluation of health aspects of sugars contained in carbohydrate sweeteners. In: Glinsmann WH, Irausquin H, Youngmee KP (eds) Report from FDA's Sugar Task Force (Reprinted from: J Nutr 116: S1–S216)

Section II

# Bibliography

# Selected References


Aas K (1984) Antigens in food. Nutr Rev 42: 85–91

Abderhalden E, Buadze S (1932) Appearance of saccharase in blood serum and urine after parenteral administration of sucrose. Fermentforsch 13: 228–243

Abderhalden E, London ES (1926) The appearance of amino acids and sugar in the blood and lymph after their introduction into the blood and intestine. Pflugers Arch Ges Physiol 212: 740

Abe Y (1963) The nature of plateau in the membrane action potential of cardiac muscle fiber of the rabbit. Kyushu J Med Sci 14: 85–97

Abelin I, Pfister H (1954) The reaction of the adrenal to the feeding of carbohydrate. Acta Endocrinol (Copenh) 16: 19–29

Abell LL, Mosbach EH, Kendall FE (1965) Influence of diet on the hypocholesterolemic action of methyltestosterone in dogs. J Nutr 87: 285–292

Abelson DC, Pergola G (1984) The effect of sucrose concentration on plaque pH in vivo. Clin Prev Dent 6: 23–26

Abrahams VC, Pickford M (1954) Simultaneous observations on the rate of urine flow and spontaneous uterine movements in the dog, and their relationship to posterior lobe activity. J Physiol (Lond) 126: 329–346

Acheson K, Flatt JP, Jéquier E (1982) Glycogen synthesis versus lipogenesis after a 500 gram carbohydrate meal in man. Metabolism 31: 1234–1240

Acheson KJ, Schutz Y, Bessard T, Ravussin E, Jeguier E (1984) Nutritional influences on lipogenesis and thermogenesis after a carbohydrate meal. Am J Physiol 246: E62–70

Ackermann, RH (1981) Auswirkungen hochdosierter parenteraler Fruktose-, Glukose- und Mannitzufuhr auf die Rattenniere. Infusionstherapie 1: 9–15

Adamkiewicz VW, Sacra PJ (1967) Histamine and sugars. Fed Proc 26: 224–230

Adamu L, Joseph PK, Augusti DT (1982) Hypolipidemic action of onion and garlic unsaturated oils in sucrose fed rats over a two-month period. Experientia 38: 899–901

Addy M, Perriam E, Sterry A (1982) Effects of sugared and sugar-free chewing gum on the accumulation of plaque and debris on the teeth. J Clin Periodontol 9: 346–354

Adelman RC, Ballard FJ, Weinhouse S (1967) Purification and properties of rat liver fructokinase. J Biol Chem 242: 3360–3365

Adlung J, Grazikowske H, Oltmanns D (1977) Uber den Einfluss von Fruktose-infusionen auf die Glukosebildung und -oxidation bei Gesunden und Kranken mit Leberzirrhose. Infusionstherapie 4: 266–270

Adrian J (1974) Nutritional and physiological consequences of the Maillard reaction. World Rev Nutr Diet 19: 71–122

Adshead VM, Parke JM, Chambers PJ, Davies RM, Cole JA (1983) An in vitro study of the role of sucrose and interactions between oral bacteria in possible mechanisms of dental plaque formation. Arch Oral Biol 28: 723–727

Affarah HB, Hall WD, Heymsfield SB, Kutner M, Wells JO, Tuttle EP (1986) High carbohydrate diet: antinatriuretic and blood pressure response in normal man. Am J Clin Nutr 44: 341–348

Aguilar JR, Arroyave G, Gallardo C (1977) Manual de supervisión y control del programa de fortificación de azucar con vitamina A. Instituto de Nutrición de Centro America y Panama, Guatemala City (Publ INCAP E-913)

AHA Committee Report (1982) Rationale of the diet–heart statement of the American Heart Association. Nutr Today Nov/Dec: 15–19

Ahrens EH Jr, Hirsch J, Oette K, Farquhar JW, Stein Y (1961) Carbohydrate-induced and fat-induced lipemia. Trans Assoc Am Physicians 73: 134–146

Ahrens RA (1974) Sucrose, hypertension, and heart disease: an historical perspective. Am J Clin Nutr 27: 403–422

Ahrens RA, Demuth P, Lee MK, Majkowski JW (1980) Moderate sucrose ingestion and blood pressure in the rat. J Nutr 110: 725–731

Ahrens RA, Garland SL, Kigutha HN, Russek E (1985) The disaccharide effect of sucrose feeding on glucoronide excretion and bile concentration of injected phenolphthalein in guinea-pigs. J Nutr 115: 288–291

Ainamo A, Ainamo J (1984) The dentition is intended to last a lifetime. Int Dent J 34: 87–92

Ainamo J, Sjoblom M, Ainamo A, Tiainen L (1977) Growth of plaque while chewing sucrose and sorbitol flavoured gum. J Clin Periodontol 4: 151–160

Akamatsu T, Ohba M, Narahara N et al. (1970) Effect of abdominal surgery on glucose tolerance, plasma levels of insulin and glycogenic amino acids. Keio J Med 19: 103–114

Akgun S, Ertel NH (1980) A comparison of carbohydrate metabolism after sucrose, sorbitol, and fructose meals in normal and diabetic subjects. Diabetes Care 3: 582–585

Akgun S, Ertel NH (1981) Plasma glucose and insulin after fructose and high-fructose corn syrup meals in subjects with non-insulin-dependent diabetes mellitus. Diabetes Care 4: 464–467

Akinyanju PA, Qureshi RU, Salter AJ, Yudkin J (1968) Effect of an atherogenic diet containing starch or sucrose on the blood lipids of young men. Nature (Lond) 218: 975–977

Albanese AA, Higgons RA, Orto L, Belmont A, DiLallo R (1954) Effect of age on the utilization of various carbohydrates by man. Metabolism 3: 154–159

Albanese AA, Orto L, Rossy J, DiLallo R, Belmont A (1955) Effect of carbohydrates on blood amino nitrogen. Metabolism 4: 160–165

Alberti KGMM, Keen H (1978) Summary and discussion of session I: the reasons for substituting or not substituting sugar. Health and sugar substitutes. In: Proceedings of the ERGOB conference, Geneva. Karger, Basel, pp 290–298

Albrink MJ, Ullrich IH (1986) Interaction of dietary sucrose and fibre on serum lipids in healthy young men fed high carbohydrate diets. Am J Clin Nutr 43: 419–428

Aldinger SM, Speer VC, Hays VW, Catron DV (1961) Effect of saccharin and sucrose on the performance of young pigs. J Anim Sci 20: 249–254

Alfin-Slater RB (1967) Carbohydrate-lipid effects on cholesterol metabolism. J Dairy Sci 50: 781–786

Alias AG (1985) Increased production and consumption of cane sugar in the Third World to prevent undernutrition and famine. N Engl J Med 313: 894

Allen RJL, Leahy JS (1966) Some effects of dietary dextrose, fructose, liquid glucose and sucrose in the adult male rat. Br J Nutr 20: 339–347

Allen RJL, Brook M, Lister RE, Sim AK, Warwick MH (1966) Metabolic differences between dietary liquid glucose and sucrose (baboon, *Papio anubis*). Nature (London) 211: 1104

Alman JE (1982) Declining caries prevalence – statistical considerations. J Dent Res 61(Sp Iss): 1361–1363

Al Nagdy S, Miller DS, Qureshi RU, Yudkin J (1966) Metabolic differences between starch and sucrose. Nature (Lond) 209: 81

Al Nagdy S, Miller DS, Yudkin J (1970) Changes in body composition and metabolism induced by sucrose in the rat. Nutr Metab 12: 193

Amano S, Ito S (1943) Glykogen und Geschwulstbildung. Gann 37: 200–203

American Council on Science and Health (1986) Sugars and your health. American Council on Science and Health, New York

American Diabetes Association (1980) Statement on sweeteners. J Am Diet Assoc 76: 549

Andersen DW, Filer LJ Jr, Wu-Rideout YC, White LB, Stegink LD (1982) Utilization of intravenously administered glucose-oligosaccharides in growing miniature pigs. Pediatr Res 16: 304–309

Andersen DW, Daabees TT, Applebaum AE, Filer LJ Jr, Stegink LD (1983) Utilization of intravenously administered b-cellobiose and maltose by young pigs. J Nutr 113: 1039–1045

Andersen DW, Filer LJ Jr, Stegink LD (1983) Utilization of intravenously infused glucose-oligosaccharides in fasted and fed pigs. J Nutr 113: 430–435

Anderson EP, Smith JK, Elvehjem CA, Phillips PH (1947) Dental caries in the cotton rat. IX.

Effect of milk rations. Proc Soc Exp Biol Med 66: 67–69
Anderson JA (1984) Non-immunologically-mediated food sensitivity. Nutr Rev 42: 109–116
Anderson JT, Grande F, Matsumoto Y, Keys A (1963) Glucose, sucrose and lactose in the diet and blood lipids in man. J Nutr 79: 349–359
Anderson JW (1977) High polysaccharide diet studies in patients with diabetes and vascular disease. Cereal Food World 22: 12–13, 22
Anderson JW, Ward K (1978) Long-term effects of high-carbohydrate, high-fiber diets on glucose and lipid metabolism: a preliminary report on patients with diabetes. Diabetes Care 1: 77–82
Anderson JW, Herman RH, Zakim D (1973) Effect of high glucose and high sucrose diets on glucose tolerance of normal men. Am J Clin Nutr 26: 600–607
Anderson K, Dobrota M, Haga HJ (1979) The effect of sucrose on assaying enzymes and protein in the subcellular fractions of the rat kidney cortex. J Biochem Biophys Methods 1: 309–311
Anderson RA Jr, Oswald C, Leto S, Zaneveld LJD (1980) Inhibition of human acrosin by fructose and other monosaccharides. Biol Reprod 22: 1079–1082
Anderson RJ, Bradnock G, Beal JF, James PMC (1982) The reduction of dental caries prevalence in English school children. J Dent Res (Sp Iss): 1311–1316
Anderson TA (1969) Effect of carbohydrate source on serum and hepatic cholesterol levels in the cholesterol fed rat. Proc Soc Exp Biol Med 130: 884–887
Anderson WAD (1941) Sucrose nephrosis and other types of renal tubular injuries. South Med J 34: 257–263
Anderson WAD, Bethea WR Jr (1940) Renal lesions following administration of hypertonic solutions: 6 cases. J Am Med Assoc 114: 1983–1987
Andersson G, Brohult J, Sterner G (1969) Case report: increasing metabolic acidosis following fructose infusion in two children. Acta Paediatr Scand 58: 301–304
Andlaw RJ (1977) Diet and dental caries – a review. J Hum Nutr 31: 45–52
Angel JF, Back DW (1981) Immediate and late effects of premature weaning of rats to diets containing starch or low levels of sucrose. J Nutr 111: 1805–1815
Anonymous (1934) Sweets in diet, especially of children. Committee on Foods and Nutrition. J Am Med Assoc 103: 110
Anonymous (1942) Some nutritional aspects of sugar, candy and sweetened carbonated beverages. Committee on Foods and Nutrition. J Am Med Assoc 120: 763–765
Anonymous (1965) Sucrose intolerance: an enzymatic defect. Nutr Rev 23: 101–102
Anonymous (1965) Sugar and starch as foods. Br Med J i: 330–331
Anonymous (1966) Sugar in food and in organisms. A bibliography. Brasil Açucareiro 67(6): 57–59
Anonymous (1969) Sugar, chromium, and serum cholesterol. Nutr Rev 27: 185–186
Anonymous (1970) Sugar and coronary heart disease. Nutr Rev 28: 228–229
Anonymous (1971) Influence of age and strain on the response of rats to dietary fructose and sucrose. Nutr Rev 29: 124–126
Anonymous (1971) Sweet mystery of life. Food Cosmet Toxicol 9: 439–442
Anonymous (1973) Statement on hypoglycemia. J Am Med Assoc 223: 682
Anonymous (1974) Frequency of eating and dental caries prevalence. Nutr Rev 32: 139–141
Anonymous (1974) The role of sugars in hyperlipidemia. Nutr Rev 32: 340–342
Anonymous (1977) Sucrose malabsortion. Br Med J i: 1558–1559
Anonymous (1978) Too much sugar? Consumer Rep 43: 136–142
Anonymous (1979) Classification and diagnosis of diabetes mellitus and other categories of glucose intolerance. Diabetes 28: 1039–1057
Anonymous (1979) Diet and dental caries. Joint report by the British Dietetic Association and the British Paedodontic Society. Proc Br Paedodont Soc 9: 11–12
Anonymous (1979) Sweeteners and dental disease in children. Br Dent J 147: 83
Anonymous (1982) Retinopathic effect of sucrose-rich diets due to fructose. Nutr Rev 40: 117–118, 128
Anonymous (1983) Statement on diet and dental caries. J Am Dent Assoc 107: 78
Anonymous (1983) Sucrose transport into *Streptococcus mutans*. Nutr Rev 41: 256–258
Anonymous (1984) Diet, nutrition, and oral health: a rational approach for the dental practice. J Am Dent Assoc 109: 20–32
Anonymous (1984) Adverse food reactions that do not involve immunologic reactions. In: Anderson JA, Sogn DD (eds) Adverse reactions to foods. American Academy of Allergy and Immunology Committee on Adverse Reactions to Foods, National Institute of Allergy and Infectious Diseases, Rockville, Maryland, pp 116–121 (NIH Publ No 84: 2442)
Anonymous (1984) Retinopathic effect of sucrose-rich diets due to fructose. Nutr Rev 42: 29–30

Anonymous (1985) Sucrose. Memorandum of conference. Cancer Assessment Committee. Center for Food Safety and Applied Nutrition, US Food and Drug Administration, Washington DC

Anonymous (1985) Dietary carbohydrate intake and catecholamine secretion in man. Nutr Rev 43: 138–140

Antar MA (1968) Studies of thrombus formation and fatty acid composition of serum lecithins and cephalins in man during high sugar and starch diets. J Atheroscl Res 8: 569–578

Antar MA, Little JA, Lucas C, Buckley GC, Csima A (1970) Interrelationship between the kinds of dietary carbohydrate and fat in hyperlipoproteinemic patients. Atherosclerosis 11: 191–201

Anton HU (1953) Glycemia: blood sugar, hypoglycemia behavior during labor, significance of uterine contractions. Zentralbl Gynakol 75: 129–138

Antonis A, Bersohn I (1961) The influence of diet on serum-triglycerides in South African white and Bantu prisoners. Lancet I: 3–9

Appleton TC, Pelc SR, Dingle JT, Fell HB (1969) Endocytosis of sugars in embryonic skeletal tissues in organ culture. III. Radiographic distribution of sucrose-$^{14}C$. J Cell Sci 4: 133–137

Aptekar SG, Minina AV (1971) The fatty acid composition of the tissue lipids of rats maintained on atherogenic diets with qualitatively different carbohydrates. Vopr Pltan 30: 16–20

Arakawa S (1956) Experimental production of sarcoma of mice with the subcutaneous high concentrated sugar solutions, especially lactose and mixture of laevulose and glucose. Gann 48: 363–364

Araujo PE, Mayer J (1979) Dietary caffeine, glucose tolerance and insulin sensitivity in mice. J Food Sci 44: 1761–1764

Araujo RL, Borges EL, Silva JDB, Palhares RD, Vieira EC (1978) Effect of the intake of vitamin A fortified sugar by pre-school children. Nutr Rep Int 18: 429

Araujo RL, Souza MSL, Mata-Machado AJ et al. (1978) Response of retinol serum levels to the intake of vitamin A-fortified sugar by pre-school children. Nutr Rep Int 17: 307

Arbogast BW, Berry DL, Newell CL (1984) Injury of arterial endothelial cells in diabetic, sucrose-fed and aged rats. Atherosclerosis 51: 31–45

Ariyama H, Takabasi K (1931) The assimilation limits of carbohydrates. Bull Agric Chem Soc Jpn 6: 5–8

Ariyama H, Takabasi K (1931) The relative nutritive value of various carbohydrates and related compounds. Bull Agric Chem Soc Jpn 6: 1–5

Armbrecht HJ, Wasserman RH (1976) Enhancement of $Ca^{++}$ uptake by lactose in the rat small intestine. J Nutr 106: 1265–1271

Armstrong BK (1979) Diet and hormones in the epidemiology of breast and endometrial cancers Nutr Cancer 1: 90–95

Armstrong B, Doll R (1975) Environmental factors and cancer incidence and mortality in different countries, with special reference to dietary practices. Int J Cancer 15: 617–631

Arnfred T (1977) Psicosuria after ingestion of honey cake (gingerbread). Ugeskr Laeger 139: 647 (in Danish)

Arnould E (1952) White sugar: dynamogenic economic food. Presse Med 60: 1391

Aron H, Hirsch H (1929) Treatment of pyuria in infants and children by means of exclusive sugar diet. Monatsschr Kinderheildk 43: 385–396

Arroyave G, Aguilar Jr, Portela IE (1975) Manual de operaciones para la fortificación de azucar con vitamina A. Instituto de Nutricion de Centro America y Panama, Guatemala City (Publ INCAP E-853)

Arroyave G, Aguilar JR, Flores M, Guzman MA (1979) Evaluation of sugar fortification with vitamin A at the national level, Pan American Health Organization, Washington DC (Sci Publ No 384)

Arroyave G, Mejia LA, Aguilar JR (1981) The effect of vitamin A fortification of sugar on the serum vitamin A levels of preschool Guatemalan children: a longitudinal evaluation. Am J Clin Nutr 34: 41–49

Arvidsson Lenner R (1976) Studies of glycemia and glucosuria in diabetics after breakfast meals of different composition. Am J Clin Nutr 29: 716–725

Arvill A, Johansson B, Jonsson O (1968) Osmotic effects on isolated vascular smooth muscle as reflected in the distribution of water, urea-$^{14}C$ and sucrose-$^{14}C$ (rat portal vein). Acta Physiol Scand 73: 8A

Ashby J, De Serres FJ, Draper M et al. (1985) Evaluation of short-term tests. Report of the International Programme on Chemical Safety's Collaborative Study on In Vitro Assays. Elsevier, Amsterdam (Progress in Mutation Research, vol 5)

Ashley BCI, Whyte HM (1967) Studies in acute under-nutrition. The effects of carbohydrate, cortisone, and an anabolic hormone on protein, water, and electrolyte metabolism. Aust J Exp Biol Med Sci 45: 249–259

Ashley DVM, Fleury MO, Golay A, Maeder E, Leathwood PD (1985) Evidence for diminished brain 5-hydroxy tryptamine biosynthesis in obese diabetic and non-diabetic humans. Am J Clin Nutr 42: 1240–1245

Astrom S, Arrhenius EK, Von der Decken A (1981) Reduced transcription activity of the rat liver chromatin after protein restriction and selective digestion of nuclei with micrococcus nuclease. J Nutr 111: 1258–1264

Atwell ME, Waterhouse C (1971) Glucose production from fructose. Diabetes 20: 193–199

Audia M (1959) Tendon water and electrolyte and saccharide spaces in normal versus adrenalectomized rats. Am J Physiol 197: 1086–1088

Auricchio S, Rubino A, Prader A et al. (1965) Intestinal glycosidase activities in congenital malabsorption of disaccharides. J Pediatr 66: 555–564

Ayers C, Kolthoff C, Durr D (1981) Identification and control of sugar sources in the diet. Gen Dent July/Aug: 312–315

Babington MA, Spadaro DC (1982) Cariogenic medications. Pediatr Nurs 8: 165–171

Babonneix L (1935) Sugars in diet. Gaz Hop 108: 591–592

Bach C, Thiriez H, Schaeffer P, Cayroche P (1962) Sucrose intolerance in an infant. Arch Fr Pediatr 19: 1138–1145

Bacigalupo FA, Couch JR, Pearson PB (1950) Effect of sucrose, lactose, galactose, and fructose on fecal excretion of biotin, riboflavin and PGA (pteroyglutamic acid) in mature cotton rat. Am J Physiol 162: 131–135

Back DW, Angel AF (1982) Effects of premature weaning on the metabolic response to dietary sucrose in adult rats. J Nutr 112: 978–985

Bacon BR, Park CH, Fowell EM, McLaren CE (1984) Hepatic steatosis in rats fed diets with varying concentrations of sucrose. Fundam Appl Toxicol 4: 819–826

Badawy AB, Evans M (1976) The regulation of rat liver tryptophan pyrrolase activity by reduced nicotinamide-adenine dinucleotide (phosphate). Experiments with glucose and nicotinamide. Biochem J 156: 381–390

Badawy AB, Punjami NF, Evans M (1980) Unsuitability of control sucrose or glucose in studies of the effects of chronic ethanol administration on brain 5-hydroxytryptamine metabolism. J Pharmacol Methods 3: 167–171

Bagley R, Ford M, Green LF (1976) Changes in plasma lipids after substituting glucose-syrup solids for table sucrose. Proc Nutr Soc 35: 68A–69A

Baglioni S, Bracaloni L, Galamini A (1927) The physiological working of alcohols. Changes in the glycemia and alcoholemia after introduction of alcoholic drink or of sucrose. Atti R Accad Lincei Roma 6(5): 34–49

Baily CS, Hsiao S, King JE (1986) Hedonic reactivity to sucrose in rats: modification by pimozide. Physiol Behav 38: 447–452

Baird JD (1973) The role of obesity in the development of clinical diabetes. In: Robertson FR, Proudfood AT (eds) Anorexia nervosa and obesity. Royal College of Physicians, Edinburgh, pp 83–99

Balint P, Harsing L, Lenner M (1948) Tubulary azotemia caused by sucrose injections. Orv Hetil 89: 434–436

Baltzell JK, Berdanier CD (1985) Effect of the interaction of dietary carbohydrate and fat on the response of rats to starvation-refeeding. J Nutr 115: 104–110

Banerjee S, Kumar KS, Bandypadhyay A (1986) Carbohydrate and lipid metabolism treated with pyrrolidinomethyl tetracycline. Proc Soc Exp Biol Med 122: 652–657

Bantle JP, Laine DC, Castle GW, Thomas JW, Hoogwerf BJ, Goetz FC (1983) Postprandial glucose and insulin response to meals containing different carbohydrates in normal and diabetic subjects. N Engl J Med 309: 7–12

Baranowski T (1935) Sucrose in blood-pancreas malfunctioning. Klin Wochenschr 14: 1719–1722

Barbas MI, Shvarts LS (1938) Carbohydrate metabolism under normal and pathological conditions. Klin Med (Mosk) 16: 509–513

Barber SA, Bull NL, Cameron AM (1986) A dietary survey of an isolated population: the islanders of Orkney. Hum Nutr Appl Nutr 40A: 462–470

Barbour AD (1929) Enzymatic hydrolyses of glycogen. J Biol Chem 85: 29–45

Bardach L, Geduldig J, Mandel ID (1980) Dietary patterns in caries resistant (CR) vs caries susceptible (CS) adults. J Dent Res 59: 508 (Abstract)

Bardiger M, Stock AL (1972) The effects of sucrose-containing diets low in protein on ocular refraction in the rat. Abstr Commun 31: 4A–5A

Barnett SA (1956) Behaviour components in the feeding of wild and laboratory rats. Behaviour 9: 24–43
Bar-on H, Stein Y (1968) Effect of glucose and fructose administration on lipid metabolism in the rat. J Nutr 94: 95–105
Bar-on H, Roheim PS, Eder HA (1976) Hyperlipoproteinemia in streptozoticin-treated rats. Diabetes 25: 509–515
Bar-on H, Chen YI, Reaven GM (1981) Evidence for a new cause of defective plasma removal of very low density lipoproteins in insulin-deficient rats. Diabetes 30: 496–499
Barr M, Kohn SR, Tice LF (1957) The effect of various sugars and polyols on the stability of vitamin B12. J Am Pharm Assoc 46: 650–652
Barry RD (1984) High-fructose corn syrup and pure fructose: potential consumption and substitution for sugar by 1990. Presented to the American Medical Association Resource Conference on HFCS and Fructose, Chicago, Illinois, 2 October. Chicago, Illinois, American Medical Association
Barter PJ, Carroll KF, Nestel PJ (1971) Diurnal fluctuations in triglyceride, free fatty acids, and insulin during sucrose consumption and insulin infusion in man. J Clin Invest 50: 583–591
Basilico MZ, Gutman R, Yommi MR, Francone O, Lombardo YB (1983) Post-heparin plasma hepatic triglyceride lipase and monoglyceride hydrolase activities in hyperlipemia induced by a sucrose rich diet. Biomed Pharmacol 37: 36–41
Basilico MZ, Chanussot F, Villaume C, Lombardo YB, Debry G (1984) Effect of carbohydrate diet type upon obesity and hyperlipemia in the Zucker fa/fa rat. Ann Nutr Metab 28: 253–260
Batchelor BR, Penner J, Hirsch J, Stern JS (1976) Effects of hypophysectomy and acute growth hormone treatment upon glucose metabolism in adipose tissues and isolated adipocytes of rats. Horm Metab Res 8: 24–33
Battistini G (1930) In nutrition of health and sick infants. Clin Pediatr 12: 311–354
Baxter JH (1947) A study of the hemorrhagic-kidney syndrome of choline deficiency: the protective effect of starch. J Nutr 34: 333–349
Beaton JR, Sangster JF (1965) Comparative effect of sucrose and cornstarch in low-protein diets fed to rats exposed to cold. Can J Physiol Pharmacol 43: 241–249
Beavers WR (1959) Hypothermia: effect of hypertonic solutions on incidence of ventricular fibrillation. Am J Physiol 196: 709–710
Beck RC (1967) Clearance of ingested sucrose solutions from the stomach and intestine of the rat. J Comp Physiol [A] 64: 243–249
Beck RC (1969) Clearance of ingested sucrose solutions from the gastrointestinal tract of the rat. Replication report. Psychol Rep 24: 753–754
Becker DC, Terrill SW (1954) Various carbohydrates in a semipurified diet for the growing pig. Arch Biochem Biophys 50: 399–403
Becker DE, Ullrey DE, Terrill SW, Notzold RA (1954) Failure of the newborn pig to utilize dietary sucrose. Science 120: 345–346
Beckmann K (1939) The meaning of sugar in nutritive therapy. Fortschr Ther 15: 309–316
Beck-Nielsen H, Pedersen O, Schwartz Sorensen N (1978) Effects of diet on the cellular insulin binding and the insulin sensitivity in young healthy subjects. Diabetes 15: 289–296
Beck-Nielsen H, Pedersen O, Lindskov HO (1980) Impaired cellular insulin binding and insulin sensitivity induced by high-fructose feeding in normal subjects. Am J Clin Nutr 33: 273–278
Beebe CG, Schemmel R, Michelson O (1976) Blood pressure of rats as affected by diet and concentration of NaCl in drinking water. Proc Soc Exp Biol Med 151: 395–399
Beerman W (1973) Directed changes in the pattern of balbiani ring puffing in chironomus: effects of a sugar treatment. Chromosoma (Berlin) 41: 297–326
Beevers H (1961) Metabolic production of sucrose from fat. Nature (Lond) 191: 433–436
Begg TB, Oliver MF (1969) Dietary sugar intake and ischaemic heart disease. Br Heart J 31: 792–793
Behall K (1982) Influence of estrogen content of oral contraceptives and consumption of sucrose on blood parameters. Diss Abstr Int[B]43: 1437
Behall KM, Moser PB, Kelsay JL, Prather ES (1980) The effect of kind of carbohydrate in the diet and use of oral contraceptives on metabolism of young women. II. Serum glucose, insulin and glycogen. Am J Clin Nutr 33: 1041–1048
Behall KM, Moser PB, Kelsay JL, Prather ES (1980) The effect of kind of carbohydrate in the diet and use of oral contraceptives on metabolism of young women. III. Serum lipid levels. Am J Clin Nutr 33: 825–831
Behar D, Rapoport SL, Adams AS, Berg CJ, Cornblath M (1984) Sugar challenge testing with children considered behaviourally "sugar reactive". Nutr Behav 1: 277–288

Belkhadir J, Rosset T, Elgrably F et al. (1985) Effect of an extra intake of carbohydrate at dinner on morning after fasting plasma glucose levels in types I and II diabetes. Br Med J 291: 1608

Bell RR, Draper HH, Bergan JG (1973) Sucrose, lactose and glucose intolerance in northern Alaskan Eskimos. Am J Clin Nutr 26: 1185–1190

Benadé AJ, Wyndham CH, Strydom NB, Rogers GG (1971) The physiological effects of a mid-shift feed of sucrose. S Afr Med J 45: 711–718

Benadé AJS, Jansen CR, Rogers GG, Wyndham CH, Strydom NB (1973) The significance of an increased RQ after sucrose ingestion during prolonged aerobic exercise. Pflugers Arch 342: 199–206

Benadé AJS, Wyndham CH, Jansen CR, Rogers GG, de Bruin EJP (1973) Plasma insulin and carbohydrate metabolism after sucrose ingestion during rest and prolonged aerobic exercise. Pflugers Arch 342: 207–218

Bender AE (1981) Refined foods. Bibl Nutr Dieta 30: 160–165

Bender AE, Damji KB (1972) Some effects of dietary sucrose. World Rev Nutr Diet 15: 104–155

Bender AE, Thadani PV (1970) Some metabolic effects of dietary sucrose. Nutr Metab 12: 22–39

Bender AE, Damji DB, Yapa CGR (1970) Effects of dietary sucrose on the metabolism in vitro of liver from rats of different strains. Biochem J 119: 351–352

Bender AE, Damji KB, Khan MA, Khan IH, McGregor L, Yudkin J (1972) Sucrose induction of hepatic hyperplasia in the rat. Nature (Lond) 238: 461–462

Bender RC, Ansbacher S, Flanigan GE, Supplee GC (1936) The influence of dextrin and sucrose on growth and dermatitis. J Nutr 11: 391–400

Benditt EP, Benditt JM (1973) Evidence for a monoclonal origin of human atherosclerotic plaques. Proc Natl Acad Sci USA 70: 1753–1756

Benedict FG, Carpenter RM (1918) Food ingestion and energy transformation with special reference to the stimulating effects of nutrients. Carnegie Research Institute, Washington DC, pp 47–250 (No 261)

Bennett AE, Doll R, Howell RW (1970) Sugar consumption and cigarette smoking. Lancet I: 1011–1014

Berdanier CD (1974) Metabolic characteristics of the carbohydrate-sensitive BHE strain of rats. J Nutr 104: 1246–1256

Berdanier CD (1975) Effect of maternal sucrose intake on the metabolic patterns of mature rat progeny. Am J Clin Nutr 28: 1416–1421

Berdanier CD, Burell BR (1980) Effect of adrenalectomy on the responses of BHE rats to either a sucrose or starch diet. J Nutr 110: 298–304

Berger M, Muller WA, Renold AE (1978) Relationship of obesity to diabetes: some facts, many questions. In: Katzen HM, Mahler RJ (eds) Diabetes, obesity and vascular disease. Advances in modern nutrition. Hemisphere, Washington DC, vol 2, pp 211–228

Bergström J, Hultman E (1967) Synthesis of muscle glycogen in man after glucose and fructose infusion. Acta Med Scand 182: 93–107

Bergström J, Hultman E, Roch-Norlund AE (1968) Lactic acid accumulation in connection with fructose infusion. Acta Med Scand 184: 359–364

Bergström J, Fürst P, Gallyas F et al. (1972) Aspects of fructose metabolism in normal man. Acta Med Scand Suppl 542: 57–64

Berman D, Sayers M, Lynch SR, Charlton RW, Bothwell TH (1977) Iron absorption from a cereal-based meal containing cane sugar fortified with ascorbic acid. Br J Nutr 38: 261–269

Bernacki RJ, Bosmann HB (1971) The effects of sucrose and various salts on the growth and lysosomal enzyme activity of L5178Y cells. J Cell Sci 8: 399–406

Bernstein AO, Gregersen MI (1938) The diffusion of glucose and sucrose from the cerebro-spinal fluid. Am J Physiol 123: 747–751

Bertino M, Beauchamp GK, Engelman K (1982) Long-term reduction in dietary sodium alters the taste of salt. Am J Clin Nutr 36: 1134–1144

Bessard T, Schutz Y, Jéquier E (1983) Energy expenditure and post-prandial thermogenesis in obese women before and after weight loss. Am J Clin Nutr 38: 680–693

Best CH, Hartroft WS, Lucas CC, Ridout JH (1949) Alcohol and sucrose in the production of liver damage in rats. Fed Proc 8: 10

Best CH, Hartroft WS, Lucas CC, Ridout JH (1949) Liver damage produced by feeding alcohol or sugar and its prevention by choline. Br Med J iv: 1001–1006

Bett DG, Morland J, Yudkin J (1967) Sugar consumption in acne vulgaris and seborrhoeic dermatitis. Br Med J iii: 153–155

Beyer P, Barrand P, Freysz H, Zawislak R (1980) Permeabilité intestinale a un polymere de fructose chez le nourrisson atteint de diarrhée benigne. Ann Pediatr (Paris) 27: 433–438

Bianco-Porro G, Panza E (1985) Smoking, sugar, and inflammatory bowel disease. Br Med J 291: 971

Bibby BG (1975) The cariogenicity of snack foods and confections. J Am Dent Assoc 90: 121–132

Bibby BG (1977) Evaluation of cariogenicity of snack foods. National Institute of Dental Research, Bethesda, Maryland (Contract No NO1 DE 22039/NIDR 72–2039)

Bibby BG (1978) Foods and dental caries. In: Food, nutrition and dental health, American Dental Association Health Foundation, Pathotox, Park Forest South, pp 257–278

Bibby BG (1983) Fruits and vegetables and dental caries. Clin Prev Dent 5: 3–11

Bibby BG (1985) Cereal foods and dental caries. Cereal Food World 30: 851–855

Bibby BG, Mundorff SA, Huang CT (1983) Enamel demineralization tests with some standard foods and candies. J Dent Res 62: 885–889

Bickel A (1935) Blood alcohol curve following drinking of malt liquor and sugar solutions. Arch Verdauungskr 57: 239–243

Bierman EL (1979) Carbohydrates, sucrose and human disease. Am J Clin Nutr 32: 2712–2722

Bierman EL (1979) Carbohydrate and sucrose intake in the causation of atherosclerotic heart disease, diabetes mellitus, and dental caries. Am J Clin Nutr 32: 2644–2647

Bierman EL, Nelson R (1975) Carbohydrates, diabetes, and blood lipids. World Rev Nutr Diet 22: 280–287

Binns NM (1981) Caries and carbohydrates – a problem for dentists and nutritionists. Dent Health (Lond) 20: 5–10

Birkhed D (1984) Sugar content, acidity and effect on plaque pH of fruit juices, fruit drinks, carbonated beverages and sport drinks. Caries Res 18: 120–127

Birkhed D, Edwardsson S (1978) Acid production from sucrose substitutes in human dental plaque. Health and sugar substitutes. In: Proceedings of the ERGOB conference, Geneva. Karger, Basel, pp 211–217

Birkhed D, Frostell G (1978) Caries in rats fed highly or slightly hydrolysed lycasin. Caries Res 12: 250–255

Birkhed D, Topitsoglou V, Edwardsson S, Frostell G (1981) Cariogenicity of invert sugar in long term rat experiments. Caries Res 15: 302–307

Bittner AS, Burritt EA, Moser J, Street JC (1982) Composition of dietary fiber: neutral and acidic sugar composition. J Food Sci 47: 1469–1471

Bittner K, Korbas J (1969) Saccharose intolerance as a cause of chronic diarrhea in an 8-month-old infant. Wiad Lek 22: 1433–1435 (in Polish)

Bjorkman O, Crump M, Phillips RW (1984) Intestinal metabolism of orally administered glucose and fructose in Yucatan miniature swine. J Nutr 114: 1413–1420

Bjorling S, Frostell G, Dahlqvist A (1971) Effects of consumption of hydrogenated saccharides and sucrose on the blood sugar concentration. Acta Odontol Scand 29: 31–41

Black AE, Ravenscroft C, Simms AJ (1984) The NACNE report: are the dietary goals realistic? Comparisons with the dietary patterns of dietitians. Hum Nutr Appl Nutr 38A: 165–179

Blackham RJ (1929) Sucrose as food. J State Med 37: 455–461

Blacklock NJ (1982) Epidemiology of urolithiasis. In: Chisholm GD, Williams DI (eds) Scientific foundations of urology, 2nd edn. Heinemann, London, pp 251–259

Blackwell IG, Fosdick LS (1955) Effect of sucrose on the solubility of human dental enamel. Proc Inst Med Chicago 20: 273–274

Blackwell RQ, Fosdick LS (1956) Studies on the physiochemical phenomena related to dental caries. III. The effect of sucrose upon the solubility of human dental enamel. J Dent Res 35: 210–219

Blackwell RQ, McMillan L, Fosdick LS (1961) Sucrose retardation of acid etching in dental enamel: an electron-microscopic study. J Dent Res 40: 16–22

Blagowestschenski A (1926) The influence of high concentrations of neutral materials on the action of peptase. Biochem Z 168: 6–13

Blair DB, Tuba J (1963) Rat intestinal sucrase. I. Intestinal distribution and reaction kinetics. Can J Biochem Physiol 41: 905–916

Blair DB, Yakimets W, Tuba J (1963) Rat intestinal sucrase. II. The effects of rat age and sex and of diet on sucrase activity. Can J Biochem Physiol 41: 917–929

Blakely SR, Hallfrisch J, Reiser S, Prather ES (1981) Long-term effects of moderate fructose feeding on glucose tolerance parameters in rats. J Nutr 111: 307–314

Blundell JE (1983) Problems and processes underlying the control of food selection and nutrient intake. In: Wurtman RJ, Wurtman JJ (eds) Nutrition and the brain. Raven Press, New York, vol 6, pp 163–221

Bocking A, Riede UN (1979) Morphometrical analysis of an ultrahistochemical demonstration of nonspecific esterases in hepatocytes of mice after fructose overload. J Histochem Cytochem 27: 967–974

Bode CH, Bode JCH (1980) Adaptive changes in activities of enzymes of the carbohydrate metabolism in rat liver and jejunal mucosa to high fructose diets. Modifying effect of the fat content of the diet. Z Gastroenterol 18: 38–44

Bode CH, Schumacher H, Goebell H, Zelder O, Pelzel H (1971) Fructose induced depletion of liver adenine nucleotides in man. Horm Metab Res 3: 289–290

Bode CH, Zelder O, Rumpelt HJ, Wittkamp U (1973) Depletion of liver adenosine phosphates and metabolic effects of intravenous infusion of fructose or sorbitol in man and in the rat. Eur J Clin Invest 3: 436–441

Bode CH, Durr HK, Bode JCH (1981) Effect of fructose feeding on the activity of enzymes of glycolysis, gluconeogenesis, and the pentose phosphate shunt in the liver and jejunal mucosa of rats. Horm Metab Res 13: 379–383

Bode CH, Eisenhardt JM, Haberich FJ, Bode JCH (1981) Influence of feeding fructose on fructose and glucose absorption in rat jejunum and ileum. Res Exp Med (Berlin) 179: 163–168

Bode JCH, Bode CH, Ohta W, Martini GA (1980) Adaptive changes of the activity of enzymes involved in fructose metabolism in the liver and jejunal mucosa of rats following fructose feeding. Res Exp Med (Berlin) 178: 55–63

Bodine JH, Lu KH, West W (1951) Endogenous oxygen uptake of intact (grasshopper) embryos, homogenates, and cellular constituents in isotonic, sucrose and Ringer solutions. Proc Soc Exp Biol Med 78: 593–595

Boecker S, Forster H (1977) Influence of carbohydrates on uric acid synthesis in the isolated perfused chicken liver. Adv Exp Med Biol 76A: 500–508

Bohannon HM (1982) The impact of decreasing caries prevalence: implications for dental education. J Dent Res 61(Sp Iss): 1369–1377

Bohannon NV, Karam JH, Forsham PH (1980) Endocrine responses to sugar ingestion in man: advantages of fructose over sucrose and glucose. J Am Diet Assoc 76: 555–560

Boll M, Brückner E, Berndt J (1982) Age-related differences in dietary regulation of lipogenic enzymes in rat liver. Hoppe-Seyler's Z Physiol Chem 363: 103–106

Bollenback (1982) Sucrose and health. In: Lineback DR, Inglett GE (eds) Food carbohydrates. AVI Publishing, Westport, Connecticut, pp 63–73

Bolzano K, Sandhofer F, Sailer S, Braunsteiner H (1972) The effect of oral administration of sucrose on the turnover rate of plasma free fatty acids and on the esterification rate of plasma free fatty acids to plasma triglycerides in normal subjects, patients with primary endogenous hypertriglyceridemia and patients with well controlled diabetes mellitus. Horm Metab Res 4: 439–446

Bonnevie-Nielsen V (1980) Experimental diets affect pancreatic insulin and glycogen differently in male and female mice. Metabolism 29: 386–391

Boogaerts JR, Malone-McNeal M, Archambault-Scheynayder J, Davis RA (1984) Dietary carbohydrate induces lipogenesis and very low density protein synthesis. Am J Physiol 246: E77–E83

Booth DA, Campbell AT, Chase A (1970) Temporal bounds of post-ingestive glucose-induced satiety in man. Nature (Lond) 228: 1104–1105

Boot-Handford R, Heath H (1980) Identification of fructose as the retinopathic agent associated with the ingestion of sucrose-rich diets in the rat. Metabolism 29: 1247–1252

Boot-Handford R, Heath H (1981) The effect of dietary fructose and diabetes on the rat kidney. Br J Exp Pathol 62: 398–406

Booyens J, de Waal VM (1970) The level of sucrose intake by three groups of Indian subjects. S Afr Med J 44: 1415–1417

Borden LW, Ostrom CA, Koulourides T (1980) Establishment of potentially cariogenic streptococci in an experimental human plaque. I. *Streptococcus mutans*. J Dent Res 59: 588–593

Bornet F, Haardt MJ, Costagliola D, Blayo A, Slama G (1985) Sucrose or honey at breakfast have no additional acute hyperglycaemic effect over an hyperglucidic amount of bread in Type 2 diabetic patients. Diabetologia 28: 213–217

Borsini F, Bendotti C, Samanin R (1985) Salbutamol, d-amphetamine and d-fenfloramine reduce sucrose intake in freely fed rats by acting on different neurochemical mechanisms. Int J Obes 9: 277–283

Bose B (1982) Honey or sugar in treatment of infected wounds? Lancet I: 963

Bossetti BM, Kocher LM, Moranz JF, Falko JM (1984) The effects of physiologic amounts of simple sugars on lipoprotein, glucose and insulin levels in normal subjects. Diabetes Care 7: 309–312

Bossmann VK (1979) In vitro Experimente zur Plaquebildung. Dtsch Zahnaerztl Z 34: 437–439

Bouillon DJ, Berdanier CD (1983) Effect of maternal carbohydrate intake on mitochondrial activity and on lipogenesis by the young and mature progeny. J Nutr 113: 2205–2216

Bourne AR, Richardson DP, Bruckdorfer KR, Yudkin J (1975) Some effects of different dietary carbohydrates on pregnancy and lactation in rats. Nutr Metab 19: 73–90

Bourne AR, Richardson DP, Bruckdorfer KR, Yudkin J (1975) The effects of feeding starch, sucrose, glucose or fructose to rats during pregnancy and early lactation. Proc Nutr Soc 34: 80A–81A

Bowen WH (1975) Role of carbohydrates in dental caries. In: Jeanes A, Hodge J (eds) Symposium on physiological effects of food carbohydrates. American Chemical Society, Washington DC, pp 150–155

Bowen WH, Amsbaugh SM, Monell-Torrens S, Brunelle J, Kuzmiak-Jones H, Cole MF (1980) A method to assess cariogenic potential of foodstuffs. J Am Dent Assoc 100: 677–681

Boyd EM, Godi I, Abel M (1965) Acute oral toxicity of sucrose. Toxicol Appl Pharmacol 7: 609–618

Boyd EM, Mulrooney DA, Pitman CA et al. (1965) Benzylpenicillin toxicity in animals on a synthetic high sucrose diet. Can J Physiol Pharmacol 43: 47–54

Boyd JD, Ilines HM, Loese CE (1925) Study on the reaction of long-term intravenous glucose injection. Am J Physiol 74: 656–671

Braham RL, Barkin PR, Morris ME, Roberts MW (1978) Nutrition and its importance in dental health. J Fam Pract 6: 49–58

Brake PJ, Wright FAC, Melton LD (1980) Concentration of sugars in commercial infant foods in New Zealand. NZ Med J 92: 320–323

Brandes JW, Korst HA, Littman KP (1982) Sugar-free diet as long-term or interval treatment in the remission phase of Crohn's disease – a prospective study. Leber Magen Darm 12: 225–228

Bras G, Ross MH (1964) Kidney disease and nutrition in the rat. Toxicol Appl Pharmacol 6: 247–262

Bray GA (1970) The myth of diet in the management of obesity. Am J Clin Nutr 23(9): 1141–1148

Bray GA (1980) Scientific perspectives for the 1980s. Dietary guidelines: the shape of things to come. J Nutr Ed 12: 97–99

Bray GA, Dahms WT, Atkinson RL, Mena I, Schwartz A (1980) Factors controlling food intake: a comparison of dieting and intestinal bypass. Am J Clin Nutr 33: 376–382

Breukink HJ (1974) Oral mono- and disaccharide tolerance tests in ponies. Am J Vet Res 35: 1523–1527

Brewer DB, Heath D (1964) Development of sucrose vacuoles from liver cell lysosomes. J Pathol Bacteriol 87: 405–408

Brice J, Coles BL, Jourdan MH, Macdonald I (1969) The influence of frequency of sucrose intake on the concentration of lipids, proteins and glucose in the plasma. Proc Nutr Soc 28: 62A (Abstract)

Briese E, Quijada M (1978) Sugar solutions taste better (positive alliesthesia) after insulin. Proc Physiol Soc 285: 20P–21P

Briese E, Quijada M (1979) Positive alliesthesia after insulin. Experientia 35: 1058–1059

Brin M (1980) Red cell transketolase as an indicator of nutritional deficiency. Am J Clin Nutr 33: 169–171

Brindley DN, Saxton J, Shahdullah H, Armstrong M (1985) Possible relationships between changes in body weight set-point and stress metabolism after treating rats chronically with D-fenfluramine. Effects of feeding rats acutely with fructose on the metabolism of corticosterone, glucose, fatty acids, glycerol and triacylglycerol. Biochem Pharmacol 34: 1265–1271

Brinkman-Engels M, Tijmstra TJ (1977) De samenhang tussen de gebitsgezondheid en enkele sociaal-wetenschappelijke variabelen. Ned Tijdschr Tandheelkd 3: 100–106

Bristol JB, Emmett PM, Heaton KW, Williamson RCN (1985) Sugar, fat and the risk of colerectal cancer. Br Med J 291: 1467–1470

British Diabetic Association (1982) Cooking with sweeteners. Balance 69: June. British Diabetic Association, London, p 19

British Diabetic Association (1983) Dietary recommendations for diabetics for the 1980s: a policy statement by the British Diabetic Association. British Diabetic Association, Medical Advisory Committee, Nutrition Subcommittee, London

Brody S, Wolitzky DL (1983) Lack of mood changes following sucrose loading. Psychosomatics 24: 155–162

Broitman SA, Gottlieb LS, Vitale JJ (1976) Augmentation of ethanol absorption by mono- and disaccharides. Gastroenterology 70: 1101–1107

Brook M, Noel P (1969) Influence of dietary liquid glucose, sucrose and fructose on body fat formation. Nature (Lond) 222: 562–563
Brooke JD, Green LF (1974) Carbohydrate availability in human recovery from physical work exhaustion. Proc Nutr Soc 33: 12A–13A
Brookes N, MacKay D (1967) Diffusion of labelled substances through isolated rat diaphragm. J Physiol (Lond) 191: 74P–75P
Brooks CC, Miyahara AY, Huck DW, Ishizaki SM (1972) Relationship of sugar-induced lesions in the heart of the pig to live weight, serum cholesterol and diet. J Anim Sci 35: 31–37
Brosh S, Boer P, Sperling O (1984) Effects of fructose on purine nucleotide metabolism in isolated rat hepatocytes. Adv Exp Mol Biol 165: 481–485
Brown AT (1975) The role of dietary carbohydrates in plaque formation and oral disease. Nutr Rev 33: 353–361
Brown DF, Daudiss K (1973) Hyperlipoproteinemia. Prevalence in a free-living population in Albany, New York. Circulation 42: 558–566
Brown JP, Bibby BG (1970) Effect on rat caries of sugars administered before tooth eruption. Caries Res 4: 56–62
Brown RH (1982) Evidence of decrease in the prevalence of dental caries in New Zealand. J Dent Res 61(Sp Iss): 1327–1330
Brown TS, Murphy H (1973) Factors affecting sucrose preference behavior in rats with hippocampal lesions. Physiol Behav 11: 833–844
Bruckdorfer KR, Yudkin J (1975) A comparison of dietary starch and dietary sucrose in the pig. Nutr Metab 19: 225–232
Bruckdorfer KR, Khan I, Yudkin J (1971) The effect of chromium on the hyperlipaemic action of sucrose. Nutr Metab 13: 36–43
Bruckdorfer KR, Khan IH, Yudkin J (1972) The lipid content of the aortas of rats given sucrose. Abstr Commun 31: 9A–11A
Bruckdorfer KR, Kang SS, Yudkin J (1974) Insulin sensitivity of adipose tissue of rats fed with various carbohydrates. Proc Nutr Soc 33(1): 4A–5A
Bruckdorfer KR, Kang SS, Khan IH, Bourne AR, Yudkin J (1974) Diurnal changes in the concentrations of plasma lipids, sugars, insulin and corticosterone in rat fed diets containing various carbohydrates. Horm Metab Res 6: 99–106
Bruckdorfer KR, Worcester N, Yudkin J (1974) The effect of dietary sucrose on plasma lipids and on the liver of the spiny mouse (*Acomys cahirinus*). Proc Nutr Soc 33: 3A–4A
Brunelle JA, Carlos JP (1982) Changes in the prevalence of dental caries in US schoolchildren, 1961–80. In: Glass RL (ed) The first international conference on the declining prevalence of dental caries. J Dent Res 61: 1346–1351
Brunzell JD (1978) Use of fructose, sorbitol, or xylitol as a sweetener in diabetes mellitus. J Am Diet Assoc 73: 499–506
Bucko A, Simko V, Kopec Z, Damy A, Deakova H, Randusova A (1969) Effect of various kinds of carbohydrates on amylase formation in the pancreas. Nutr Dieta 11: 203–213
Buerger M, Baur M (1927) Convulsive and fatal effect of osmotic deprivation of water by hypertonic sugar solutions. IV. Physiological principles of somotherapy. Z Ges Exp Med 56: 1–30
Buergi E (1943) Presumable significance of binding of sugar and phosphate with medicaments. Schweiz Med Wochenschr 73: 1176
Bujard E (1937) Nuclear reactions studied in relation to mitoses produced by intravenous injections of ordinary sugar. Rev Med Suisse Romande 57: 511–519
Bulbuk GA (1968) Role of glucocorticoids and mineralocorticoids in the regulation of sucrose hydrolysis and enzymic synthesis in the small intestine. Fiziol Zh SSSR 54: 607–613
Bullock LT, Gregersen MI, Kinney R (1935) The use of hypertonic sucrose solution intravenously to reduce cerebro-spinal fluid pressure without a secondary rise. Am J Physiol 112: 82–96
Buñag RD, Tomita T, Sasaki S (1983) Chronic sucrose ingestion induces mild hypertension and tachycardia in rats. Hypertension 5: 218–225
Bundy KT, Morgan KJ, Zabik ME (1982) Nutritional adequacy of snacks and sources of total sugar intake among US adolescents. J Can Diet Assoc 43: 358–365, 374
Burch HB, Max P Jr, Chuy K, Lowry OH (1969) Metabolic intermediates in liver of rats given large amounts of fructose or dihydroxyacetone. Biochem Biophys Res Commun 34: 619–626
Burch HB, Choi S, Dence CN, Alvey TR, Coley BR, Lowry OH (1980) Metabolic effects of large fructose loads in different parts of the rat nephron. J Biol Chem 255: 8239–8244
Burch HB, Cole B, Choi S, Alvey TR, Dence C (1980) Diversity of effects of fructose loads on

different parts of nephron. Int J Biochem 12: 37–40

Burgess EA, Levin B, Mahalanabis D, Tonge RE (1964) Hereditary sucrose intolerance: levels of sucrase activity in jejunal mucosa. Arch Dis Child 39: 431–443

Burns-Cox CJ, Doll R, Ball KP (1969) Sugar intake and myocardial infarction. Br Heart J 31: 485–490

Buss RW, Kansal PC, Roddam RF, Pino J, Boshell BR (1982) Mixed meal tolerance test and reactive hypoglycemia. Horm Metab Res 14: 281–283

Butterfield WJH, Whichelow M (1968) Effect of diet, sulfonylureas and phenoformin on peripheral glucose uptake in diabetes and obesity. Lancet II: 785–788

Butterfield WJH, Sargeant BM, Whichelow MJ (1964) The metabolism of human forearm tissues after ingestion of glucose, fructose, sucrose, or liquid glucose. Lancet I: 574–577

Butterfield WJH, Hanley T, Whichelow MJ (1965) Peripheral metabolism of glucose and free fatty acids during oral glucose tolerance test. Metabolism 14: 851–866

Buttolph ML, Newberne PM (1982) Modified food starch ingestion in hamster and rat. In: Gawthorne JM, Howell JMcC, White CL (eds) Trace element metabolism in man and animals. Proceedings of the fourth international symposium on trace element metabolism in man and animals. Springer, Berlin Heidelberg New York

Cabanac M, Duclaux R (1970) Obesity: absence of satiety aversion to sucrose. Science 168: 496–497

Cahill GF Jr, Soeldner JS (1974) A non-editorial on non-hypoglycemia. N Engl J Med 291: 905–906

Cahill GF Jr, Ashmore J, Earle AS, Zottu S (1958) Glucose penetration into liver. Am J Physiol 192: 491–496

Cahlin E, Jönsson J, Persson B et al. (1973) Sucrose feeding in man. Effects on substrate incorporation into hepatic triglycerides and phosphoglycerides in vitro and on removal of intravenous fat in patients with hyperlipo-proteinaemia. Scand J Clin Lab Invest 32: 21–33

Cakala S, Borkowski T, Albrycht A (1974) Rumen acidosis in sheep fed various doses of sucrose. Pol Arch Weter 17: 117–130 (in Polish)

Caldwell RC (1959) A method of measuring the adhesion of foodstuffs to tooth surfaces. J Dent Res 18: 188–196

Caltabiano D (1925) Action of parenteral injections of saccharose, maltose, and lactose on insulin hypoglycemia. Policlinico 32: 489–502

Cameron AT (1943) The relative sweetness of sucrose, glucose, and fructose. Trans Roy Soc Can Sect 5 Ser 3(37): 11–27

Cameron IR, Davson H, Segal MB (1969) The effect of hypercapnia on the blood–brain barrier to sucrose in the rabbit. Yale J Biol Med 42: 241–247

Cameron-Clarke A, Manchester KL (1985) Effect of dietary carbohydrate and lipid on very low density lipoprotein secretion by rat hepatocytes. Nutr Rep Int 32: 1451–1459

Campbell GD (1978) Sucrose in human nutrition – importance of delineation of safe ingestant levels – a current survey. Health and sugar substitutes. In: Proceedings of the ERGOB conference, Geneva. Karger, Basel, pp 2–9

Campbell GD, Goldberg MD (1966) Sugar orgy – a preliminary report. S Afr Med J 40: 27–28

Campbell GD, Jackson WPU (1980) Sugar for energy – misleading advertising. S Afr Med J 57: 767

Campbell RG, Zinner DD (1970) Effect of certain dietary sugars on hamster caries. J Nutr 100: 11–20

Campbell WR, Hanna MI (1926) Estimation of fructose, sucrose, and insulin. J Biol Chem 69: 703–711

Campus A (1926) The injection of saccharose and milk secretion in sheep. Arch Farmacol Sper Sci Affini 41(2): 39–46

Cannon G (1985) The sugar lobby. Lancet I: 224

Capozzi L (1967) Cariogenicity of sugars. Ann Fac Med Chir Univ Stud Perugia 58: 585–600

Cappellato M (1942) Sui sarcomi sperimentali da glucosio nel ratto bianco. Tumori 16: 38–54

Cardini CE, Leloir LF, Chiriboga J (1955) The biosynthesis of sucrose. J Biol Chem 214: 149–155

Carlos JP (1982) The prevention of dental caries: ten years later. J Am Diet Assoc 104: 193–197

Carlos JP (1983) A debate over the role of sugars in the etiology of dental caries. J Pedod 1: 330–332

Carlson LA, Böttiger LE (1972) Ischaemic heart disease in relation to fasting values of plasma triglycerides and cholesterol. Stockholm prospective study. Lancet I: 865–868

Carlsson J, Sundstrom B (1968) Variations in the composition of early dental plaque following ingestion of sucrose and glucose. Odont Rev 19: 161–169

Carney-Crane S, Lachance PA (1983) Effects of sucrose, glucose and fructose on spontaneous activity and brain monoamines in rat pups. Nutr Rep Int 28: 991–997

Carpenter TM, Lee RC (1938) Action of muscle working on the metabolism of men after taking cane sugar and galactose. Arbeits-Physiol 10: 172–187

Carroll C (1963) Influences of dietary carbohydrate. Fat combinations on various functions associated with glycolysis and lipogenesis in rats. 1. Effects of substituting sucrose for rice starch with unsaturated and saturated fat. J Nutr 79: 93–100

Carroll C (1964) Influences of dietary carbohydrate-fat combinations on various functions associated with glycolysis and lipogenesis in rats. II. Glucose vs sucrose with corn oil and two hydrogenated oils. J Nutr 82: 163–172

Carroll C, Williams L (1982) Choline deficiency in rats as influenced by dietary energy sources. Nutr Rep Int 25: 773–782

CAS (1978) Chemical abstracts ninth collective index (9CI), 1972–76. Chemical Abstracts Services, Columbus, Ohio, vols 76–85

Cassamassimo PS, Entwwistle B, Ernest A, McCabe E (1984) Dental health in children with phenylketonuria (PKU) and other inborn errors of amino acid metabolism managed by diet. US Department of Health and Human Services, Public Health Service, Health Resources and Services Administration, Bureau of Health Care Delivery and Assistance, Division of Maternal and Child Health, Bethesda, Maryland, 1–28 (Publ No HRS-D-MC 84-1)

Cassidy GJ, Dworkin S, Finney WH (1926) Effect of various sugars (and of adrenalin and pituitrin) in restoring the shivering reflex. Am J Physiol 77: 211–218

Castaigne P, Cambier J, Schuller E (1963) Study of blood sugar regulation by continuous measurement. Presse Med 71: 2555–2557

Caster WO, Parthemos MD (1976) Growth, hemoglobin, cholesterol, and blood pressure observed in rats fed common breakfast cereals. Am J Clin Nutr 29: 529–534

Castonguay TW, Applegate EA, Upton DE, Stern JS (1984) Hunger and appetite. Old concepts/new distinctions. In: Present knowledge in nutrition, 5th edn. The Nutrition Foundation, Washington DC, pp 19–34

Carthcart EP, Markowitz J (1927) Influence of various sugars on the respiratory quotient. J Physiol (Lond) 63: 309–324

Cauderay M (1985) Effect of glycohydrolase inhibitors on sucrose metabolism in normal men. Université de Lausanne, Faculté de Médicine, Département de Médicine Interne, Lausanne (PhD Dissertation)

Cerquiglini S, Marchetti M (1961) Reaction time of man under the influence of alcohol with and without sugar. Boll Soc Ital Biol Sper 37: 1647–1649

Cezard JP, Broyart JP, Cuisinier-Bleizes P, Mathieu H (1983) Sucrase–isomaltase regulation by dietary sucrose in the rat. Gastroenterology 84: 18–25

Cha CJ, Randall HT (1982) Effects of substitution of glucose-oligosaccharides by sucrose in a defined formula diet or interstitial dissacharides, hepatic lipogenic enzymes and carbohydrate metabolism in young rats. Metabolism 31: 57–66

Chalvardjian AM, Stephens S (1970) Lipotropic effect of dextrin versus sucrose in choline-deficient rats. J Nutr 100: 397–403

Chan RIM, Stich HF, Rosin MP, Powrie WD (1982) Antimutagenic activity of browning reaction products. Cancer Lett 15: 27–33

Chang MLW, Johnson MA (1980) Effect of garlic on carbohydrate metabolism and lipid synthesis in rats. J Nutr 110: 931–936

Chang RH (1974) Sucrose chemicals and their industrial uses. In: Inglett GE (ed) Symposium on sweeteners, AVI Publishing Westport, Connecticut, pp 74–77

Chantelau EA, Gosseringer G, Sonnenberg GE, Berger M (1985) Moderate intake of sucrose does not impair metabolic control in pump-treated diabetic out-patients. Diabetologia 28: 204–207

Charlton G, Fitzgerald DB, Keyes PH (1971) Hydrogen ion activity in dental plaques of hamsters during metabolism of sucrose, glucose and fructose. Arch Oral Biol 16: 655–661

Chatterjee A, Jalan KN, Agarwal SK et al. (1977) Evaluation of a sucrose electrolyte solution for oral rehydration in acute infantile diarrhoea. Lancet I: 1333–1335

Chauchard P, Mazoue H, Lecoq R (1945) Inhibition by sugars of the stimulation of the nervous system by xanthine bases. C R Soc Biol (Paris) 139: 12–13

Chauchard P, Mazoue H, Lecoq R (1945) New studies on the variations in nervous excitability under the influence of different sugars. C R Acad Sci 221: 543–645

Chaudhuri M, Squibb RL, Solotorovsky M (1980) Effects of glucose and fructose loading on glycogenesis in chicks infected with avian tuberculosis. Poultry Sci 59: 1736–1741

Cheraskin E, Ringdorf WM Jr, Setyaadmadja AT (1965) Periodontal pathosis in man. XIII. Effect of sucrose drinks upon sulcus depth. J Oral Ther 2: 195–202

Cheraskin E, Ringsdorf WM Jr, Setyaadmadja AT (1965) Periodontal pathosis in man. XIV. Effect of sucrose drinks upon clinical tooth mobility. J Dent Med 20: 91–96

Chesley LC, Uichanco L, Lenobel A, Kellerman H (1964) Placental permeability to sucrose: a source of error in measuring volumes of sucrose distribution in gravidas. Obstet Gynecol 23: 795–798

Chevalier M, Wiley JH, Leveille GA (1972) The age-dependent response of serum triglycerides to dietary fructose. Proc Soc Exp Biol Med 139: 220–223

Chiel HJ, Wurtman RJ (1981) Short term variations in diet composition change the pattern of spontaneous motor activity in rats. Science 213: 676–678

Chirife J, Herszage L, Joseph A, Kohn ES (1983) In vitro study of bacterial growth inhibition in concentrated sugar solutions: microbiological basis for the use of sugar in treating infected wounds. Antimicrob Agric Chemother 23: 766–773

Chlouverakis C, Schnatz JD (1974) Incorporation of dietary fat, sucrose and protein into plasma and tissues triglycerides of streptozotocin diabetic rats. Horm Res 5: 293–303

Chludzinski Am, Germaine GR, Schacetele CF (1974) Purification and properties of dextransucrase from *Streptococcus mutans*. J Bacteriol 11: 1–17

Choh H (1939) Permeability of lymphatic vessels of the skin. I. The effect of various stimulants. J Chosen Med Assoc 29: 2469–2488

Chung-Ja M, Randall HT (1982) Effects of substitution of glucose-oligosaccharides by sucrose in a defined formula diet on intestinal disaccharidases, hepatic lipogenic enzymes and carbohydrate metabolism in young rats. Metabolism 31: 57–66

Ciaccio C, Racciusa S (1927) Influence of saccharose administered in various quantities upon dextrose hyperglycemia. Boll Soc Ital Biol Sper 2: 309–311

Clancy KL, Bibby BG, Goldberg HJV, Ripa LW, Barenie J (1977) Snack food intake of adolescents and caries development. J Dent Res 45: 568–573

Clancy KL, Goldberg HJV, Ritz A (1978) Snack food consumption of 12-year-old inner city children and its relationship to oral health. J Public Health Dent 38: 227–234

Clark DG, Storer GB, Topping DL (1980) Inhibition of the substrate cycle glucose : glucose-6-phosphate by physiological concentrations of fructose in perfused rat liver. Biochem Biophys Res Commun 93: 155–161

Cleaton-Jones P, Walker ARP, Retief DH (1975) What is the role of sugar (sucrose) in dental caries today? Tydskr Tandheelkd Ver S Afr 30: 637

Cleave TL (1968) Sucrose intake and coronary heart disease. Lancet II: 1187

Cleek JL, Phillips RW, Johnson BD (1979) Availability of oral carbohydrates to neonatal calves. J Am Vet Med Assoc 174: 373–377

Clement R (1967) Saccharose intolerance. Presse Med 75: 969–970

Clevidence BA, Srinivasan SR, Webber LS, Radhakrishnamurthy B, Dalferes E, Berenson GS (1981) Serum lipoprotein and blood pressure levels in rhesus monkeys fed sucrose diets. Biochem Med 25: 186–197

Coarse JF, Cardoni AA (1975) Use of fructose in the treatment of acute alcoholic intoxication. Am J Hosp Pharm 32: 518–519

Coburn FF, Annegers J (1950) Effect of dietary substances on cholate synthesis in the dog. Am J Physiol 163: 48–53

Cogan DG, Kinoshita JH, Kador PF et al. (1984) Aldose reductase and complications of diabetes. Ann Intern Med 101: 82–91

Cohen AM (1972) Effect of sucrose feeding on glucose tolerance. Acta Med Scand 542(Suppl): 173–179

Cohen AM (1978) Genetically determined response to different ingested carbohydrates in the production of diabetes. Horm Metab Res 10: 86–92

Cohen AM (1986) Metabolic responses to dietary carbohydrates: interactions of dietary and hereditary factors. Prog Biochem Pharmacol 21: 74–103

Cohen AM, Rosenmann E (1971) Diffuse intercapillary glomerulosclerosis in sucrose-fed rats. Diabetes 7: 25–28

Cohen AM, Shoshan S (1968) Effect of glucose, fructose, and starch on aorta collagen in the rat. Preliminary note. J Atheroscler Res 8: 371–375

Cohen AM, Teitelbaum A (1964) Effect of dietary sucrose and starch on oral glucose tolerance and insulin-like activity. Am J Physiol 206: 105–108

Cohen AM, Teitelbaum A (1966) Effect of different levels of protein in sucrose and starch diets on lipid synthesis in the rat. Isr J Med Sci 2: 727–731

Cohen AM, Teitelbaum A (1968) Effect of glucose, fructose, sucrose, and starch on lipogenesis in rats. Life Sci 7(2 pt 2): 23–29

Cohen AM, Yudkin J (1967) The effect of dietary sucrose upon the response to sodium tolbutamide in the rat. Biochim Biophys Acta 141: 637–638

Cohen AM, Bavly S, Poznanski R (1961) Change of diet of Yemenite Jews in relation to diabetes and ischaemic heart disease. Lancet II: 1399–1401

Cohen AM, Kaufmann NA, Poznanski R, Blondheim SH, Stein Y (1966) Effect of starch and sucrose on carbohydrate-induced hyperlipaemia. Br Med J i: 339–340

Cohen AM, Teitelbaum A, Balogh M, Groen JJ (1966) Effect of interchanging bread and sucrose as main source of carbohydrate in a low fat diet on the glucose intolerance curve of healthy volunteer subjects. Am J Clin Nutr 19: 59–62

Cohen AM, Teitelbaum A, Cohen B (1970) Effect of glucose, fructose, sucrose, and starch on nitrogen utilization in rats fed high and low protein diets. Isr J Med Sci 6: 86–89

Cohen AM, Briller S, Shafrir E (1972) Effect of longterm sucrose feeding on the activity of some enzymes regulating glycolysis, lipogenesis and gluconeogenesis in rat liver adipose tissue. Biochim Biophys Acta 279: 129–138

Cohen AM, Teitelbaum A, Saliternick R (1972) Genetics and diet as factors in development of diabetes mellitus. Metabolism 21: 235–240

Cohen AM, Teitelbaum A, Briller S, Rosenmann E, Yanio L, Shafrir E (1974) Role of diet and genetics in the development of angiopathy in diabetes. Horm Metab Res 4: 117–123

Cohen AM, Teitelbaum A, Rosenmann E (1977) Diabetes induced by a high fructose diet. Metabolism 26: 17–24

Cohen BL (1978) Relative risks of saccharin and calorie ingestion. Science 199: 983

Cohen L, Perkins EG, Dobrilovic L (1965) The influence of different dietary carbohydrates on serum and liver lipids. In: Proceedings of the meat industry research conference, Chicago, pp 70–73

Cohen MR, Cohen RM, Pickar D, Murphy KL (1985) Maloxone reduces food intake in humans. Psychosom Med 47: 132–138

Cohen PS, Tokieda FK (1972) Sucrose–water preference reversal in the water-deprived rat. J Comp Physiol [A] 79: 254–258

Cohen SG, Sapp TM (1964) Polysaccharide effects simulating hypersensitivity responses in the rabbit. Am J Physiol 207: 389–392

Cohn PF, Gabbay SI, Weglicki WG (1976) Serum lipid levels in angiographically defined coronary artery disease. Ann Intern Med 84: 241–245

Cole MF, Bowden GH, Korts DC, Bowen WH (1978) The effect of pyridoxine, phytate and invert sugar on production of plaque acids in situ in the monkey (*M. fascicularis*). Caries Res 12: 190–201

Coleman DL (1978) Diabetes and obesity: thrifty mutants? Nutr Rev 36: 129–132

Collier G, Bolles R (1968) Some determinants of intake of sucrose solutions. J Comp Physiol [A] 65: 379–383

Collins RG, Densen AL, Becks H (1942) Study of caries-free individuals. II. Is an optimum diet or a reduced carbohydrate intake required to arrest dental caries. J Am Dent Assoc 29: 1169–1178

Colman G, Bowen WH, Cole MF (1977) The effects of sucrose, fructose and a mixture of glucose and fructose on the incidence of dental caries in monkeys (*M. fascicularis*). Br Dent J 142: 217–221

Colmenares JL, Wurtman RJ, Fernstrom JD (1975) Effects of ingestion of a carbohydrate fat meal on the levels and synthesis of 5-hydroxyindoles in various regions of the rat central nervous system. J Neurochem 25: 825–829

Coltart TM (1969) Changes in serum phospholipids in male and female baboons on a sucrose diet. Nature (Lond) 222: 575–576

Coltart TM, Crossley JN (1970) Influence of dietary sucrose on glucose and fructose tolerance and triglyceride synthesis in the baboon. Clin Sci 38: 427–437

Coltart TM, Macdonald I (1971) Effect of sex hormones on fasting serum triglycerides in baboons given high-sucrose diets. Br J Nutr 25: 323–331

Conard V (1955) Mesure de l'assimilation du glucose: bases théoretiques et application cliniques. Acta Gastroenterol Belg 18: 655–705

Conners CK (1984) Nutritional therapy in children. In: Galler JR (ed) Nutrition and behavior. Plenum, New York, pp 159–192

Conners CK, Blouin AG (1982) Nutritional effects on behavior of children. J Psychiatr Res 17: 193–201

Conners CK, Goyette CH, Southwick DA, Lees JM, Andrulonis PA (1976) Food additives and hyperkinesis: a controlled double-blind experiment. Pediatrics 58: 154–166

Consolazio CF, Matoush LO, Johnson HL et al. (1968) Effects of high-carbohydrate diets on performance and clinical symptomatology after rapid ascent to high altitude. Fed Proc 28: 937–943

Constant MA, Phillips PH (1954) The occurrence of a calcinosis syndrome in cotton rats. IV. The effect of diet and the age of the animals on the development of the disease and on the urinary excretion of various metabolites. J Nutr 52: 165–186

Constantinides SM, Bedford CL (1965) Sucrose determination in sugar beets using paper chromatography and spectrophotometry. J Am Soc Sugar Beet Technol 13: 185–191

Constantopoulos G, Boyd EM (1968) Maximal tolerated amounts of sucrose given by daily intragastric administration to albino rats. Food Cosmet Toxicol 6: 712–727

Constantopoulos G, Boyd EM (1968) Factors affecting sucrose toxicity. Int Z Klin Pharmakol Toxikol 1: 539–544

Cook GC (1969) Absorption productions of D(–)fructose in man. Clin Sci 37: 675–687

Cook GC (1970) Comparison of the absorption and metabolic products of sucrose and its monosaccharides in man. Clin Sci 38: 687–697

Cook GC (1971) Absorption and metabolism of D(–)fructose in man. Am J Clin Sci 24: 1302–1307

Cooper PL, Wahlqvist ML, Simpson RW (1986) Sucrose in the diabetic rat. Rec Adv Clin Nutr 2: 271–283

Coquet B, Rondot G, Mary MC (1980) Thirteen-week oral toxicity study in the rat with Lycasin 80/55 (unpublished report from the Institut Français de Recherches et Essai Biologiques, l'Arbresle, France; submitted to the World Health Organization by Roquette Freres SA, France. In: Joint FAO/WHO Expert Committee on Food Additives (ed) Toxicological evaluation of certain food additives and contaminants. Cambridge University Press, Cambridge, pp 189–190 (WHO Food Additive Series No 20, 1987)

Corbella RG (1926) Sucrose as possible antidote for cyanide poisoning. Semana Med 2: 1437

Cori CF (1925) The fate of sugar in the animal body. I. The rate of absorption of hexoses and pentoses from the intestinal tract. J Biol Chem 66: 691–715

Cori GT, Ochoa S, Stein MW, Cori CF (1951) The metabolism of fructose in liver. Isolation of fructose-1-phosphate and inorganic pyrophosphate. Biochim Biophys Acta 7: 304–317

Cornblath M, Schwartz R (1976) Disorders of carbohydrate metabolism in infancy, 2nd edn. Saunders, Philadelphia, Pennsylvania, p 322

Corvilain J (1956) Reactivity of the adrenal dependent on the osmolarity of the blood. Livre Jubilaire du Prof Paul Govaerts. Impr médical et scientifique, Brussels, pp 480–500

Costella JC, Virgo BB (1980) Is dieldrin-induced congenital inviability mediated by central nervous system hyperstimulation or by altered carbohydrate metabolism. Can J Physiol Pharmacol 58: 633–637

Costill DL (1985) Carbohydrate nutrition before, during and after exercise. Fed Proc 44: 364–368

Cotlove E (1954) Mechanism and extent of distribution of insulin and sucrose in chloride space of tissues. Am J Physiol 176: 396–410

Cottet-Emard JM, Peyrin L, Bonnod J (1980) Dietary induced changes in catecholamine metabolites in rat urine. J Neurol Transmission 48: 189–201

Coulston AM, Swislocki ALM (1985) Metabolic effects of high carbohydrate, moderate sucrose diets in patients with non-insulin-dependent diabetes mellitus (NIDDM). Diabetes 34(Suppl): 34A (Abstract No 133)

Coulston AM, Hollenbeck CB, Donner C, Williams R, Chiou YM, Reaven GM (1985) Metabolic effects of added dietary sucrose in individuals with non-insulin-dependent diabetes mellitus. Metabolism 34: 962–966

Council on Foods and Nutrition (1942) Some nutritional aspects of sugar, candy and sweetened carbonated beverages: report. J Am Med Assoc 120: 763–765

Councils on Dental Health and Dental Therapeutics of American Dental Association (1953) Sugar and dental caries: effect on teeth of sweetened beverages and other sugar-containing substances (joint report). J Am Dent Assoc 47: 387–415

Coussons HS (1984) Diet and hyperactivity. J Okla State Med Assoc 77: 169–173

Cox RP, Gesner BM (1965) Effect of simple sugars on the morphology and growth pattern of mammalian cell cultures (mouse, monkey kidney, human skin, HeLa cells). Proc Natl Acad Sci USA 54: 1571–1579

Craig JT (1977) Xylitol – non-cariogenic sweetener. Food Technol NZ 12: 27, 29–30

Craighead JE (1978) Current views on the etiology of insulin-dependent diabetes mellitus. N Engl J Med 299: 1439–1445

Craik JD, Elliott KRF (1980) Transport of D-glucose and D-galactose into isolated rat hepatocytes. Biochem J 192: 373–375

Crandall DL, Goldstein BM, Lizzo FH, Lozito RJ, Cervoni P (1986) Myocardial adaptations to dietary sucrose in spontaneously hypertensive rats. Res Commun Chem Pathol Pharmacol 53: 237–240

Crane RK (1960) Intestinal absorption of sugars. Physiol Rev 40: 789–825

Crane RK (1968) Absorption of sugars. Aliment Canal 3: 1323–1351

Crane RK (1979) Digestion and absorption of water soluble organics. In: Crane RK (ed) International review of physiology. II. Gastrointestinal physiology. University Park Press, Baltimore, vol 12, p 325

Crane SC, Lachance PA (1983) Effects of sucrose, glucose and fructose on spontaneous activity and brain monoamines in rat pups. Nutr Rep Intern 28: 991–997

Crapo PA (1985) Simple versus complex carbohydrate use in the diabetic diet. Ann Rev Nutr 5: 95–114

Crapo PA, Kolterman OG (1984) The metabolic effects of a 2-week fructose feeding in normal subjects. Am J Clin Nutr 39: 525–534

Crapo PA, Olefsky JM (1980) Fructose – its characteristics, physiology, and metabolism. Nutr Today 15: 2–7

Crapo PA, Reaven G, Olefsky J (1976) Plasma glucose and insulin responses to orally administered single and complex carbohydrates. Diabetes 25: 741–747

Crapo PA, Reaven G, Olefsky (1977) Post-prandial plasma-glucose and insulin responses to different complex carbohydrates. Diabetes 26: 1178–1183

Crapo PA, Kolterman OG, Waldeck N, Reaven GM, Olefsky JM (1980) Post-prandial hormonal responses to different types of complex carbohydrate in individuals with impaired glucose tolerance. Am J Clin Nutr 33: 1723–1728

Crapo PA, Insel J, Sperling M, Kolterman OG (1981) Comparison of serum glucose, insulin, and glucagon responses to different types of complex carbohydrate in noninsulin-dependent diabetic patients. Am J Clin Nutr 34: 184–190

Crapo PA, Reaven GM, Olefsky JM (1981) Hormonal and substrate responses to a standard meal in normal and hypertriglyceridemic subjects. Metabolism 30: 331–334

Crapo P, Scarlett JA, Kolterman OG (1982) Comparison of the metabolic responses to fructose and sucrose sweetened foods. Am J Clin Nutr 36: 256–261

Crapo PA, Scarlett JA, Kolterman OG, Sanders LR, Hofeldt FD, Olefsky JM (1982) The effects of oral fructose, sucrose, and glucose in subjects with reactive hypoglycemia. Diabetes Care 5: 512–517

Crawford PJM (1981) Sweetened medicines and caries. Pharm J 226: 668–669

Crone C, Garlick D (1970) The penetration of insulin, sucrose, mannitol, and tritiated water from the interstitial space in muscle into the vascular system. J Physiol 210: 387–404

Crook WG (1981) Can what a child eats make him dull, stupid, or hyperactive? Ann Allergy 47: 123

Crossley JN, Macdonald I (1970) The influence in male baboons of a high sucrose diet on the portal and arterial levels of glucose and fructose following a sucrose meal. Nutr Metabol 12: 171–178

Crow KE, Newland KM, Batt RD (1981) The fructose effect. NZ Med J 93: 232–234

Csáky TZ, Fisher E (1981) Intestinal sugar transport in experimental diabetes. Diabetes 30: 568–574

Cudworth AG, Festenstein H (1978) HLA genetic heterogeneity in diabetes mellitus. Br Med Bull 34: 285–289

Curtbiryh MG, Chu GW, Lemke RP, Flanary CM, Rugh JD (1984) Bite force and muscle activity in long and short-face individuals. Int Assoc Dent Res Abstr Gen Meet No. 1471

Cutler HH (1939) Effects of sucrose on the kidney. Proc Staff Meet Mayo Clin 14: 318–320

Cybulska B, Naruszewicz M (1982) The effect of short-term and prolonged fructose intake on VLDL-TG and relative properties on apo CIII and apo CII in the VLDL fraction in type IV hyperlipoproteinaemia. Nahrung 26: 253–261

Dahlqvist A (1967) Localization of the small-intestinal disaccharidases. Am J Clin Nutr 20: 81–88

Dahlqvist A (1969) Intestinal absorption of sucrose. Acta Med Scand 542(Suppl): 13–18

Dahlqvist A (1974) Physiology of carbohydrate digestion and absorption. In: Sipple HL, McNutt KW (eds) Sugars in nutrition. Academic Press, New York, p 189

Dahlqvist A, Borgström B (1961) Digestion and absorption of disaccharides in man. Biochem J 81: 411–418

Dahlqvist A, Lindberg T (1966) Development of the intestinal disaccharidase and alkaline

phosphatase activities in the human fetus. Clin Sci 30: 517–528
Dahlqvist A, Thomson DL (1963) The digestion and absorption of sucrose by the intact rat. J Physiol (Lond) 167: 193–209
Dahlqvist A, Auricchio S, Semenza G, Prader A (1963) Human intestinal disaccharidases and hereditary disaccharide intolerance. The hydrolysis of sucrose, isomaltose, palatinose (isomaltulose), and 1,6-alpha-oligosaccharide (isomaltooligosaccharide) preparation. J Clin Invest 42: 556–562
Dako DY, Trautner K, Somogyi JC (1970) Studies on the glucose, fructose, and saccharose content of various fruits. Bibl Nutr Dieta 15: 184–198
Dalderup LM (1967) Nutrition and caries. III. The influence of the major calorie-providing food components. A: Carbohydrates. World Rev Nutr Diet 7: 84–89
Dalderup LM (1969) Cariogenicity of various sugars in rats. Voeding 30: 443–446
Dalderup LM, Visser W (1969) Influence of extra sucrose daily food on the life-span of Wistar albino rats. Nature (Lond) 222: 1050–1052
Dalderup LM, Visser W (1971) Influence of extra sucrose, fats, protein and of cyclamate in the daily food on the life-span of rats. Experientia 27: 519–521
Dalderup LM, Visser W, Keller GHM (1968) Studies with adult animals on rations with sucrose, glucose and fructose. Voeding 29: 245–252
Dall'Aglio E, Chang F, Chang H, Stern J, Reaven GM (1983) Effect of exercise and diet on triglyceride metabolism in rats with moderate insulin deficiency. Diabetes 32: 46–50
Dalton LD, Grossman SP (1982) Responses to dietary adulterations in rats with zona incerta lesions. Physiol Behav 29: 51–60
Damon CE, Pettitt BC (1980) High performance liquid chromatographic determination of fructose, glucose, and sucrose in molasses. J Assoc Off Anal Chem 63: 476–480
Dandrifosse G, Schoffeniels E (1964) Action of certain sugars on the secretion of chitinase by isolated gastric mucosa. Arch Int Physiol Biochim 72: 517–519
Danforth E Jr (1985) Diet and obesity. Am J Clin Nutr 41: 1132–1145
Daniloff AA, Krestrownikoff AN (1932) Action of stimulants (sugar, chocolate and cocoa) on urinopoietic processes under influence of muscular work. Arbeitsphysiol 5: 537–548
Danowski TS, Nolan S, Stephan T (1975) Hypoglycemia. World Rev Nutr Diet 22: 288–303
Das DK, Steinberg H (1980) The effect of fructose on purine metabolism in the lung. Life Sci 26: 1–10
Das DK, Neogi A, Steinberg H (1985) Nutritional and hormonal control of glucose and fructose utilization by lung. Clin Physiol Biochem 3: 240–248
Davidenkov S (1940) Inherited inability to eat sugar (aglycophagia). J Hered 31: 5–7
Davies JN (1966) Occurrence of sucrose in the fruit of some species of *Lycopersion*. Nature (Lond) 209: 640–641
Davson H, Matchett PA (1953) The kinetics of penetration of the blood–aqueous barrier. J Physiol (Lond) 122: 11–32
Dawes CA (1983) A mathematical model of salivary clearance of sugar from the oral cavity. Caries Res 17: 321–334
Day CDM, Daggs RG, Sedwick HJ (1935) High sugar diets and dental caries in white rat. J Am Dent Assoc 22: 913–925
De AK, Aiyar AS, Sreenivasan A (1969) Biochemical effects of irradiated sucrose solutions in the rat. Radiat Res 37: 202–215
De HN, Guha SR (1951) Biosynthesis of B-vitamins. IV. Effect of different carbohydrates and some inorganic and organic nitrogenous compounds on the biosynthesis of nictonic acid in albino rats. Indian J Med Res 39: 337–342
de Almeida MO (1924) Action of sugars on lung ventilation. C R Soc Biol (Paris) 91: 1122–1124
Deane HW (1942) Study of hepatic-cell mitochondria in fatty liver produced by high-sugar diet. Anat Rec 84: 171–191
Deane HW, Deane N, Smith HW (1955) Fate of insulin and sucrose in normal subjects as determined by a urine reinfusion technique. J Clin Invest 34: 681–684
Debure A, Gachot B, Lacour B, Kreis H (1987) Acute renal failure after use of granulated sugar in deep infected wounds. Lancet I: 1034–1035
DeFeudis FV, Elliot KA (1967) Delay or inhibition of convulsions by intraperitoneal injections of diverse substances. Can J Physiol Pharmacol 45: 857–865
DeFronzo RA, Goldberg M, Agus ZS (1976) The effects of glucose and insulin on renal electrolyte transport. J Clin Invest 58: 83–90
De Kalbermatten N, Ravussin E, Maeder E, Geser C, Jequier E, Felber JP (1980) Comparison

of glucose, fructose, sorbitol, and xylitol utilization in humans during insulin suppression. Metabolism 29: 62–67

Delaitre M, Fonty M (1962) Sugar intolerance in infants. Gaz Med France 69: 213–220

Delaitre M, Fonty M, Varlet, Fourrier (1961) Chronic diarrhea in a nursing infant caused by intolerance to sucrose. Arch Fr Pediatr 18: 1202–1210

Delak M, Adamic S (1959) Sucrose intoxication in sheep. Vet Archiv 29: 214–223

Dell'Acqua G (1935) Experimental influence of liver lipase. Z Klin Med 128: 95–97

Delthil P (1953) Intolerance of saccharose as cause of nutritional disorders in infants. Presse Med 61: 1643–1645

Demyanovskii SYa, Galitsova R, Rozhdestvenskaya V (1940) The effect of carbohydrates on proteolysis. Uchenye Zapiski Moscow Gosudarst Pedagog Inst 21(4): 73–99

Dennison CI, Randolph PM (1981) Diet and dental caries. In: Randolph PM, Dennison CI (eds) Diet, nutrition and dentistry. CV Mosby, St Louis, Missouri, pp 200–203

DePaola DP (1982) The influence of food carbohydrates on dental caries. In: Lineback DR, Inglett GE (eds) Food carbohydrates. AVI Publishing, Westport, Connecticut, pp 134–152

DePaola PF, Soparkar PM, Tavares M, Allukian M Jr, Peterson H (1982) A dental survey of Massachusetts school children. J Dent Res 61(Sp Iss): 1356–1360

Deren JJ, Bloitman SA, Zamcheck N (1967) Effect of diet upon intestinal disaccharidases and disaccharide absorption. J Clin Invest 46: 186–195

Derman D, Sayers M, Lynch SR, Charlton R, Bothwell TH, Mayet F (1977) Iron absorption from a cereal based meal containing cane sugar fortified with ascorbic acid. Br J Nutr 38: 261–269

Deshaies Y, Vallerand AL, Bukowiecki LJ (1983) Serum lipids and cholesterol distribution in lipoproteins of exercise-trained female rats fed sucrose. Life Sci 33: 75–82

De Silva Abiaka M (1969) Dietary sucrose and cardiovascular disease: a review. J Nutr Diet 6: 343–355

DeVries JW, Heroff JC, Egberg DC (1979) High pressure liquid chromatographic determination of carbohydrates in food products: evaluation of method. J Assoc Off Anal Chem 62: 1292–1296

De Zeeuw-Van Gerven M, Thoden Van Velzen SK (1978) Floride caries bij zeer jonge kinderen. Ned Tijdschr Tandheelkd 85: 434–437

Diehl JF, Bissett JK (1963) Determination of insulin and sucrose space in tissues of vitamin E-deficient rabbits. J Arkansas Med Soc 60: 41–42

Diehl JF, Bissett JK (1963) Insulin and sucrose distribution in tissues of vitamin E-deficient and control rabbits. Proc Soc Exp Biol Med 112: 173–176

Dienhart GB, Tumbleson ME, Hicklin KW, Hutcheson DP (1975) Plasma lactic acid and pyruvic acid concentrations following intragastric infusion of ethanol in adult miniature swine (*Sus scrofa*). Int J Biochem 6: 211–218

Diet and Behavior: A Multidisciplinary Evaluation (1986) Sponsored by the American Medical Association International Life Science Institute and Nutrition Foundation, Arlington, Virginia (1984). Nutr Rev 44(Suppl): 1–254

Dillmann WH (1984) Fructose feeding increases $Ca^{++}$-activated myosin ATPase activity and changes myosin isoenzyme distribution in the diabetic rat heart. Endocrinology 114: 1678–1685

Dirks B (1985) Nutrition, sugars, dental caries. Riv Ital Sci Alim 14: 307–316

Dische Z, Goldhammer H (1932) Action of various carbohydrates on the blood sugar and blood phosphorus by the working of muscles. Biochem Z 247: 8–34

Dixon CH (1973) In vitro effects of sodium and calcium cyclamates, cyclohexylamine and sucrose on growth rate and chromosomes of Chinese hamster fibroblasts. Diss Abstr Int [B] 33: 5933

Dodds C, Miller AL, Rose CFM (1960) Blood pyruvate and lactate response of normal subjects to dextrose, sucrose, and liquid glucose. Lancet II: 178–180

Dodds MWJ, Edgar WM (1986) Effects of dietary sucrose levels on pH fall and acid-anion profile in human dental plaque after a starch mouth-rinse. Arch Oral Biol 31: 509–512

Doff RS, Rosen S, App G (1977) Root surface caries in the molar teeth of rice rats. II. Quantitation of lesions induced by high sucrose diet. J Dent Res 56: 1111–1114

Douwes AC, Fernandes J, Jongbloed AA (1980) Diagnostic value of sucrose tolerance test in children evaluated by breath hydrogen measurement. Acta Paediatr Scand 69: 79–82

Downer MC (1982) Secular changes in caries experience in Scotland. J Dent Res 61(Sp Iss): 1336–1339

Draper HH (1976) A review of recent nutritional research in the Arctic. In: Shephard RJ (ed) International symposium on circumpolar health. University of Toronto Press, Toronto, Buffalo, pp 120–129

Dreizen S, Spies TD (1952) Incidence of dental caries in habitual sugar cane chewers. J Am Dent Assoc 45: 193–200

Drewnowski A, Greenwood MRC (1983) Cream and sugar: human preferences for high-fat foods. Physiol Behav 30: 629–633

Drewnowski A, Gruen RK, Grinker J (1982) Carbohydrates, sweet taste, and obesity: changing consumption patterns and health implications. In: Lineback DR, Inglett GE (eds) Food carbohydrates. AVI Publishing, Westport, Connecticut, pp 153–169

Drewnowski A, Brunzell JD, Sande K, Iverius PH, Greenwood MR (1985) Sweet tooth reconsidered: taste responsiveness in human obesity. Physiol Behav 35: 617–622

Drucker DB (1983) The role of sugar in the aetiology of dental caries. The microbiological evidence. J Dent 11: 205–207

Duggan JM (1979) Ischaemic heart disease – an hypothesis to integrate the role of insulin, fibre and sucrose. Med Hypotheses 5: 209–219

Duguid R (1985) In vitro acid production by the oral bacterium *Streptococcus mutans* 10449 in various concentrations of glucose, fructose and sucrose. Arch Oral Biol 30: 319–324

Dulloo AG, Eisa OA, Miller DS, Yudkin J (1985) A comparative study of the effects of white sugar, unrefined sugar and starch on the efficiency of food utilization and thermogenesis. Am J Clin Nutr 42: 214–219

Dunkley LC, Mettrick DF (1977) Hymenolepos diminuta: migration, and the rat host's intestinal and blood plasma glucose levels following dietary carbohydrate intake. Exp Parasitol 41: 213–228

Dunlop M, Court JM, Larkins RG (1981) The effects of maternal carbohydrate (sucrose) supplementation on the growth of offspring of pregnancies with habitual consumption. Biol Neonate 40: 196–198

Dunmire DL, Otto SE (1979) High pressure liquid chromatographic determination of sugars in various food products. J Assoc Off Anal Chem 62: 176–184

Durand AMA, Fischer M, Adams M (1968) The influence of type of dietary carbohydrate: effect on histological findings in two strains of rats. Arch Pathol 85: 318–324

Dyerberg J (1986) Linoleate-derived polyunsaturated fatty acids and prevention of atherosclerosis. Nutr Rev 44: 125–134

Dziezak JD (1986) Special report: sweeteners and product development. J Food Technol 40: 11–130

Eacho PI, Sweeny D, Weiner M (1981) Effects of glucose and fructose on conjugation of *p*-nitrophenol in hepatocytes of normal and streptozotocin-diabetic rats. Biochem Pharmacol 30: 2616–2619

Eaton RP, Kipnis DM (1969) Effects of high carbohydrate diets on lipid and carbohydrate metabolism in the rat. Am J Physiol 217: 1160–1165

Ebeling WW (1933) Resorption of dextrose by the colon. Arch Surg 26: 134–152

Ebell S (1978) Caries and sugar substitutes. Ernaehrungsforschung 23: 171–174 (in German)

Edgar WM (1983) The role of sugar in the aetiology of dental caries. The physicochemical evidence. J Dent Res 11: 199–205

Edgar WM, Bibby BG, Mundorff S, Rowley J (1975) Acid production in plaques after eating snacks: modifying factors in foods. J Am Dent Assoc 90: 409–425

Edgar WM, Geddes DAM, Jenkins ON, Rugg-Gunn AJ, Howell R (1978) Effects of calcium glycerophosphate and sodium fluoride on the induction in vivo of caries-like changes in human dental enamel. Arch Oral Biol 23: 655–661

Eggelton P (1963) A comparative investigation of the sugars and hexitols in foetal fluids and foetal and maternal bloods. J Physiol (Lond) 166: 28P–29P

Eggermont E (1968) The biochemical defects of sucrose intolerance and in glucose–galactose malabsorption. University of Louvain, Belgium, 71 pp (Thesis)

Eggermont E (1969) Hydrolysis of the naturally occurring alpha-glucosides by the human intestinal mucosa. Eur J Biochem 9: 483–487

Eisa OA, Yudkin J (1985) Some nutritional properties of unrefined sugar and its promotion of the survival of new-born rat. Br J Nutr 54: 593–603

Eisenstein R, Battifora H, Ellis H, Rosato J, Hobart J (1969) Effects of sucrose and actinomycin on hepatic acid hydrolases. Arch Pathol 87: 514–520

Elias A, Delgado A (1976) Intensive beef production from sugar cane. XII. Effect of maize–wheat supplementation to the molasses/urea diet in bull fattening. Cuban J Agric Sci 10: 153–160

Ellestad-Sayed JJ, Haworth JC, Hildes JA (1978) Disaccharide malabsorption and dietary patterns in two Canadian Eskimo communities. Am J Clin Nutr 31: 1473–1478

Elliot KAC (1946) Swelling of brain slices and the permeability of brain cells to glucose. Proc Soc Exp Biol Med 63: 234–236
Ellis FM, Krantz JC Jr (1941) Sugar alcohols: metabolism and toxicity studies with mannitol and sorbitol in man and animals. J Biol Chem 141: 147–154
Ellis L, Faulkner JM (1939) Circulatory effects of intravenous injection of fifty percent dextrose and sucrose solutions in patients with heart disease. Am Heart J 17: 542–551
Ellwood KC, Michaelis OE IV, Hallfrisch JG, O'Dorisio TM, Cataland S (1983) Blood insulin, glucose, fructose and gastric inhibitor polypeptide levels in carbohydrate-sensitive and normal men given a sucrose or invert sugar tolerance test. J Nutr 113: 1732–1736
Elwood PC, Moore S, Waters WE, Sweetnam P (1970) Sucrose consumption and ischaemic heart disease in the community. Lancet I: 1014–1016
Elwood PC, Baiton D, Moore F, Davies DF, Wakley EJ, Langman M (1971) Cardiovascular surveys in areas with different water supplies. Br Med J ii: 362–363
Empey EL (1960) Occurrence of kidney calcification in female albino rats fed a diet of milk, alkali and sucrose: the effect of adding magnesium to the diet. Diss Abstr Int [B] 20: 3454
Engleman EG, Lankton J, Lankton B (1971) Granulated sugar as treatment for hiccups in conscious patient. N Engl J Med 285: 1489
Erbersdobler HF, Brandt A, Scharrer E, Von Wangenheim B (1981) Transport and metabolism studies with fructose amino acids. Prog Food Nutr Sci 5: 257–263
Ericsson Y (1951) Salivary amylase and its significance in the caries process. Acta Odontol Scand 9: 89–110
Ericsson Y, Hellstrom I, Jared B, Stjernstrom L (1954) Investigations into the relationship between saliva and dental caries. Acta Odontol Scand 11: 179–194
Eriksson C (ed) (1981) Maillard reactions in food. Pergamon, Stockholm
Erlander SR (1967) Explanation of the common relation of diabetes mellitus and atherosclerosis with consumption of carbohydrates, saturated fats, alcohol, and with smoking. Staerke 19: 179–188
Eroz K, Terzioglu M (1949) The effect of glucose injections on the blood sugar level, alkali reserve, and blood and urine pH values of starved rabbits. Arch Int Pharmacodyn Ther 80: 255–268
Ershoff BH (1952) Effect of dietary carbohydrate on the growth-promoting properties of fat in the hyperthyroid rat. Exp Med Surg 10: 21–25
Ershoff BH (1977) Effects of dietary carbohydrate on sodium cyclamate toxicity in rats fed a purified, low-fiber diet. Proc Soc Exp Biol Med 154: 65–68
Ershoff BH, Deuel HJ Jr (1947) A comparison of the nutritive value of fats when fed alone or when fed with sucrose or lactose. Am J Physiol 143: 45–50
Ershoff BH, Geiger E, Bittner E, Graham T (1953) Comparative effects of glucose, fructose, and sucrose on the thiamine requirement of the rat. Exp Med Surg 11: 293–296
Es AJH van, Groot L de, Vogt JE (1986) Energy balances of eight volunteers fed on diets supplemented with either lactitol or saccharose. Br J Nutr 56: 545–554
Escandon JC, Cunningham J, Felig P (1984) The plasma amino acid response to cafeteria feeding in the rat: influence of hyperphagia, sucrose intake, and exercise. Metabolism 33: 364–368
Esh GC, Bandyopadhyaya P (1968) Influence of different types of carbohydrates on the progress of protein depletion and repletion in rats. Indian J Biochem 5: 178–182
Esposito EJ (1971) Effects of sodium chloride and sucrose on caries activity in rats. J Dent Res 50: 850–853
Evans DP (1932) New method for using cane sugar in infant feeding: clinical report. J Med Soc N J 29: 856–859
Evers WD, Chenoweth WL, Bennink MR (1977) Effect on offspring of feeding a sucrose diet during gestation and lactation in rats. Nutr Rep Int 15: 391–396
Exton JH, Park CR (1967) Control of gluconeogenesis in liver. I. General features of gluconeogenesis in the perfused livers of rats. J Biol Chem 242: 2622–2636
Fabio U (1986) Ecosistema microbico del cavo orale e prevenzione della carie. Fed Med 39: 201–214
Fabry P, Poledne R, Kazdová L (1968) The effect of feeding frequency and type of dietary carbohydrate on hepatic lipogenesis in the albino rat. Nutr Diet (Basel) 10: 81–90
Fajans SS (1981) Etiologic aspects of types of diabetes. Diabetes Care 4: 69–75
Fanelli ZF (1930) Modifications of phagocytosis in vivo from effects of dextrose and saccharose on serum and leukocytes. Folia Med 16: 1604–1630
Fanning EA (1982) Sugar in breakfast cereals. Aust Dent J 27: 134
Fare G, Sammons DCH, Seabourne IA, Woodhouse DL (1967) Lethal action of sugars on (rat)

ascites tumour cells in vitro (liver tumour). Nature (Lond) 213: 308–309
Farman AG (1975) Sugar and dental caries. S Afr Med J 49: 998
Feather MS (1982) Sugar dehydration reactions. In: Lineback DR, Inglett GE (eds) Food carbohydrates. AVI Publishing, Westport, Connecticut, pp 113–133
Feher G, Rodbard S, Katz LN (1942) The renal response to hypertonic sucrose solutions. Surgery 12: 705–771
Fehilly AM, Phillips KM, Sweetman PM (1984) A weighed dietary survey of men in Caerphilly, South Wales. Hum Nutr Appl Nutr 38A: 270–276
Feigal RJ, Jensen ME (1982) The cariogenic potential of liquid medications: a concern for the handicapped patient. Special Care Dent 2: 1, 20–24
Feingold DS, Avigad G (1956) Isolation of sucrose and other related oligosaccharides from a partial acid hydrolysate of inulin. Biochim Biophys Acta 22: 196–197
Feingold KR, Moser AH (1985) Effect of glucose or fructose feeding on cholesterol synthesis in diabetic animals. Am J Physiol 249: G634–G641
Fejerskov O, Antoft P, Gadegaard E (1982) Decrease in caries experience in Danish children and young adults in the 1970s. J Dent Res 61(Sp Iss): 1305–1310
Feldman SA, Rubenstein AH, Ho KJ, Taylor CB, Lewis LA, Mikkelson B (1975) Carbohydrate and lipid metabolism in the Alaskan Arctic Eskimo. Am J Clin Nutr 28: 588–594
Felici W, Boscherini B, Cardi et al. (1964) A case of chronic diarrhea from saccharose intolerance. Arch Ital Pediatr 23: 220–222
Felig P, Cherif A, Minagawa A, Wahren J (1982) Hypoglycemia during prolonged exercise in normal men. N Engl J Med 306: 895–900
Fell HG, Dingle JT (1969) Endocytosis of sugars in embryonic skeletal tissues in organ culture. I. General introduction and histological effects. J Cell Sci 4: 89–103
Felt V (1967) Different action of glucose and sucrose on non-esterified fatty acids and cholesterol in healthy women. Cesk Gastroenterol Vyziva 21: 35–39
Fenstermacher JC, Bartlett MO (1967) Sucrose space measurements in the rabbit central nervous system. Am J Physiol 212: 1268–1272
Ferguson HB, Stoddard C, Simeon JG (1986) Double blind challenge studies of behavioural and cognitive effects of sucrose-aspartame ingestion in normal children. Nutr Rev 44: 144–150
Ferguson RK, Woodbury DM (1969) Penetration of $^{14}$C-insulin and $^{14}$C-sucrose into brain, cerebrospinal fluid, and skeletal muscle of developing rats. Exp Brain Res 7: 181–194
Ferlito S, Maugeri D, Lo Furno R, Calafato M (1978) Effetti di un carico orale di glucosio, di fruttosio e di saccarosio sui livelli glicemici, insulinemici e lipidemici in soggetti normali. Arch Sci Med 135: 447–460
Ferlito S, Nolfo G, Caruso L, Calafato M, Lo Furno R (1979) Comportamento della glicemia, della insulinemia e del quadro lipidemico dopo carico orale di glucosio, saccarosio e fruttosio in donne gravide normali. Minerva Ginecol 31: 805–814
Ferlito S, Nolfo G, Caruso L, Lo Furno R Calafato M, (1979) Comportamento della gastrinemia in gravide normali dopo carico orale di glucosio, fruttosio e saccarosio. Minerva Ginecol 31: 871–878
Fernandes J (1974) The effect of disaccharides on the hyperlactacidaemia of glucose-6-phosphatase deficient children. Acta Paediatr Scand 63: 695–698
Fernstrom J (1979) Nutritional influences on brain function. In: Gryder RM, Frankos VH (eds) Effects of foods and drugs on the development and function of the nervous system: methods for predicting toxicity. Proceedings of the fifth FDA science symposium, 10–12 October, Arlington, Virginia. US Food and Drug Administration, Washington DC, pp 21–39
Fernstrom JD, Wurtman RJ (1974) Control of brain serotonin levels by the diet. Adv Biochem Psychopharmacol 11: 133–142
Fernstrom JD, Wurtman RJ, Hammarstrom-Wiklund B, Rand WM, Munro HN, Davidson CS (1979) Diurnal variations in plasma concentrations of tryptophan, tyrosine, and other neutral amino acids: effect of dietary protein intake. Am J Clin Nutr 32: 1912–1922
Fernstrom JD, Wurtman RJ, Hammarstrom-Wiklund B, Rand WM, Munro HN, Davidson CS (1979) Diurnal variations in plasma neutral amino acid concentrations among patients with cirrhosis: effect of dietary protein. Am J Clin Nutr 32: 1923–1933
Fewkes DW, Parker KJ, Vlitos AJ (1971) Sucrose. Sci Prog 59: 25–39
Feyder S (1935) Fat formation from sucrose and glucose. J Nutr 9: 457–468
Fidanza F (1973) Diet and incidence of coronary heart disease. Nutr Diet 19: 93–107
Fields M, Feretti RJ, Smith JC Jr, Reiser S (1983) Effect of copper deficiency on metabolism and mortality in rats fed sucrose or starch diets. J Nutr 113: 1335–1345
Fields M, Michaels OE IV, Hallfrisch J, Reiser S, Smith JC Jr (1983) Effect of copper deficiency

on intestinal hexose uptake and hepatic enzyme activity in the rat. Nutr Rep Int 28: 123–131
Fields M, Reiser S, Smith JC Jr (1983) Effect of copper and zinc on insulin binding and glucose transport by isolated rat adipocytes. Nutr Rep Int 28: 163–169
Fields M, Reiser S, Smith JC Jr (1983) Effect of copper on insulin in diabetic copper-deficient rats. Proc Soc Exp Biol Med 172: 137–139
Fields M, Ferretti RJ, Smith JC Jr, Reiser S (1984) Impairment of glucose tolerance in copper deficient rats: dependency on the type of dietary carbohydrate. J Nutr 114: 393–397
Fields M, Ferretti RJ, Smith JC Jr, Reiser S (1984) The interaction of type of dietary carbohydrates with copper deficiency. Am J Clin Nutr 39: 289–295
Fields M, Ferretti RJ, Smith JC Jr, Reiser S (1984) Interaction between dietary carbohydrate and copper nutriture on lipid peroxidation in rat tissues. Biol Trace Elem Res 6: 379–391
Fields M, Ferretti RJ, Smith JC Jr, Reiser S (1984) Effect of copper deficiency on metabolism and mortality in rats fed sucrose or starch diets. J Nutr 113: 1335–1345
Fields M, Ferretti RJ, Smith JC Jr, Reiser S (1984) Effect of dietary carbohydrates and copper status on blood pressure of rats. Life Sci 34: 763–769
Fields M, Ferretti RJ, Reiser S, Smith JC Jr (1984) The severity of copper deficiency in rats is determined by the type of dietary carbohydrate. Proc Soc Exp Biol Med 175: 530–537
Fillios LC, Naito C, Andrus SB, Portwan OW, Martin RS (1958) Variations in cardiovascular sudanophilia with changes in the dietary level of protein. Am J Physiol 194: 275–279
Finegan A, Hickey N, Maurer B, Mulcahy R (1968) Diet and coronary heart disease: dietary analysis on 100 male patients. Am J Clin Nutr 21: 143–148
Finer N (1985) Sugar substitutes in the treatment of obesity and diabetes mellitus. Clin Nutr 4: 207–214
Finn SB, Glass RB (1975) Sugar and dental decay. World Rev Nutr Diet 22: 304–326
Finn SB, Jamison HC (1980) The relative effects of three dietary supplements on dental caries. ASDC J Dent Child 47: 109–113
Fiorica V, Iampietro PF (1966) Tolerance to sucrose in adapted and non-adapted rats. Proc Soc Exp Biol Med 122: 647–652
Fiorica V, Iampietro PF, de Steiguer D, Moses R (1964) Sucrose excretion in the rat following repeated intra-arterial loading. Fed Proc 23(2 Pt 1): 467 (Abstract)
Fiorica V, Iampietro PF, de Steiguer D, Morrison RD (1965) Modification of sucrose adaptation in the rat. Proc Soc Exp Biol Med 118: 594–596
Fiorica V, Iampietro PF, Moses R (1966) Adaptation to sucrose loading in the rat. Can J Physiol Pharmacol 44: 389–400
Firestone AR (1982) Effect of increasing contact time of sucrose solution or powdered sucrose on plaque pH in vivo. J Dent Res 61: 1243–1244
Firestone AR, Navia JM (1984) In situ changes in plaque pH in rats after application of sugars and sugar alcohols. International Association of Dental Research, Washington DC (Abstract of General Meeting No 1467)
Firestone AR, Schmid R, Muhlemann HR (1982) Cariogenic effects of cooked wheat starch alone or with sucrose and frequency-controlled feedings in rats. Arch Oral Biol 27: 759–763
Firestone AR, Schmid R, Muhlemann HR (1984) Effect of the length and number of intervals between meals on caries in rats. Caries Res 18: 128–133
Fisher KD (1982) Quick response No. 21: evidence concerning hypersensitivity to nutritive sweeteners including table syrups. Federation of American Societies for Experimental Biology (FASEB), Rockville, Maryland (unpublished memorandum to Dr VP Frattali, FDA, 1 February)
Fishman RA (1963) Studies of the transport of sugars between blood and cerebrospinal fluid in normal states and in meningeal carcinomatosis. Trans Am Neurol Assoc 88: 114–118
Fitzgerald DB, Fitzgerald RJ (1972) Intermittent sucrose feeding and caries in hamsters. Arch Oral Biol 17: 215–217
Flaherty CF, McCurdy ML, Becker HC, D'Alessio J (1983) Incentive relativity effects reduced by exogenous insulin. Physiol Behav 30: 639–642
Fleig WE, Noether-Fleig G, Roeben H, Ditschuneit H (1984) Hormonal regulation of key gluconeogenic enzymes and glucose release in cultured hepatocytes: effects of dexamethasone and gastrointestinal hormones on glucagon action. Arch Biochem Biophys 229: 368–378
Fleming SE, GrootWassink JWD (1979) Preparation of high-fructose syrup from the tubers of the Jerusalem artichoke (*Helianthus tuberosusl*). CRC Crit Rev Food Sci Nutr 12: 1–28
Flier JS, Usher P, Deluise M (1981) Effect of sucrose overfeeding on Na,K-APase-mediated 86Rb uptake in normal and ob/ob mice. Diabetes 30: 975–978
Florence E, Quarterman J (1972) The effects of age, feeding pattern and sucrose on glucose

tolerance, and plasma free fatty acids and insulin concentrations in the rat. Br J Nutr 28: 63–74

Foerster A, Forster A (1925) Blood sugar levels and carcinogens. Klin Wochenschr 4: 1540–1543

Foerster H, Mehnert H, Beck J (1967) Influence of various orally administered carbohydrates on blood glucose level. Klin Wochenschr 45: 50–51

Foglia VG (1938) Effects of prolonged fasting or a sugar diet on the endocrine function of the pancreas. C R Soc Biol (Paris) 127: 694–695

Folkart GR, Dancis J, Money WL (1960) Transfer of carbohydrates across guinea pig placenta. Am J Obstet Gynecol 80: 221–223

Folke LE, Gawronski TH, Staat RH, Harris RS (1972) Effect of dietary sucrose on quantity and quality of plaque. Scand J Dent Res 80: 529–533

Fölsch VR, Ebert R, Creutzfeldt W (1981) Response of serum levels of gastric inhibitory polypeptide and insulin to sucrose ingestion during long-term application of acarbose. Scand J Gastroenterol 16: 629–632

Fontes G, Yovanovitch A (1925) Influence of inanition and ingestion of saccharose on urea content of blood. C R Soc Biol (Paris) 93: 690–691

Food and Drug Laboratories (1973) Teratologic evaluation of FDA 71–76 (pure cane sugar; sucrose) in mice and rats. Food and Drug Research Laboratories Inc, Waverley, New York (Prepared under DHEW Contract No FDA 21–260)

Foppiani E (1986) Caratteristiche dei prodotti dietetici per nutrizione enterale. In: Alimentazione enterale specializzata, Milano Instituto Scotti Bassani per le Ricerche e l'Informazione Scientifica e Nutrizionale, Milan, pp 45–64

Ford WCL, Harrison A (1983) D-[I-$^{14}$C]Mannitol and [U-$^{14}$C]sucrose as extracellular space markers for human spermatozoa and the uptake of 2-deoxyglucose. J Reprod Fertil 69: 479–487

Fordtran JS, Ingelfinger FJ (1968) Absorption of water, electrolytes, and sugars from the human gut. Aliment Canal 3: 1457–1490

Forscher BK, Fosdick LS (1954) Physicochemical phenomena related to dental caries: effect of sugar on rate of diffusion of hydrogen ion. J Dent Res 33: 206–211

Forster H (1974) Comparative metabolism of xylitol, sorbitol and fructose. In: Sipple HL, McNutt KW (eds) Sugars in nutrition. Academic Press, New York, pp 257–260

Forster H (1976) Possible side effects of glucose, fructose, sorbitol and xylitol in man. Int Z Vitamin Ernahrungforsch 15: 116–130

Forster H (1976) Metabolism of glucose substitutes compared to that of glucose. Int Z Vitamin Ernahrungforsch 15: 68–74

Foucard T (1984) Developmental aspects of food sensitivity in childhood. Nutr Rev 42: 98–104

Fosdick LS (1952) New concepts concerning role of sugar in dental caries. Oral Surg 5: 615–624

Fosdick LS, Forscher BK (1954) Physicochemical phenomena related to dental caries: effect of sugar on rates of diffusion of calcium and phosphate ions. J Dent Res 33: 212–217

Fournier P (1954) Differential activity of sugars with respect to the utilization of calcium. C R Acad Sci 239: 718–720

Fournier P, Dupuis Y, Fournier A (1972) Mechanisms of action of sugars on calcium absorption and retention. C R Acad Sci (Paris) 275: 85–88

Francois R, Frederich A, Vicens-Calvert E, Bertrand M, Ruitton-Ucliengo (1963) Isolated intolerance to saccharose. Pediatrie 18: 563–579

Francois R, Moreau P, David M et al. (1966) Saccharose intolerance. Rein Fole 9: 147–157

Frandsen EK, Grunnet N (1971) Kinetic properties of triokinase from rat liver. Eur J Biochem 23: 588–592

Frederick CA, Guerrant NB, Dutcher RA, Knight CA (1940) Urinary excretion of ascorbic acid by the rat as influenced by ingestion of certain carbohydrates. Proc Soc Exp Biol Med 44: 203–207

Freier S (1962) The relation between sucrosuria, hiatus hernia and mental defect. Arch Dis Child 37: 74–77

Freycon MT (1961) The intolerance of saccharose. Pediatrie 16: 749–750

Fridhandler L, Quastel JH (1955) Absorption of sugars from isolated surviving intestine. Arch Biochem Biophys 56: 412–423

Friedenwald JS (1930) Permeability of lens capsule to sugars. Arch Ophth 4: 350–360

Friedhoff R, Simon JA, Friedhoff AJ (1971) Sucrose solution vs no-calorie sweetener vs water in weight gain. J Am Diet Assoc 59: 485–486

Friedman L, Richardson HL, Richardson ME, Lethco EJ, Wallace WC, Sauro FM (1972) Toxic response of rats to cyclamates in chow and synthetic diets. J Natl Cancer Inst 49: 751–760

Fritz H, Hess R (1968) Prenatal development in the rat following administration of cyclamate,

saccharin and sucrose. Experientia 24: 1140–1141
Froesch ER (1972) Fructose metabolism in adipose tissue. Acta Med Scand 542(Suppl): 37–42
Froesch ER (1972) Essential fructosuria and hereditary fructose intolerance. In: Stanbury JB, Wyngaarden JB, Frederickson DS (eds) The metabolic basis of inherited disease. McGraw-Hill, New York
Froesch ER, Ginsberg JL (1962) Fructose metabolism of adipose tissue. I. Comparison of fructose and glucose metabolism in epididymal adipose tissue of normal rats. J Biol Chem 237: 3317–3324
Froesch ER, Keller U (1971) Metabolism and oxidation of U-$^{14}$C-glucose, xylitol, fructose and sorbitol in the fasted and in the streptozotocin-diabetic rat. Diabetologia 7: 349–356
Frostell G (1969) Dental plaque pH in relation to intake of carbohydrate products. Acta Odontol Scand 27: 3–29
Frostell G (1971) The caries-inducing properties of lycasin in comparison to sucrose. Dtsch Zahnaerztl Z 26: 1181–1187
Frostell G, Keyes PH, Larson RH (1967) Effect of various sugars and sugar substitutes on dental caries in hamsters and rats. J Nutr 93: 65–76
Frostell G, Blomlof L, Blomqvist T et al. (1974) Substitution of sucrose by lycasin in candy. "The Roslagen Study" Acta Odontol Scand 32: 235–254
Fry AJ (1972) The effect of a "sucrose-free" diet on oral glucose tolerance in man. Nutr Metab 14: 313–323
Fry AJ, Grenby TH (1972) The effects of reduced sucrose intake on the formation and composition of dental plaque in a group of men in the Antarctic. Arch Oral Biol 17: 873–882
Fujimaki Y (1934) Sucrose in diet: effects on growth of animals. Trans Soc Pathol Jap 23: 279
Fullerton DT, Getto CJ, Swift WJ, Carlson IH (1985) Sugar opioids and binge eating. Brain Res Bull 14: 673–680
Furuno K, Shimakawa K, Suzuoki Z (1975) Effects of nutritional factors on the development of ethanol-induced fatty liver in KK and KK-A mice. J Nutr 105: 1263–1268
Gabrielson IW, Gryboski JD, Yannet H (1962) Urinary excretion of lactose and sucrose by mentally retarded patients. Lancet I: 361
Gabris J, Cajkova A (1965) Changes in the lactic acid content of the meat of pigs fed on sugar before slaughtering. Folia Vet 9: 25–30
Gachev ER, Spirirova PS (1982) Influence of some alimentary factors (sacharosa, ethanol) on the triglyceride content in blood serum and liver. Sci Work High Med Inst 4: 18–22
Gallagher IHC, Pearce EIF (1983) The fermentation of sucrose, sorbitol, and xylitol by *Propionibacterium avidum*, resulting in the formation of caries-like lesions in enamel. NZ Dent J 79: 75–79
Gamble JL Jr, Robertson JS, Hannigan A, Foster CG, Farr LF (1953) Chloride, bromide, sodium, and sucrose spaces in man. J Clin Invest 32: 483–489
Garcia-Palmieri MR, Tillotson J, Cordero E et al. (1977) Nutrient intake and serum lipids in urban and rural Puerto Rican men. Am J Clin Nutr 30: 2092–2100
Garcia-Palmieri MR, Sorlie P, Tillotson J, Costas R, Cordero E, Rodriquez M (1980) Relationship of dietary intake to subsequent coronary heart disease incidence: the Puerto Rico Heart Health Program. Am J Clin Nutr 33: 1818–1827
Gardner LB, Reiser S (1982) Effects of dietary carbohydrate on fasting levels of human growth hormone and cortisol. Proc Soc Exp Biol Med 169: 36–40
Gardner LB, Spannhake EB, Keeney M (1977) Effect of dietary carbohydrate on serum insulin and glucagon in two strains of rats. Nutr Rep Int 15: 361–366
Garg VK, Agrawal VP (1980) Effect of fructose, leucine and prefeeding of sucrose on transport of nutrients. Rev Esp Fisiol 36: 231–236
Garn SM, Cole PE, Solomon MA, Schaefer AE (1980) Relationships between sugar-foods and the DMFT in 1968–1970. Ecol Food Nutr 9: 135–138
Garn SM, Solomon MA, Cole PE (1980) Sugar-food intake of obese and lean adolescents. Ecol Food Nutr 9: 219–222
Garn SM, Solomon MA, Schaefer A (1980) Internal validation of sugar-food intakes in obese adolescents. Am J Clin Nutr 33: 1890–1891
Gasch AT, Michaelis OE, Douglass LW, Moser PB (1984) Blood zinc, copper, insulin and glucose levels in carbohydrate-sensitive and normal men given a sucrose or invert sugar tolerance test. Nutr Res 4: 967–976
Gawecki J, Urbanowicz M, Jeszka J, Mazur B (1976) Feeding habits and their physiological determinants. Relationships between consumers preference for sweetness of drinks and certain physiological parameters. Acta Physiol Pol 27: 455–460

Geddes DAM, Edgar WM, Jenkins GN, Rugg-Gunn AJ (1977) Apples, salted peanuts and plaque pH. Br Dent J 142: 317–319

Geddes DAM, Cooke JA, Edgar WM, Jenkins GN (1978) The effect of frequent sucrose mouth rinsing on the induction in vivo of caries-like changes in human dental enamel. Arch Oral Biol 23: 663–665

Geiselman PJ (1985) Appetite, hunger and obesity as a function of dietary sugar intake. Appetite 6: 64–79

Geiselman PJ, Novin D (1982) The role of carbohydrates in appetite, hunger and obesity. J Intake Res 3: 203–223

Gergely P, Toth B, Farkas H, Bot G (1985) Effect of fructose 1-phosphate on the activation of liver glycogen synthase. Biochem J 232: 133–137

Germaine GR, Schachtele CF (1976) *Streptococcus mutans* dextransucrase. Mode of interaction with high-molecular weight dextran and role in cellular aggregation. Infect Immun 13: 365–372

Germaine GR, Chludzinski AM, Schachtele CF (1974) *Streptococcus mutans* dextransucrase: requirement for primer dextran. J Bacteriol 120: 287–294

Germaine GR, Harlander SK, Leung WS, Schachtele CF (1977) *Streptococcus mutans* dextransucrase: functioning of primer dextran and endogenous dextranase in water-soluble and water-insoluble glucan synthesis. Infect Immun 16: 637–648

Gerritsen GC (1976) The role of nutrition to diabetes in relation to age. In: Rockstein M, Sassmano ML (eds) Nutrition, longevity and aging. Academic Press, New York, pp 229–252

Giboudeau J, Roland J, Orcel L (1969) Influence de la dose de glucose injectée par voie veineuse sur l'aspect histologique et l'activité fonctionnelle du foie chez le rat. Soc Biol 8 Juillet: 1511–1517

Giebisch G, Kraupp O, Pillat B, Stormann H (1957) Replacement of extra-cellular sodium chloride by sodium sulfate or sucrose and the effect on the isolated perfused mammalian muscle. Arch Ges Physiol 265: 220–236

Gigon A (1924) The fluctuation of hydrogen ion concentration in blood under different conditions. Z Ges Exp Med 44: 95–106

Gill HS, Toor HS (1976) Evaluation of bioassays of the toxicity to fish of sugar factory effluents. Acta Hydrobiol 18: 323–329

Ginsberg H, Olefsky JM, Kimmerling G, Crapo P, Reaven GM (1976) Induction of hypertriglyceridemia by a low-fat diet. J Clin Endocrinol Metab 42: 729–735

Ginsburg V, Hers HG (1960) On the conversion of fructose to glucose by guinea pig intestine. Biochim Biophys Acta 38: 427–434

Gitzelmann R, Steinmann B, Van den Berghe G (1983) Essential fructosuria, hereditary fructose intolerance, and fructose 1,6-diphosphatase deficiency. In: Stanbury JB, Wyngaarden JB, Fredrickson DS, Goldstein JL, Brown MS (eds) The metabolic basis of inherited disease, 5th edn. McGraw Hill, New York, pp 118–140

Glass RL (1979) Effects on caries of sugar and stannous chelate in chewing gums. Caries Res 13: 89

Glass RL (1982) Secular changes in caries prevalance in two Massachusetts towns. J Dent Res 61(Sp Iss): 1352–1355

Glass RL, Fleisch S (1974) Diet and dental caries: dental caries incidence and the consumption of ready-to-eat cereals. J Am Diet Assoc 88: 807–813

Glatzel H (1977) Fructose und andere Saccharide in der Ernaehrung des Menschen. Eine kritische Literaturubersicht. Ther Gegenw 116: 1100–1108

Glatzel H (1977) Fructose und andere Saccharide in der Ernaehrung des Menschen. Eine kritische Literaturubersicht. Ther Gegenw 116: 1327–1329

Glavin GB, Mikhail AA (1976) Stress and ulcer etiology in the rat. Physiol Behav 16: 135–139

Glickman I (1971) Periodontal disease. N Engl Med J 284: 1071–1077

Glueck CJ, Connor WE (1978) Diet–coronary heart disease relationships reconnoitered. Am J Clin Nutr 31: 727–737

Goda T, Yamada K, Sugiyama M, Moriuchi S, Hosoya N (1982) Effect of sucrose and acarbose feeding on the development of streptozotocin-induced diabetes in the rat. J Nutr Sci Vitaminol (Tokyo) 28: 41–56

Goetz PW (1985) The new encyclopaedia Britannica 15th edn. Encyclopaedia Britannica, Chicago, vol 2, pp 355–356

Goetzl FR, Goldschmidt M, Wheeler P, Stone F (1949) Influence of sugar upon olfactory acuity and upon the sensation complex of appetite and satiety. Gastroenterology 12: 252–257

Goetzl FR, Ahokas AJ, Payne JG (1950) Occurrence in normal individuals of diurnal variations in acuity of sense of taste for sucrose. J Appl Physiol 2: 619–626

Goffart H, Weischer B (1963) Nematode control with sugar? Anz Schadlingskunde 36: 57–60
Goldblatt MW (1925) Action of various carbohydrates on the ketosis of hunger in humans. Biochem J 19: 948–957
Goldstein JL, Hazzard WR, Schrott HG et al. (1973) Hyperlipidemia in coronary disease. I. Lipid levels in 500 survivors of myocardial infarction. J Clin Invest 52: 1533–1543
Gollnick D (1971) Sugar in relation to maximum sport performance. Proceedings of a symposium on sugar and health, Netherlands, September, pp 15–20
Gomez MI, Markakis P, (1974) Mercury content of some foods. J Food Sci 39: 673–675
Gonnermann B, Schaffer-Spiegel R, Laube H, Schatz H (1982) The effect of a saccharose rich diet on carbohydrate and lipid metabolism of streptozotocin diabetic rats and genetically determined 'diabetic' mice (gg-diab). Int J Obes 6: 41–48
Gonzalez-Calvin JL, Munoz JR, Valdecasas JG et al. (1981) Efectos de la ingestion de sacarosa sobre la diuresis, calciuria y otros constituyentes urinarios en sujectos sanos. Rev Clin Esp 160: 293–297
Gonzalez-Calvin JL, Gomez AZ, Valdecasas JG et al. (1982) Efectos de la ingestion de sacarosa sobre la calciuria y otros constituyentes urinarios en litiasicos renales y familiares asintomaticos. Rev Clin Esp 160: 325–327
Goodrich KP, Zaretsky H (1962) Running speed as a function of concentration of sucrose incentive during pretraining. Psychol Rep 11: 463–468
Gordon H, Middleton K, Seal D, Sullens K (1985) Sugar and wound healing. Lancet II: 663–664
Gordon T, Castelli WP, Hjortland MC, Kannel WB, Dawber TR (1977) High density lipoprotein as a protective factor against coronary heart disease. Am J Med 62: 707–714
Gordon T, Kagan A, Garcia-Palmieri M et al. (1981) Diet and its relation to coronary heart disease and death in three populations. Circulation 63: 500–515
Gorodetskii VK, Shchors EI (1969) Determination of keto sugars in biological fluids with the aid of beta-indoleacetic acid. Vopr Med Khim 15: 310–314
Gorodetskii VK, Mikhailov VI, Rosenfeld EL (1971) On the occurrence of sucrose in urine of man and animals. Vopr Med Khim 17: 206–211 (in Russian)
Gorodetskii VK, Lando LI, Babenko GI (1971) Blood glucose and urine sucrose levels under the effects of a reducing diet. Vopr Pitan 30: 15–20
Gorostiza E, Marche C, Broyart JP, Balmain N, Cezard JP (1984) Influence of starvation on sucrase regulation by dietary sucrose in the rat. Am J Clin Nutr 40: 1017–1022
Gorouben J-C, Bedu J, Le Balle J-C et al. (1963) Sucrose intolerance. A clinical and biological study of 5 cases. Arch Fr Pediatr 20: 253–283
Gotto AM Jr, Bierman EL, Connor WE et al. (1984) Recommendations for treatment of hyperlipidemia in adults. Circulation 69: 1067A–1090A
Gottschall EG, McMillan DB (1985) Structural changes in rat hepatocytes following ingestion of sugar solutions. Acta Anat (Basel) 123: 178–188
Graber CD, Boltjes BH (1968) Blood cholesterol levels, autochthonous intestinal flora and *Clostridium perfringens* recovery in rhesus monkeys fed a high sucrose diet. J Chronic Dis 21: 255–264
Grabfield GP, Swanson D (1942) The uricosuric effects of certain polyhydric alcohols and saccharides. J Pharmacol 74: 106–113
Gracey M, Burke V (1973) Sugar induced diarrhoea in children. Arch Dis Child 48: 331–336
Graf H (1983) Potential cariogenicity of low and high sucrose dietary patterns. J Clin Periodontol 10: 636–642
Graf RJ, Halter JB, Pfeifer MA, Halar E, Brozovich F, Porte D Jr (1981) Glycemic control and nerve conduction abnormalities in noninsulin-dependent diabetic subjects. Ann Intern Med 94: 307–311
Grafe E, Meythaler F (1928) The regulation of insulin production. Part II. The action of carbohydrates (also grape sugar) on the insulin delivery. Arch Exp Pathol Pharmakol 131: 90–91
Graham S, Marshall J, Mettlin C et al. (1982) Diet in the epidemiology of breast cancer. Am J Epidemiol 116: 68–75
Grande F (1967) Dietary carbohydrates and serum cholesterol. Am J Clin Nutr 20: 176–184
Grande F (1974) Sugars in cardiovascular disease. In: Sipple HL, McNutt KW (ed) Sugars in nutrition. Academic Press, New York, pp 401–437
Grande F (1975) Sugar and cardiovascular disease. World Rev Nutr Diet 22: 248–269
Grande F (1984) Body weight, composition and energy balance. In: Present knowledge in nutrition 5th edn. The Nutrition Foundation, Washington DC, pp 7–18
Grande F, Prigge WF (1974) Serum lipid changes produced in dogs by substituting coconut oil

for either sucrose or protein in the diet. J Nutr 104: 613–618
Grande F, Anderson JT, Keys A (1965) Effect of carbohydrates of leguminous seeds, wheat, and potatoes on serum cholesterol concentration in man. J Nutr 86: 313–317
Granneman JG, Campbell RG (1984) Effects of sucrose feeding and denervation on lipogenesis in brown adipose tissue. Metabolism 33: 257–261
Granneman JG, Wade GN (1983) Effect of sucrose overfeeding on brown adipose tissue lipogenesis and lipoprotein lipase activity in rats. Metabolism 32: 202–207
Grant EC (1979) Food allergies and migraine. Lancet I: 966–969
Grant NC, Fahrenbach MJ (1959) Effect of dietary sucrose and glucose on plasma cholesterol in chicks and rabbits. Proc Soc Exp Biol Med 100: 250–252
Grasso P, Golberg L (1966) Early changes at the site of repeated subcutaneous injection of food colorings. Food Cosmet Toxicol 4: 269–282
Grasso P, Golberg L (1966) Subcutaneous sarcoma as an index of carcinogenicity potency. Food Cosmet Toxicol 4: 297–320
Grau R (1955) The use of sugar in meat processing. Die Fleischwirtsch 7: 182–184
Gray GE (1986) Diet, crime and delinquency: a critique. Nutr Rev 44: 89–93
Gray GE, Gray LK (1983) Diet and juvenile delinquency. Nutr Today May–June: 14–22
Gray GM, Ingelfinger FJ (1964) Sucrose digestion and absorption in man: jejunal and ileal studies. Gastroenterology 46: 774–775
Gray GM, Ingelfinger FJ (1964) Sucrose absorption in man: differential absorption of hydrolysis products. J Clin Invest 43: 1305
Gray GM, Ingelfinger FJ (1965) Intestinal absorption of sucrose in man: the site of hydrolysis and absorption. J Clin Invest 44: 390–398
Gray GM, Ingelfinger FJ (1966) Intestinal absorption of sucrose in man: interrelation of hydrolysis and monosaccharide product absorption. J Clin Invest 45: 388–398
Gray RS, Olefsky JM (1982) Effect of a glucosidase inhibitor on the metabolic response of diabetic rats to a high carbohydrate diet, consisting of starch and sucrose, or glucose. Metabolism 31: 88–92
Green K, Otori T (1970) Studies on cornea: physiology in vitro. Exp Eye Res 9(2): 268–280
Green LF, Dale TLC, Ford MA, Bagley R (1971) The effect of a glucose syrup drink on plasma triglyceride concentrations after a high-fat meal and a low-fat meal. Proc Nutr Soc 30: 92A
Green RM, Hartles RL (1967) The effect of uncooked and roll-dried wheat starch, alone and mixed in equal quantity with sucrose, on dental caries in the albino rat. Br J Nutr 21: 921–924
Green RM, Hartles RL (1969) Effect of diets containing different mono- and disaccharides on the incidence of dental caries in the albino rat. Arch Oral Biol 14: 235–241
Green RM, Hartles RL (1970) The effects of diets containing varying percentages of sucrose and maize starch on caries in the albino rat. Caries Res 4: 188–192
Greenwald HM, Pennell S (1930) The carbohydrate metabolism of the normal new-born infant. II. The effect on the concentration of the blood sugar of feeding various sugars to new-born infants. Am J Dis Child 39: 493–503
Greenwald P, Ershow AG, Novelli WD, Benton CM (1985) Cancer, diet and nutrition. A comprehensive source book, 1st edn. Marquis WHO's WHO, Chicago, Illinois
Gregersen MI, Wright L (1935) Sucrose effect of intravenous injection of sucrose and glucose upon reducing power of cerebrospinal fluid, before and after hydrolysis. Am J Physiol 112: 97–108
Greisheimer EM, Johnson OH (1930) Glycogen formation in rat. I. Food forms with 60% the same calorie content in the form of starch or cane sugar or fat or casein. Am J Physiol 94: 11–12
Grenby TH (1965) The influence of cooked and raw wheat starch on dental caries in the rat. Arch Oral Biol 10: 433–438
Grenby TH (1967) A comparison of dental caries in two strains of rats fed on high-sugar diets. Helv Odontol Acta 11: 43–49
Grenby TH (1967) Investigations in experimental animals on the cariogenicity of diets containing sucrose and/or starch. Caries Res 1: 208–221
Grenby TH (1968) Dental plaque formation on a chocolate diet. Proc Nutr Soc 27: 42A–43A
Grenby TH (1970) Replacement of sucrose by dried glucose syrup in the diet of rats susceptible to dental caries. Proc Nutr Soc 29: 63A (Abstract)
Grenby TH (1974) Dental plaque formation in relation to type of carbohydrate. Proc Nutr Soc 33: 24A–25A
Grenby TH (1975) Dental plaque, dental caries and sugar intake. The effects on the plaque of a low-calorie sweetener used in beverages in place of ordinary sugar. Br Dent J 139: 129–134

Grenby TH, Bull JM (1972) Changes in the dental plaque after eating sweets containing starch hydrolysates instead of sucrose. Abstr Commum 32: 39A
Grenby TH, Eikrem G (1974) Dental plaque formation in relation to type of carbohydrate. Proc Nutr Soc 33: 24A–25A
Grenby TH, Hutchinson JB (1969) The effects of diets containing sucrose, glucose or fructose on experimental dental caries in two strains of rats. Arch Oral Biol 14: 373–380
Grenby TH, Powell JM, Gleeson MJ (1974) Effect of sweets made with and without sucrose on the dental plaque, and the correlation between the extent of plaque and human dental caries experience. Arch Oral Biol 19: 217–224
Grevenstuk A (1929) Protein sugar in metabolism of carbohydrates. Ned Tijdschr Geneeskd 1: 495–500
Grewal T, Gopaldas T, Hartenberger P, Ramakrishnan I, Ramachandran G (1973) Influence of sugar and flavour on the acceptibility of instant CSM. Trials on young children from an urban orphanage. J Food Sci Technol (India) 10: 149–152
Gridley DS, Kettering JD, Slater JM, Nutter RL (1983) Modification of spontaneous mammary tumours in mice fed different sources of protein, fat and carbohydrate. Cancer Lett 19: 133–146
Griffiths JR, Stevens AN, Gadian DG, Iles RA, Porteous R (1980) Hepatic fructose metabolism studied by $^{31}P$ nuclear magnetic resonance in the anaesthetized rat. Biochem Soc Trans 8: 641
Grigoresco C, Halfon JJP, Bros M et al. (1985) Effet metabolique de la prise quotidienne de 30g fructose pendant 2 mois. Diabetes Metab 11: 327–328
Grinker J (1978) Obesity and sweet taste. Am J Clin Nutr 31: 1078–1087
Grippaudo G (1985) Cariogenicità degli zuccheri e dolcificanti non cariogenici. Riv Soc Ital Sci Alim 14: 293–298
Groen JJ, Balogh M, Yaron E, Cohen AM (1966) Effect of interchanging bread and surcrose as main source of carbohydrate in a low fat diet on the serum cholesterol levels of health volunteer subjects. Am J Clin Nutr 19: 46–58
Gropp VJ (1973) Studien uber die verwertung von nahrungskohlenhydraten beim mastkalb 2. mitteilung lactose, saccharose und fructose. Z Tierphysiol 32: 144–163
Gross MD (1984) Effect of sucrose on hyperkinetic children. Pediatrics 74: 876–878
Groth W (1963) Effect of sugar feeding and fasting on blood sugar and on adrenal and pancreatic islet morphology of chickens. In: Proceedings of the 3rd international symposium on structural metabolism of pancreatic islets, Uppsala. Stockholm, pp 419–427
Growdon JH, Wurtman RJ (1979) Dietary influences on the synthesis of neurotransmitters in the brain. Nutr Rev 37: 129–136
Gruebner I (1970) The effect of potassium and chloride ions on the resting potential of myelinated nerve fibres under the influence of sucrose and 2,4-dinitrophenol. Acta Biol Med Ger 25: 367–372
Grundy SM, Blackburn H, Mattson F (1982) Rationale of the diet–heart statement of the American Heart Association. Nutr Today September–October: 16–20
Gryder RM, Frankos VH (eds) (1979) The effects of foods and drugs on the development and function of the nervous system: methods for predicting toxicity. In: Proceedings of the fifth FDA science symposium. Office of Health Affairs, US Food and Drug Administration, Arlington, Virginia
Gudman-Hoyer E (1985) Sucrose malabsorption in children. A report of thirty-one Greenlanders. J Pediatr Gastroenterol Nutr 4: 873–877
Gudman-Hoyer E, Krasilnikoff PA (1977) The effect of sucrose malabsorption on the growth pattern in children. Scand J Gastroenterol 12: 103–107
Guenther T, Coecke C (1966) Permeability changes during cell isolation. Z Naturforsch 21: 1171–1174
Guggenheim B (ed) (1979) Health and sugar substitutes. Proceedings of the ERGOB conference, Geneva. Karger, Basel, p 315
Guggenheim B, Muhlemann HR (1978) Summary and discussion of session V: substitutes in oral health – metabolic criteria indicative of cariogenicity. In: Health and sugar substitutes. Proceedings of the ERGOB conference, Geneva. Karger, Basel, pp 336–347
Gurtler H, Stohr J, Brenner KV (1980) Altersabhängigikeit der Fruktose- und Galaktosetoleranz nach intravenoser Stobbelastung von Ferkeln. Arch Exp Vet Med (Leipzig) 34: 767–775
Gustafsson BE, Quensel CE, Lanke LS et al. (1954) The Vipeholm dental caries study. The effect of different levels of carbohydrate intake on caries activity in 436 individuals observed for five years. Acta Odontol Scand 11: 232–365
Guttman N (1953) Operant conditioning, extinction, and periodic reinforcement in relation to concentration of sucrose used as reinforcing agent. J Exp Psychol 46: 213–224

Guyton AC (1981) Transport through the cell membrane. In: Textbook of medical physiology. Saunders, Philadelphia, Pennsylvania, pp 840–841

Hackett AF, Rugg-Gunn AJ, Appleton DR, Allison M, Eastoe JE (1984) Sugar eating habits of 405 11 to 14-year-old English children. Br J Nutr 51: 347–356

Hackett AF, Appleton DR, Rugg-Gunn AJ, Eastoe JE (1985) Some influences on the measurement of food intake during a dietary survey of adolescents. Hum Nutr Appl Nutr 39A: 167–177

Haga K (1980) Effect of coupling sugar as compared with sucrose on plaque accumulation in monkeys. Nippon Shishubyo Gakkaikaishi 22: 55–71

Hahn EW, Hays RL, Kendall KA (1962) Evidence for decreased ovarian steroidogenesis in pregnant albino rats fed on a sucrose diet. In: 46th annual meeting, Atlantic City, New Jersey, April. Fed Proc 21: 436

Halbhuber KJ, Linss W, Geyer G (1977) Zur saccharoseinduzierten Aggutinabilität menschlicher Erythrozyten. Acta Biol Med Ger 36: 885–886

Hald PM, Heinsen AJ, Peters JP (1948) Effect of isotonic solutions of sulfates and phosphates on the distribution of water and electrolytes in human blood. Am J Physiol 152: 77–85

Haldi J, Wynn W (1947) Blood-sugar levels and the behaviour pattern of young healthy adults several hours after the ingestion of large amounts of sucrose. Am J Physiol 150: 263–266

Haldi J, Wynn W (1952) Cariogenicity of sugar-containing diets. J Nutr 46: 425–431

Haldi J, Bachmann G, Ensor C, Wynn W (1938) Comparative effects of a high-glucose and a high-fructose diet on activity, body weight and various constituents of the liver and body of the albino rat exercising at will. J Nutr 16: 239–248

Haldi J, Wynn W, Ensor C (1947) Modification of the action of caffeine on the spontaneous activity of the white rat by the concomitant administration of various food materials. J Nutr 34: 389–399

Haldi J, Wynn W, Shaw JH; Sognnaes RF (1953) The relative cariogenicity of sucrose when ingested in the solid form and in solution by the albino rat. J Nutr 49: 295–305

Haldi J, Law ML, John K (1967) The role of saliva in the differential cariogenic properties of glucose and sucrose. J Dent Res 46: 739–741

Hall CE, Hall O (1966) Comparative ability of certain sugars and honey to enhance saline polydipsia and salt hypertension. Proc Soc Exp Biol Med 122: 362–365

Hall CE, Hall O (1966) Comparative effectiveness of glucose and sucrose in enhancement of hypersalimentation and salt hypertension. Proc Soc Exp Biol Med 123: 370–374

Hall CE, Hall O (1966) Salt hypertension induced by drinking saline, and the effect of different concentrations of sucrose and maltose upon its development. Tex Rep Biol Med 24: 445–456

Hall JF Jr, McClure GS (1936) Insensible water loss in relation to water ingestion in man. Am J Physiol 115: 670–678

Haller JK, Milner JA (1979) Effect of refined carbohydrates on colon cancer in the albino rat. Fed Proc 38: 714

Hallfrisch J, Lazar R, Jorgenson C, Reiser S (1979) Insulin and glucose responses in rats fed sucrose or starch. Am J Clin Nutr 32: 787–793

Hallfrisch J, Lazar R, Reiser S (1979) Effect of feeding sucrose or starch to rats made diabetic with streptozotocin. J Nutr 109: 1909–1915

Hallfrisch J, Cohen L, Reiser S (1981) Effects of feeding rats sucrose in a high fat diet. J Nutr 111: 531–536

Hallfrisch J, Ellwood KC, Michaelis OE, Reiser S, O'Dorisio TM, Prather EM (1983) Effects of dietary fructose on plasma glucose and hormone responses in normal and hyperinsulinaemic men. J Nutr 113: 1819–1826

Hallfrisch J, Lyon R, Michaelis OE, Reiser S (1983) Effects of a combination of common snack foods on some risk factors in heart disease and diabetes in rats. J Am Coll Nutr 2: 141–148

Hallfrisch J, Reiser S, Prather ES (1983) Blood lipid distribution of hyperinsulinemic men consuming three levels of fructose. Am J Clin Nutr 37: 740–748

Hallfrisch J, Reiser S, Prather ES, Canary JJ (1985) Relationships of glucoregulatory hormones in normal and hyperinsulinemic men consuming fructose. Nutr Res 5: 585–594

Halperin ML, Cheema-Dhadli S (1982) Comparison of glucose and fructose transport into adipocytes of the rat. Biochem J 202: 717–721

Halperin-Walega ES, Shively CA, Griffith JW, Greene FE (1985) Adverse effect of a sucrose-based semipurified diet on development and postnatal growth of Fischer rats. Toxicol Appl Pharmacol 80: 284–292

Ham WE, Scott KW (1953) Intestinal synthesis of biotin in the rat. Effect of deficiencies of certain B vitamins and of sulfasuxadine and terramycin. J Nutr 51: 423–433

Hamalainen MM, Makinen KK (1982) Metabolism of glucose, fructose and xylitol in normal and streptozotocin-diabetic rats. J Nutr 112: 1369–1378

Hamilton CL (1972) An observation of long-term experimental obesity and diabetes mellitus in the monkey. J Med Primatol 1: 247–255

Hamilton CL, Kuo PT, Feng L (1974) The effects of high carbohydrate diets on hyperinsulinemic monkeys (*Macaca mulatta*). J Med Primatol 3: 276–284

Hamilton LW (1971) Starvation induced by sucrose ingestion in the rat: partial protection by septal lesions. J Comp Physiol [A] 77: 59–69

Hanson LA, James P, Johansson SGO, Olson RE, White P (1984) General discussion. Summary. Nutr Rev 43: 117–139

Hanzlik PJ, Karsner HT (1924) The anaphylactic phenomena of different intravenously injected substances. J Pharm Exp Ther 23: 173–235

Hanzlik PJ, Karsner HT (1924) Handling of anaphylactic phenomena caused by materials due to hypertonic sugar and salt solution. J Pharm Exp Ther 23: 237–242

Hara E, Saito M (1981) Impaired insulin secretion after oral sucrose and fructose in rats. Endocrinology 109: 966–970

Harding TS (1942) Facts about sugar. Am J Pharm 114: 134–137

Harper AE (1978) Dietary goals – a skeptical view. Am J Clin Nutr 31: 310–321

Harper AE, Gans DA (1986) Claims of antisocial behaviour from consumption of sugar. Food Tech 40: 142–149

Harper AE, Katayama MC (1953) The influence of various carbohydrates on the utilization of low-protein rations by the white rat. I. Comparison of sucrose and cornstarch in 9% casein rations. J Nutr 49: 261–275

Harper AE, Katayama MC, Jelinek B (1952) Influence of dietary carbohydrate on levels of amino acids in the feces of the white rat. Can J Med Sci 30: 578–584

Harper AE, Monson WJ, Arata DA, Benton DA, Elvehjem CA (1953) Influence of various carbohydrates on utilization of low protein rations by white rat, comparison of several proteins and carbohydrates. J Nutr 51: 523–537

Harper AE, Monson WT, Benton DA, Elvehjem CA (1953) The influence of protein and certain amino acids particularly threonine on the deposition of fat in the liver of the rat. J Nutr 50: 383–393

Harper DS, Osborn JC, Clayton RC, Apiess SA, Hefferren JJ (1984) Comparison of nutrient balanced and unbalanced supplementation in rat cariogenicity testing. International Association of Dental Research, Washington DC (Abstract of General Meeting No 1469)

Harper KH, Worden AN (1964) Comparative toxicity studies on glucose, fructose, and sucrose. Toxicol Appl Pharmacol 6: 365 (Abstract No 70)

Harris R (1977) Sweeteners or sugar. Aust Dent J 22: 403–405

Hartles RL (1951) The effect of a high-sucrose diet on the calcium and phosphorus content of the enamel and dentine of rat incisors. Biochem J 48: 245–249

Hartles RL (1951) Influence of a high-sucrose diet on the calcium and phosphorus percentage of the rat femur, and a comparison with its effect on the enamel and dentine of the rat incisor teeth. Biochem J 49: 574–577

Hartles RL (1965) Dietary and environmental factors influencing caries resistance. In: Caries resistant teeth. Ciba Foundation Symposium. Little, Brown, New York, pp 289–312

Hartles RL (1967) Carbohydrate consumption and dental caries. Am J Clin Nutr 20: 152–156

Hartles RL, Wasdell MR (1954) Metabolism of the oral flora. II. Oxidation of some sugars by mixed human saliva. Biochem J 56: 353–355

Hartles RL, Lawton FE, Slack GL (1956) Experimental dental caries in albino rat: production of carious lesions in animals maintained on finely powdered purified diet containing 67 percent sucrose. Br J Nutr 10: 234–240

Hashimoto Y, Yasu T (1966) Protective effect of some neutral solutes on the inactivation of myosin A-adenosinetriphosphatase. Nature (Lond) 211: 194–195

Hassinger W, Sauer G, Cordes U, Krause U, Beyer J, Baessler KH (1981) The effects of equal caloric amounts of xylitol, sucrose and starch on insulin requirements and blood glucose levels in insulin-dependent diabetics. Diabetes 21: 37–40

Haug A, Hostmark AT, Spydevold O (1985) Plasma lipoprotein distribution, faecal cholesterol excretion, and activities of lipoprotein lipase, hepatic lipase and lecithin: cholesterol acyltransferase in rats fed diets rich in sucrose or sunflower oil. Acta Physiol Scand 125: 609–617

Hausmann W (1925) Zur Kenntnis der toxischen Wirkung konzentrierter Zuckerlosungen. Wien Klin Wochenschr 38: 332

Hawker JS (1967) Inhibition of sucrose phosphatase by sucrose. Biochem J 102: 401–460

Haworth JC (1960) Sugars in the blood and urine of children following the ingestion of disaccharides. Arch Dis Child 35: 552–560

Hayashi Y, Nagao M, Sugimura T et al. (eds) (1986) Diet, nutrition and cancer. Japan Scientific Societies Press, Tokyo; VNU Science Press BV, Utrecht

Hayford JT, Danney MM, Thompson RG (1979) Triglyceride-integrated concentration: relationship to insulin-integrated concentration. Metabolism 28: 1078–1085

Hayford JT, Danney MM, Wiebe D, Roberts S, Thompson RG (1979) Triglyceride integrated concentration: effect of variation of source and amount of dietary carbohydrate. Am J Clin Nutr 32: 1670–1678

Hays RL, Kendall KA (1961) Maintenance of pregnancy with prolactin or progesterone in rats on a sucrose diet. Endocrinology 68: 177–178

Hays RL, Hahn EW, Kendall KA (1965) Evidence for decreased steroidogenesis in pregnant rats fed a sucrose diet. Endocrinology 76: 771–772

Hayward SJ, Loeb L (1937) Effects of sugar, glycerol and urea on hormones of cattle anterior pituitary glands. Proc Soc Exp Biol Med 36: 250–253

Heath H, Hamlett VC (1976) The sorbitol pathway. Effect of streptozotocin induced diabetes and the feeding of a sucrose-rich diet on glucose, sorbitol and fructose in the retina, blood and liver of rats. Diabetologia 12: 43–46

Heath H, Kang SS, Phillippou D (1975) Glucose, glucose-6-phosphate, lactate and pyruvate content of the retina, blood and liver of streptozotocin-diabetic rats fed sucrose or starch-rich diets. Diabetologia 11: 57–62

Heaton KW (1976) Fiber, blood lipids, and heart disease. Am J Clin Nutr 29: 125–126

Heaton KW (1984) The sweet road to gall stones. Br Med J 288: 1103–1104

Heaton KW, Emmett PM, Henry CL, Thornton JR, Manhire A, Hartog M (1983) Not just fibre. The nutritional consequences of refined carbohydrate foods. Hum Nutr Clin Nutr 37C: 31–35

Hefti A, Schmid R (1979) Effect on caries incidence in rats of increasing dietary sucrose levels. Caries Res 13: 298–300

Hegsted DM, Stare FJ (1971) Carbohydrates in relation to cardiovascular disease. In: Proceedings of the symposium on sugar and health, Netherlands, pp 36–41

Hei TK, Sudilovsky O (1985) Effects of a high sucrose diet on the development of enzyme altered foci in chemical hepatocarcinogenesis in rats. Cancer Res 45: 2700–2705

Heidecker HA, Borgman RF, Maurice DV (1985) Influence of dietary carbohydrates upon the metabolism of lipids and minerals in rabbits. Nutr Res 5: 983–992

Heinz F, Lamprecht W, Kirsch J (1968) Enzymes of fructose metabolism in human liver. J Clin Invest 47: 1826–1832

Heinz F, Schlegel F, Krause PH (1975) Enzymes of fructose metabolism in human kidney. Enzyme 19: 85–92

Heinz F, Schlegel F, Krause PH (1975) Enzymes of fructose metabolism in human small intestine mucosa. Enzyme 19: 93–101

Heller L (1976) Diskussionsbemerkung zu nebenwirkungen von zuckern und zucker-austausch-stoffen. Int Z Vit Ernahrungforsch 15: 150–152

Helmholz HF (1933) Renal changes in rabbit resulting from intravenous injection of hypertonic solution. J Pediatr 3: 144–157

Helmholz HF, Bollman JL (1939) The diuretic action of sucrose and other solutions. Proc Staff Meet Mayo Clin 14: 567–569

Helmholz HF, Bollman JL (1940) The intravenous administration of sucrose solutions as a means of producing intense diuresis. J Lab Clin Med 25: 1180–1187

Helmholz HF, Bollman JL (1942) The intravenous use of a sucrose-ringer solution to produce maximal diuresis. J Lab Clin Med 27: 606–615

Hems G (1978) The contribution of diet and childbearing to breast cancer rates. Br J Cancer 37: 974–982

Hendler RC, Walesky M, Sherwin RS (1983) Isocaloric sucrose substitution prevents or reverses the diet-induced fall in metabolic rate. Clin Res 31: 465A

Henry LD, Marshall MS (1927) The stability of carbohydrate solutions. J Lab Clin Med 12: 474–477

Herbert V (1981) Will questionable nutrition overwhelm nutrition science? Am J Clin Nutr 34: 2848–2853

Herman RH (1974) Hydrolysis and absorption of carbohydrates and adaptive responses of the jejunum. In: Sipple HL, McNutt KW (eds) Sugars in nutrition. Academic Press, New York, pp 145–172

Herman RH, Zakim D, Stifel FB (1970) Effect of diet on lipid metabolism in experimental

animals and man. Fed Proc Fed Am Soc Exp Biol 29: 1302–1307
Hermansen L, Hultman E, Saltin B (1967) Muscle glycogen during prolonged severe exercise. Acta Physiol Scand 71: 129–139
Herold PM, Kinsella JE (1986) Fish oil consumption and decreased risk of cardiovascular disease: a comparison of findings from animal and human feeding trials. Am J Clin Nutr 43: 566–598
Hers HG (1952) La fructokinase du foie. Biochim Biophys Acta 8: 416–423
Hers HG (1962) Triokinase. In: Colowich SP, Kaplan NO (eds) Methods in enzymology. Academic Press, New York, vol 5, pp 362–364
Hers HG, Ginsburg V (1960) On the conversion of fructose to glucose by guinea pig intestine. Biochim Biophys Acta 38: 427–434
Hers HG, Kusaka T (1953) Le Metabolisme du fructose-1-phosphate dans le foie. Biochim Biophys Acta 11: 427–437
Hersage L, Montenegro JR (1984) Treatment of suppurating wounds by application of sucrose. Gaz Med 91: 59–62
Hershgold EJ, Riley MB (1959) Diet induced variations in tolerance to altitude hypoxia in the mouse. Proc Soc Exp Biol Med 100: 831–834
Herxheimer A, Woodbury DM (1960) The effect of deoxycorticosterone on salt and sucrose taste preference thresholds and drinking behaviour in rats. J Physiol (Lond) 141: 253–260
Herzberg GR, Rogerson M (1981) Dietary corn oil does not suppress the fructose induced increase in hepatic fatty acid synthesis. Nutr Res 1: 73–82
Hess WC, Dhariwal A, Chambliss JF, Alba ZC (1961) Effect of dietary casein–sucrose ratios on amino acid composition of dentinal protein. J Dent Res 40: 87–89
Hessov I (1974) Effects of fructose and glucose infusions on blood acid–base equilibrium in the post-operative period. Acta Chir Scand 140: 347–351
Heukenkamp PU, Zollner N (1972) The comparative metabolism of carbohydrates administered intravenously. Nutr Metab 14(Suppl): 58–73
Heymann W, Hartman ME (1948) Hyperlipemia following intravenous infusion of hypertonic solution of sucrose. Am J Dis Child 75: 68–75
Heywood R, Chesterman H, Allen TR et al. (1977) Sorbitol toxicity studies in the beagle dog (unpublished report from the Huntingdon Research Center; submitted to the World Health Organization by Hoffmann La Roche and Co, Ltd, Basel). In: Joint FAO/WHO Expert Committee on Food Additives (ed) Summary of toxicological data of certain food additives and contaminants. World Health Organization, Geneva, pp 30–31 (WHO Food Additive Series No 13, 1978)
Hill P (1970) Effect of fructose on rat lipids. Lipids 5: 621–627
Hill R, Baker N, Chaikoff IL (1954) Altered metabolic patterns induced in the normal rat by feeding an adequate diet containing fructose as the sole carbohydrate. J Biol Chem 209: 705–716
Hill RM, Lewis HB (1922) Hydrolysis of sucrose in human stomach. Am J Physiol 59: 413–420
Hilton HW (1966) Pesticides and food additives in sugar cane and sugar products. Residue Rev 15: 1–30
Himms-Hagen J (1984) Thermogenesis in brown adipose tissue as an energy buffer. N Engl J Med 311: 1549–1558
Himsworth HP (1935) Diet and the incidence of diabetes mellitus. Clin Sci 2: 117–148
Hinkle MM (1979) Prohibiting the sale of confections in schools – Why? How? When? Dent Assist 48: 23–26
Hippchen LJ (1981) Some possible biochemical aspects of criminal behavior. Int J Biosocial Res 2: 37–42
Hirashima M (1939) Effect of sugar diets on development of bones of rabbits. Sei-i-Kai Med J 58: 1
Hirsch E, Walsh M (1982) Effect of limited access to sucrose on over-eating and patterns of eating. Physiol Behav 29: 129–134
Hirsch E, Ball E, Godkin L (1982) Sex differences in the effects of voluntary activity on sucrose induced obesity. Physiol Behav 29: 253–262
Hiruma K (1923) Permeability changes in solution of nonconductors. Pflugers Arch Physiol 200: 497–510
Ho RS, Aranda CG (1983) Influence of acarbose on hyperglycemia induced by various carbohydrates in rats and oral starch tolerance in monkeys. Arch Int Pharmacodyn Ther 261: 147–156
Hobson P (1978) The role of dietary counselling in prevention of dental caries. Dent Health (Lond) 17: 17–22

Hockett RG (1949) Sugar: a must in the diet. I. Natl Bottlers Gaz 68: 98–104
Hockett RG (1949) Sugar: a must in the diet. II. Natl Bottlers Gaz 68: 116–121
Hodges RE, Rebello T (1983) Carbohydrates and blood pressure. Ann Intern Med 98: 838–841
Hodges RE, Krehl WA, Stone DB, Lopez AS (1967) Dietary carbohydrates and low cholesterol diets: effects on serum lipids of man. Am J Clin Nutr 20: 198–208
Hoehn SK, Carroll KK (1979) Effects of dietary carbohydrate on the incidence of mammary tumors induced in rats by 7,12-dimethylbenz(a)anthracene. Nutr Cancer 1: 27–30
Hoelscher HA (1950) Demonstration of dehydrogenases in tumor cells with tetrazolium salts. Z Krebsforsch 56: 587–595
Hoffman O, Gudernatsch F (1936) Sugar and hormonal regulators of sugar metabolism in relation to the activity of thyroxine. Endokrinologie 18: 96–114
Hollenbeck CB, Coulston AM, Donner C, Williams RA, Reaven GM (1985) The effects of variations in percent of naturally occurring complex and simple carbohydrates on plasma glucose and insulin response in individuals with non-insulin-dependent diabetes mellitus. Diabetes 34: 151–155
Holmes JH, Cizek LJ (1951) Observations on sodium chloride depletion in the dog. Am J Physiol 164: 407–414
Holmes R, Clune S, Cacciarelli A, Yang SS (1983) Chronic diarrhea and failure to thrive. J Pediatr 103: 491–495
Holsinger VH, Sutton CS, Vettel HE, Allen C, Talley FB (1977) Acceptability of whey-soy drink mix prepared with cottage cheese whey. J Dairy Sci 60: 1841–1845
Holt PR, Dominguez AA, Kwartier J (1979) Effect of sucrose feeding upon intestinal and hepatic lipid synthesis. Am J Clin Nutr 32: 1792–1798
Holt S, York DA, Fitzsimons TR (1983) The effects of corticosterone cold exposure and overfeeding with sucrose on brown adipose tissue of obese Zucker rats (fa/fa). Biochem J 214: 215–223
Hooper RH, Short AH (1977) The hepatocellular uptake of glucose, galactose, and fructose in conscious sheep. J Physiol (Lond) 264: 523–539
Hopkins FB (1940) Has it been proved that sugar is significant factor in dental caries? J Am Dent Assoc 27: 1801–1803
Horiuchi M (1932) Tissue respiration. II. The respiration of the diaphragm. Mitteil Med Ges Tokjo 46: 223–235
Horowitz AM (1983) Health education and promotion to prevent dental caries. The opportunity and responsibility of dental hygienists. Dent Hyg (Chicago) 57: 8, 10–15
Horowitz AM, Suomi JD, Peterson JK, Lyman BA (1977) Effects of supervised daily dental plaque removal by children. II. 24 months' results. J Public Health Dent 37: 180–187
Hosoya N, Goda T, Moriuchi S (1982) Effects of sucrose and acarbose feeding on the development of streptozotocin-induced diabetes in the rat. Int Cong Ser Excerpta Med 594: 183–185
Hostmark AT, Blom PCS (1985) Previous exercise nullifies the plasma triacylglycerol response to repeated fructose ingestion in young men. Acta Physiol Scan 125: 553–554
Hostmark AT, Glattre E (1975) Plasma lipid concentration and lipoprotein distribution in exercising and nonexercising rats fed on a high sucrose diet. Experientia 35: 627–628
Hostmark AT, Spydevold O, Lystad E, Kristensen E, Goffeng Bay I (1982) Plasma high density lipoprotein subgroup distribution in rats fed diets with varying amounts of sucrose and sunflower oil. Lipids 17: 489–499
Hostmark AT ,Spydevold O, Lystad E, Eilertsen E (1982) Plasma lipoproteins in rats fed starch, sucrose, glucose or fructose. Nutr Rep Int 25: 161–167
Hotta K (1939) Comparison between disease similar to Kaschin-Beck's osteo-arthritis produced by introduction of excessive iron into drinking water and osteopathia acidosa produced by administration of sugar: experiments on rabbits. J Orient Med 31: 78
Hotta K (1939) Comparative studies between Kaschin-Beck disease produced by excess feeding of iron and osteopathia acidosa produced by excess feeding of sucrose. J Orient Med 31: 1191–1197
Hotta N, Kakuta H, Fukasawa H et al. (1985) Effects of a fructose rich diet and the aldose reductase inhibitor ONO-2235 on the development of diabetic neuropathy in streptozotocin treated rats. Diabetologia 28: 176–180
House RF (1964) Resistance to extinction as a function of sucrose concentration and number of acquisition trials. Diss Abstr Int [B] 24: 5567–5568
Howell RW, Wilson DG (1969) Dietary sugar and ischaemic heart disease. Br Med J iii: 145–148
Howell RW, Wilson DG (1969) Sugar and ischaemic heart disease. Br Med J iv: 559–560
Hrdlicka J, Janicek G, Cuda P (1967) Study of changes during thermal and hydrothermal processes.

XI. Changes in the nutritive value of soybean meal with the addition of certain sugars. Sb Vys Sk Chem Technol Praze Potraviny 18: 29–54

Huang CT, Bibby BG (1977) Must we stick with sugar? NY State Dent J 43: 278–281

Hubbard RS, Anderson RK (1942) The distribution of sucrose in body fluids following intravenous injections. Am J Physiol 137: 722–726

Hubbard RS, Zoll J (1949) Entrance of intravenously injected sucrose into the cerebrospinal fluid. Proc Soc Exp Biol Med 70: 394–396

Hubner D, Schafer R, Pietsch HP (1979) Zur enzymatischen Bestimmung von Glucose und Saccharose in ausgewählten diatetischen Lebensmitteln. Nahrung 23: 723–729

Hue L (1974) The metabolism and toxic effects of fructose. In: Sipple HL, McNutt KW (eds) Sugars in nutrition. Academic Press, New York, pp 357–371

Hueper WC (1965) Are sugars carcinogens? An experimental study. Cancer Res 25: 440–443

Hughes AF, Freeman RB, Fadem T (1974) The teratogenic effects of sugars on the chick embryo. J Embryol Exp Morphol 32: 661–674

Hugill JAC (1983) Sugars and dental decay. Lancet I: 598–599

Hulley SB, Rosenman RH, Bawol RD, Brand RJ (1980) Epidemiology as a guide to clinical decisions: the association between triglyceride and coronary heart disease. N Engl J Med 302: 1383–1389

Hummel H (1924) The significance of grape sugar for detoxification processes in organisms. Klin Wochenschr 3: 1573–1575

Hummel H, Pueschel J (1927) The action of sugar in guanidine poisoning and its significance for the permeability theory of muscles. Pflugers Arch Physiol 217: 441–455

Hummler H (1976) Chronische Verträglichkeitsprüfung mit Xylit an Hunden (unpublished report submitted to the World Health Organization by Hoffmann La Roche and Co, Ltd, Basel). In: Joint FAO/WHO Expert Committee on Food Additives (ed) Summary of toxicological data of certain food additives and contaminants. World Health Organization, Geneva pp 131–132 (WHO Food Additive Series No 12, 1977)

Hummler H (1978) Reproduction study in rabbits in oral administration of Ro-06-7045-xylitol phase II teratology study (unpublished report submitted to the World Health Organization by Hoffmann La Roche and Co, Ltd, Basel). In: Joint FAO/WHO Expert Committee on Food Additives (ed) Summary of toxicological data of certain food additives and contaminants. World Health Organization, Geneva pp 32–33 (WHO Food Additive Series No 13, 1978)

Hundley JM (1949) Influence of fructose and other carbohydrates on the niacin requirement of the rat. J Biol Chem 181: 1–9

Hunt JN (1954) The inhibitory action of sucrose on gastric digestive activity in patients with peptic ulcer. Guy's Hosp Rep 103: 161–173

Hunt JN (1956) Some properties of an alimentary osmoreceptor mechanism. J Physiol (Lond) 132: 267–288

Hunter B, Grahan C, Heywood R, Prentice D, Roe R, Noakes D (1978) Tumorigenicity and carcinogenicity study with xylitol in long-term dietary administration to mice (unpublished report from the Huntingdon Research Center submitted to the World Health Organization by Hoffmann La Roche and Co, Ltd, Basel). In: Joint FAO/WHO Expert Committee on Food Additives (ed) Summary of toxicological data of certain food additives and contaminants. World Health Organization, Geneva pp 28–29 (WHO Food Additive Series No 13, 1978)

Hunter B, Colley J, Street A, Heywood R, Prentice D, Magnusson G (1978) Xylitol tumorigenicity and toxicity study in long-term dietary administration to rats (unpublished report submitted to the World Health Organization by Hoffmann La Roche and Co, Ltd, Basel). In: Joint FAO/WHO Expert Committee on Food Additives (ed) Summary of toxicological data of certain food additives and contaminants. World Health Organization, Geneva pp 29–30 (WHO Food Additive Series No 13, 1978)

Hurst WJ, Martin RA (1980) High performance liquid chromatographic determination of carbohydrates in chocolate. Collaborative study. J Assoc Off Anal Chem 63: 595–599

Hurst WJ, Martin RA, Zoumas BL (1979) Application of HPLC on characterization of individual carbohydrates in foods. J Food Sci 44: 892–895

Hurst WJ, Martin RA, Zoumas BL (1983) Carbohydrate composition of candy bars. J Am Diet Assoc 83: 53–54

Hurttia H, Veli-Matti Multanen, Makinen KK, Tenovuo J, Paunio K (1984) Effects on oral health of mouthrinses containing xylitol, sodium cyclamate and sucrose sweeteners in the absence of oral hygiene. III. Composition and bone resorbing potential of dental plaque. Proc Finn Dent Soc 80: 20–27

Hutteroth TH, Wagner R, Knolle J (1977) Schwere toxische Leberschädigung nach Überdisierung von parenteral verabreichten Kohlenhydraten. Med Klin 72: 703–707

Hutton J, Koulourides T, Borden L (1982) Evaluation of cariostatic disciplines for postradiation caries. Caries Res 16: 390–397

Huttunen JK (1976) Serum lipids, uric acids and glucose during chronic consumption of fructose and xylitol in healthy human subjects. Int J Nutr Res Vitaminol 15: 105–115

Huttunen JK, Ehnholm C, Kekki M, Nikkila EA (1976) Post-heparin plasma lipoprotein lipase and hepatic lipase in normal subjects and patients with hypertriglyceridemia: correlations to sex, age and various parameters of triglyceride metabolism. Clin Sci Mol Med 50: 249–260

Huttunen JK, Makinen KK, Scheinin A (1976) Turku sugar studies. XI. Effects of sucrose, fructose and xylitol diets on glucose lipid and urate metabolism. Acta Odontol Scand 33(Suppl 70): 230–245 reprinted 34: 345–351

Huxley HG (1970) A comparison in experimental animals between diets containing sucrose and glucose. NZ Dent J 66: 6–166

Huxley HG (1971) The cariogenicity of various percentages of dietary sucrose and glucose in experimental animals. NZ Dent J 67: 85–98

Huxley HG (1977) The cariogenicity of dietary sucrose at various levels in two strains of rat under unrestricted and controlled frequency feeding conditions. Caries Res 11: 237–242

Iampietro PF (1955) Fluid, sucrose, and electrolyte distributions after intraperitoneal injections of sucrose solution in the rat. Am J Physiol 182: 148–154

Iampietro PF (1956) Adaptations to intraperitoneal sucrose solutions in the rat. Am J Physiol 184: 390–395

Ichiara A, Greenberg DM (1959) Studies on the purification and properties of D-glycemic acid kinase. J Biol Chem 225: 949–958

Ikeda T (1982) Sugar substitutes: reasons and indications for their use. Int Dent J 32: 33–43

Ikeno N, Honma H, Ishikawa J (1976) Experimental gingivitis by feeding soft food containing sucrose – repeated observation in a *Macaca irus* monkey. Nippon Shishubyo Gakkaikaishi 18: 350–361

Ikkos K (1955) Metabolism, comparison of volume of distribution of sucrose, insulin and thiosulfate in human subjects. Metabolism 4: 19–28

Imfeld Th, Hirsch RS, Muhlemann HR (1978) Telemetric recordings of interdental plaque pH during different meal patterns. Br Dent J 144: 40–45

INCAP report (English translation 1975) Fortification of sugar with vitamin A in Central America and Panama. Institute of Central America and Panama, Guatemala City (INCAP V-36)

Ingersoll CD (1927) Hydrolysis by sucrase inversion of saccharose in very concentrated solution. Bull Soc Chim Biol 9: 838–840

Inglett BE (1981) Sweeteners: a review. Food Tech 35: 37–41

Isabel IM, Staton AJ (1969) Dietary wheat starch and sucrose. Effect on levels of five enzymes in blood serum of young adults. Am J Clin Nutr 22: 701–709

Israel KD, Michaelis OE, Reiser S, Keeney M (1983) Serum uric acid, inorganic phosphorus, and glutamic oxaloacetic transaminase and blood pressure in carbohydrate sensitive adults consuming three different levels of sucrose. Ann Nutr Metab 27: 425–435

Ivanoff NM (1929) Effect of cane sugar on general and residual nitrogen of blood. Klin Med 7: 503–509

Iverson JL, Bueno MP (1981) Evaluation of high pressure liquid chromatography and gas–liquid chromatography for quantitative determination of sugars in foods. J Assoc Off Anal Chem 64: 139–143

Ivy AC, McIlvain GB (1923) The stimulation of stomach secretion by application of substances from the duodenal and jejunal mucous membranes. Am J Physiol 67: 124–140

Ivy AC, Futcher PH, Consolazio WV, Pace N, Gerrard EJ (1944) A tablet emergency ration for lifeboats and rafts. US Naval Med Bull 42: 841–847

Iwao H, Takai Y, Kenmoku A (1959) On dextrin and sucrose in the rat diet. I. The effect of carbohydrates on thiamine and riboflavin excretion of adult rats. II. The effect of carbohydrates on thiamine and riboflavin excretion of young rats. National Institute of Nutrition, Tokyo, pp 78–80 (Annual Report)

Izar G, Fortuna S (1924) The influence of small sugar doses on the complementability and agglutinating action of blood serum. Klin Wochenschr 3: 2196

Jackson CM (1930) Effects of high sugar diets on growth and structure of rat. J Nutr 3: 61–77

Jackson D (1974) Caries experience in English children and young adults during the years 1947–1972. Br Dent J 137: 91–98

Jackson D (1978) Sugar and dental caries: myth and fact. Probe 19: 388–394

Jackson D (1978) Sugar and dental caries: myth and fact. Probe 20: 49–54

Jackson D (1979) Caries experience in 15-year-old English children: 1950–1977. Probe 20: 455–457

Jackson D (1979) Caries experience in deciduous teeth of five-year-old English children: 1974–1977. Probe 20: 404–406

Jacobs HL (1963) Effect of intragastric hypertonic glucose loads on alternating current brain resistance in rats. Proc Soc Exp Biol Med 114: 657–659

Jacobs MH, Parpart AK (1933) Osmotic properties of the erythrocyte. VI. The influence of the escape of salts on hemolysis by hypotonic solutions. Biol Bull 65: 512–528

Jacobs MH, Willis M (1947) Observations on the antihemolytic action of sucrose. Fed Proc 6: 136 (Abstract)

Jacobs P (1966) Sucrosuria or hysteria? Clin Chim Acta 13: 113–116

James WPT (1970) Sugar absorption and intestinal motility in children when malnourished and after treatment. Clin Sci 39: 305–318

James WPT (1983) Energy requirements and obesity. Lancet II: 387–389

Janigan DT, Santamaria A (1961) A histochemical study of swelling and vacuolation of proximal tubular cells in sucrose nephrosis in the rat. Am J Pathol 39: 175–193

Janigan DT, Santamaria A, Trump BF (1960) Sucrose nephrosis: electron microscopic and histochemical observations. J Histochem Cytochem 8: 385–386

Jannulis GE (1934) Sugar as practical hemostatic agent. Monatsschr Ohrenh 68: 1234

Janowitz HD, Grossman MI (1951) Effect of prefeeding alcohol and bitters on food intake of dogs. Am J Physiol 164: 182–186

Jay P (1940) Role in etiology of dental caries. J Am Dent Assoc 27: 393–396

Jeffreys DB, White IR (1974) Influence of a combined oral contraceptive upon rats fed low protein/high carbohydrate diets. Nutr Metab 16: 155–162

Jefremow WW (1935) The influence of carbohydrates on fundamental nutrition: the dynamics of skin phenomena, the development of cataracts, and the phenomena of refaction due to vitamin B2 deficiency in the white rat. Probl Nutr 4: 126–128

Jenkins DJA, Taylor RH (1982) Dosage and interaction with sugars, starch and fibre. Int Cong Ser Excerpta Med 594: 86–96

Jenkins DJA, Wolever TMS, Taylor RH et al. (1981) Glycemic index of foods: a physiological basis for carbohydrate exchange. Am J Clin Nutr 34: 362–366

Jenkins GN (1981) Nutrition and caries. Proc Finn Dent Soc 77: 183–197

Jenkins GN, Tatevossian A (1971) Sucrose and the role of saccharates in enamel caries. Caries Res 5: 28 (Abstract)

Jensen BF, Norman BE (1984) *Bacillus acidopullulyticus Pullulanase*. Application and regulatory aspects for use in the food industry. Proc Biochem 19: 129–134

Jensen ME, Schachtelle CF (1983) The acidogenic potential of reference foods and snacks at interproximal sites in the human dentition. J Dent Res 62: 889–892

Jensen ME, Schachtele CF (1983) Plaque pH measurements by different methods on the buccal surfaces of human teeth after a sucrose rinse. J Dent Res 62: 1058–1061

Jodibauer A (1935) The behavior of various neutral salts on hypertonic hemolysis and the formation of acetic hematin, e.g., the influence of sugar on these two processes. Arch Exp Path Pharmakol 178: 719–723

Johansson KR, Saries WB, Shapiro SK (1948) The intestinal microflora of hens as influenced by various carbohydrates in a biotin-deficient ration. J Bacteriol 56: 619–634

Johansson SGO (1984) Immunological mechanisms of food sensitivity. Nutr Rev 42: 79–84

John HJ (1924) Time relation of fall of blood sugar with insulin. J Metab Res 4: 121–133

Johnson D, Lardy H (1958) Substrate-selective inhibition of mitochondrial oxidations by enhanced toxicity. Nature (Lond) 181: 701–702

Johnson DE, Dorr KE, Swenson WM, Service J (1980) Reactive hypoglycemia. J Am Med Assoc 243: 1151–1156

Johnson MA, Flagg EW (1986) Effects of sucrose and cornstarch on the development of copper deficiency in rats fed high levels of zinc. Nutr Res 6: 1307–1319

Johnson MA, Hove SS (1986) Development of anemia in copper-deficient rats fed high levels of dietary iron and sucrose. J Nutr 116: 1225–1238

Johnson PR, Miller TB (1982) Adverse effects of fructose in perfused livers of diabetic rats. Metabolism 31: 12–125

Jolliffe N, Shannon JA, Smith HW (1932) The excretion of urine in the dog. Am J Physiol 101: 625–646

Jones AW, Goldberg L, Neri A (1979) Effects of a sugar mixture on blood alcohol parameters and impairment in the intact rat. Acta Pharmacol Toxicol 45: 345–351

Jones DP, Sosa FR, Skroma K (1972) Effects of glucose, sucrose, and lactose on intestinal disaccharidases in the rat. J Lab Clin Med 79: 19–30

Jones PJH, Leichter J, Lee M (1981) Uptake of zinc, folate and analogs of glucose and amino acid by the rat fetus exposed to alchohol *in utero*. Nutr Rep Int 24: 75–83

Jones VA, Dickinson RJ, Workman E, Wilson AJ, Freeman AH, Hunter JO (1985) Crohn's disease: maintenance of remission by diet. Lancet I: 177–180

Jong MA, Kieboom CWA, Likassen JAM, Hoeven JS (1985) Effects of dietary carbohydrates on the numbers of *Streptococcus mutans* and *Actinomyces viscosus* in dental plaque of mono-infected gnotobiotic rats. J Dent Res 64: 1134–1137

Jordan HV, Keyes PH, Lim S (1969) Plaque formation and implantation of *Odontomyces viscosus* in hamsters fed different carbohydrates. J Dent Res 48: 824–831

Joszt A, Molinski S (1935) The caramelization of saccharidose. Biochem Z 282: 269–276

Jourdan MH (1970) The effect of 10 weeks consumption of a 75 percent sucrose diet on the incorporation of (U-$^{14}$C) fructose into the serum glycerides of male and female baboons. Proc Nutr Soc 29(Suppl): 18A

Jourdan MN (1974) Endogenous glyceride synthesis rate in baboons fed diets containing sucrose or glucose. Nutr Metab 16: 129–139

Judzewitsch RG, Jaspan JB, Polonsky KS et al. (1983) Aldose reductase inhibition improves nerve conduction velocity in diabetic patients. N Engl J Med 308: 119–125

Juhl M (1963) Control of some nematodes with sucrose. Tidsskr Planteavl 67: 474–481

Jukes TH (1985) Nutrition for mountaineers. Proc Yosemite Inst Mount Med pp 21–26

Jukes TH (1986) Sugar and health. World Rev Nutr Diet 48: 137–194

Jursons KK, Puzaka JJ, Aalmans AR, Vitolin SP (1974) Morphological changes in liver parenchyma of white rats kept on a diet containing xylite and sorbite. Izv Akad Nauk Latviiskoi SSR 9: 16–22

Kaganov IN (1949) Sucrose as food. Sakharnaya Prom 23(8): 16–17

Kahn HA, Herman JB, Medalie JH, Neufeld HN, Riss E, Goldbourt U (1971) Factors related to diabetes incidence: a multivariate analysis of two years' observation on 10 000 men. The Israel ischaemic heart disease study. J Chron Dis 23: 617–629

Kaiser FE, Mariash CN, Schwartz HL, Oppenheimer JH (1980) Inhibition of malic enzyme induction by triiodothyronine in the diabetic rat: reversal by fructose feeding. Metabolism 29: 767–772

Kalsbeek H (1982) Evidence of decrease in prevalence of dental caries in The Netherlands: an evaluation of epidemiological caries surveys on 4–6 and 11–15 year old children, performed between 1965 and 1980. J Dent Res 61(Sp Iss): 1321–1326

Kamikawa Y, Kawamura M (1930) Influence of the abnormal function of internal secretion on the production of epitheliomatous proliferation in the rabbit's ear due to tarring. Trans Jap Pathol Soc 20: 670–673

Kanarek RB, Marks-Kaufman R (1979) Developmental aspects of sucrose-induced obesity in rats. Physiol Behav 23: 881–885

Kanarek RB, Orthen-Gambill N (1982) Differential effects of sucrose, fructose and glucose on carbohydrate-induced obesity in rats. J Nutr 112: 1546–1554

Kanatani H (1959) Protective action of glucose and sucrose against retardation caused by demecolcine in the planarian *Dugesia gonocephala*. J Fac Sci Tokyo U IV Zool 8: 467–471

Kang SS (1979) The effects of dietary sucrose and streptozotocin diabetes on blood and liver constituents. Nutr Metab 23: 327–334

Kang SS, Price RG, Bruckdorfer KR, Worcester NA, Yudkin J (1976) Dietary induced renal damage in the rat. Proc Nutr Soc 36: 27A

Kang SS, Price RG, Bruckdorfer NA, Worcester NA, Yudkin J (1977) Renal damage in rats caused by dietary sucrose. Biochem Soc Trans 5: 235–236

Kang SS, Bruckdorfer KR, Yudkin J (1979) Influence of different dietary carbohydrates on liver and plasma constituents in rats adapted to meal feeding. Nutr Metab 23: 301–315

Kang SS, Price RG, Yudkin J, Worcester NA, Bruckdorfer KR (1979) The influence of dietary carbohydrate and fat on kidney calcification and the urinary excretion of *N*-acetyl-β-glucosaminidase (EC 3.2.1.30). Br J Nutr 41: 65–71

Kang SS, Noirot S, Yudkin J (1980) Renal lipogenesis and gluconeogenesis in streptozotocin-diabetic rats: effect of dietary sucrose. Proc Nutr Soc 39: 8A

Kang SS, Fears R, Noirot S, Mbanya JN, Yudkin J (1982) Changes in metabolism of rat kidney and liver caused by experimental diabetes and by dietary sucrose. Diabetes 22: 285–288

Kannell WB, Gordon T (1970) The Framingham diet study. Diet and the regulation of serum

cholesterol. US Government Printing Office, Washington DC (Public Health Service Monograph No 24)
Kannell WB, Castelli WP, Gordon T (1971) Serum cholesterol, lipoproteins, and the risk of coronary heart disease. The Framingham study. Ann Intern Med 74: 1–12
Kapoor SC, Nath R (1981) Effect of diet on the induction of cadmium induced hypertension in rats. Indian J Biochem Biophys 18(Suppl): 165 (Abstract No 544)
Kapran SK, Slinko OF (1936) Effect of prolonged ingestion of sugar on opsonic index of animal organism infected with purulent staphylococci: causes of development of pyodermitis in workers at sugar factories. Med Zhur 6: 403–410
Karam JH, Grodsky GM, Forsham PH (1973) Excessive insulin response to glucose in obese subjects as measured by immunochemical assay. Diabetes 12: 197–204
Karanja N, McCarron D (1986) Effects of dietary carbohydrates on blood pressure. Prog Biochem Pharmacol 21: 248–265
Karatygin WM, Hefter AJ (1929) Effect of food stimulants on sugar content of bile. Z Ges Exp Med 65: 183–197
Karle E, Büttner W (1971) Caries following administration of sorbitol, xylitol, lycasin and calcium saccharose phosphate in animal experiments. Dtsch Zahnaerztl Z 26: 1097–1108
Karle VE, Gehring F (1975) Wirkung der Zuckeraustauschstoffe Fruktose, Sorbit und Xylit auf Kariesbefall und Plaqueflora der Ratte. Dtsch Zahnaerztl Z 30: 356–363
Kaspar L, Irsigler K (1980) Vergleich des Blutglukoseanstiegs und des Insulin Bedarfs nach oralen Gaben von Saccharose Fruktose Sorbit oder einer Mischung Fructose-Sorbit. Wien Klin Wochenschr 92: 683–687
Katoaka K, Nakae T (1970) Chemical composition and properties of the constituents of rat milk. Rakuno Kagaku No Kenkyu 19: A1–A6
Katz J, McGarry JD (1984) The glucose paradox – is glucose a substrate for liver metabolism? J Clin Invest 74: 1901–1909
Katz S (1981) A diet counseling program. J Am Diet Assoc 102: 840–845
Katz S, Olson BL, Park KC (1975) Factors related to the cariogenic potential of breakfast cereals. Pharmacol Ther Dent 2: 109–131
Kaufmann NA, Kapitulnik J (1972) The significance of sucrose in production of hypertriglyceridemia. Acta Med Scand 542: 229–235
Kaufmann NA, Poznanski R, Blondheim SH, Stein Y (1966) Changes in serum lipid levels of hyperlipemic patients following the feeding of starch, sucrose and glucose. Am J Clin Nutr 18: 261–269
Kaufmann NA, Poznanski R, Blondheim SH, Stein Y (1966) Effect of fructose, glucose, sucrose and starch on serum lipids in carbohydrate induced hypertriglyceridemia and in normal subjects. Isr J Med Sci 2: 715–726
Kaufmann NA, Poznanski R, Blondheim SH, Stein Y (1967) Comparison of effects of fructose, sucrose, glucose, and starch on serum lipids in patients with hypertriglyceridemia and normal subjects. Am J Clin Nutr 20: 131–132
Kawamura M, Kamikawa Y (1930) Relation between the production of tar-tumour and the organs of internal secretion which have a special relation to the metabolism of carbohydrate. Trans Jap Pathol Soc 20: 668–670
Kawamura Y, Okamoto J, Funakoshi M (1968) A role of oral afferents in aversion to taste solutions. Physiol Behav 3: 537–542
Kawasaki C, Okada K (1965) Influence of sugars on the growth of *Lactobacillus fermenti* 36. J Vitamin (Kyoto) 11: 139–144
Kazumi T, Vranic M, Steiner G (1985) Changes in very low density lipoprotein particle size and production in response to sucrose feeding and hyperinsulinaemia. Endocrinology 117: 1145–1150
Keen H (1974) Diabetes and sugar consumption. Eighth International Sugar Research Symposium. International Sugar Research Foundation, Washington DC, 15 pp
Keen H, Thomas BJ, Jarrett RJ, Fuller JH (1977) Nutritional factors in diabetes mellitus. In: Yudkin J (ed) Diet of man: needs and wants. Applied Science Publishers, London, pp 89–108
Keen H, Thomas BJ, Jarrett RJ, Fuller JH (1979) Nutrient intake, adiposity and diabetes in man. Br Med J i: 655–658
Keil HL, Keil HH, Nelson VE (1934) The effect of addition of minerals and sucrose to milk as shown by growth, fertility and lactation of the rat. Am J Physiol 108: 215–220
Keith NM, Power MH (1937) The urinary excretion of sucrose and its distribution in the blood after intravenous injection into normal men. Am J Physiol 120: 203–211
Keith NM, Power MH, Peterson RD (1934) The separation of sucrose, xylose, urea and inorganic sulfates by the kidney in normal man. Am J Physiol 108: 221–228

Keller U, Froesch ER (1971) Metabolism and oxidation of U-$^{14}$C-glucose, xylitol, fructose and sorbitol in the fasted and in the streptozotocin-diabetic rat. Diabetologia 7: 349–356

Kellogg DW (1968) Relation of sucrose to lactation and ruminal fermentation. Diss Abstr Int [B] 29: 424–425

Kellogg DW (1969) Influence of sucrose on rumen fermentation pattern and milk fat content of cows fed a high-grain ration. J Dairy Sci 52: 1601–1604

Kelly TJ, Holt PR, Ai-Lien WU (1980) Effect of sucrose on intestinal very low density lipoprotein production. Am J Clin Nutr 33: 1033–1040

Kelsay JL, Behall KM, Clark WM (1974) Glucose, fructose, lactate and pyruvate in blood and lactate and pyruvate in parotid saliva in response to sugars with and without other foods. Am J Clin Nutr 27: 819–825

Kelsay JL, Behall KM, Holden JM, Prather ES (1974) Diets high in glucose or sucrose in young women. Am J Clin Nutr 27: 926–936

Kennedy PM (1980) The effects of dietary sucrose and the concentrations of plasma urea and rumen ammonia on the degradation of urea in the gastrointestinal tract of cattle. Br J Nutr 43: 125–140

Kennedy PM, Clarke RTJ, Milligan LP (1981) Influences of dietary sucrose and urea on transfer of endogenous urea to the rumen of sheep and numbers of epithelial bacteria. Br J Nutr 46: 533–541

Kergoat M, Bailbe D, Portha B (1987) Effect of high sucrose diet on insulin secretion and insulin action: a study in the normal rat. Diabetologia 30: 252–258

Kerry KR, Anderson CM (1964) A ward test for sugar in faeces. Lancet I: 981–982

Kershner J, Hawke W (1979) Megavitamins and learning disorders: a controlled double-blind experiment. J Nutr 109: 819–826

Keup U, Krause HP, Puls W, Thomas G (1982) Pharmacological studies on acarbose. I. Antihyperglycaemic effects. Int Congr Ser Excerpta Med 594: 147–150

Keys A (1971) Sucrose in the diet and coronary heart disease. Atherosclerosis 14: 193–202

Keys A (1975) Coronary heart disease. The global picture. Atherosclerosis 22: 149–192

Khachatryan GS (1956) Absorption by the brain and muscular tissues of glucose and pyruvic acid by nutritional and conditional nutrition stimulation and internal inhibition. Izv Akad Nauk Armyan SSR Biol Sel'skokhoz Nauki 9(11): 13–26

Khavari KA (1970) Some parameters of sucrose and saline ingestion. Physiol Behav 5: 633–666

Kidder DE, Manners MJ, McCrea MR (1963) The digestion of sucrose by the piglet. Res Vet Sci 4: 131–144

Kiehm TG, Anderson JW, Ward K (1976) Beneficial effects of a high carbohydrate, high fiber diet on hyperglycemic diabetic men. Am J Clin Nutr 29: 895–899

Kim J, Goldfischer S, Biempica L (1976) Changes produced by clofibrate in hepatocytes of rats with sucrose-induced hyperlipidemia. Exp Mol Pathol 25: 263–278

Kimura T, Matsumoto Y, Yoshida A (1980) Dietary sucrose-mediated changes in jejunal sucrase activity of rats. J Nutr Sci Vitaminol (Tokyo) 26: 585–597

Kinoshita JH (1974) Mechanisms initiating cataract formation. Invest Ophthalmol 13: 719–724

Kinoshita S, Schait A, Breboum M et al. (1966) Effects of sucrose on early dental calculus and plaque. Helv Odontol Acta 10: 134–137

Kinuwaki T (1932) Muscular movement and amount of carbon dioxide in blood: especially effect of sugar for restoration of fatigue. J Kumamoto Med Soc 8: 80

Kisch B (1933) Influence of carbohydrates on tissue respiration. Biochem Z 267: 32–42

Kitagawa T (1941) Effect of various sugars on abstinence reaction of tissue cultures habituated to morphine. Folia Pharmacol Jpn 32: 21–22

Kitagawa T (1941) The influence of sugar media on the abstinence of reaction of accustomed morphine tissue cultures. Folia Pharmacol Jpn 32: 219–241

Kiyasu JY, Chaikoff IL (1957) On the manner of transport of absorbed fructose. J Biol Chem 224: 935–939

Kjerulf-Jensen K (1942) The hexosemonophosphoric acids formed within the intestinal mucosa during absorption of fructose, glucose, and galactose. Acta Physiol Scand 4: 225–248

Klacho DM, Burns TW (1977) Observations on glucose homeostasis using continuous monitoring. Horm Metab Res 7: 64–71

Klasing SA, Pilch SA (1986) Review of epidemiological and clinical evidence on the role of omega-3 fatty acids in health and disease. Quick response report. Life Sciences Research Office, Federation of American Societies for Experimental Biology, Rockville, Maryland (Contract No FDA-223-84-2059)

Klatskin G, Krehl WA, Conn HO (1954) The effect of alcohol on the choline requirement. I.

Changes in the rat's liver following prolonged ingestion of alcohol. J Exp Med 100: 605–614

Kleber CJ, Schimmele RG, Putt MS, Muhler JC (1979) The effect of tablets composed of various mixtures of sugar alcohols and sugars upon plaque pH in children. J Dent Res 58: 614–618

Kelber W, Kloepfer J (1968) Rapid method for estimation of sugars in worts, beers, alcohol-free drinks and enzymic hydrolyzates by using thin-layer chromatography. Brauwissenschaft 21: 81–85

Klein H (1949) The influence of cane sugar on the blood alcohol curve. Dtsch Z Ges Gerichtl Med 39: 704–714

Klein H (1951) Blood alcohol curves after sugar-containing beverages. Dtsch Z Ges Gerichtl Med 40: 455–467

Kleinberg I (1985) Oral effects of sugar and sweeteners. Int Dent J 35: 180–189

Klodt W, Stieb B (1938) Effects of ethyl alcohol, dextrose and saccharose on stability of vitamin C. Arch Exp Pathol Pharmakol 190: 652–657

Kloppers PJ, Meiring L, Campbell GD (1980) Sugar-cane juice has no protecting factor with regard to glucose tolerance curves. S Afr Med J 57: 781–782 (in Afrikaans)

Klotzsche C (1969) Zur Frage der teratogenen und embriotoxischen Wirkung von Cyclamat Saccharin und Saccharose (Teratogenic and embryotoxic effects of cyclamate, saccharin and sucrose). Arzneimittelforschung 19: 925–928

Klurfeld DM, Weber MM, Kritchevsky D (1984) Comparison of dietary carbohydrates for promotion of DMBA induced mammary tumorigenesis in rats. Carcinogenesis 5: 423–425

Knebel R (1950) Alimentary sugar loading and circulation. Z Klin Med 146: 75–90

Knuston RA, Merbitz LA, Creekmore AA, Snipes HG (1981) Use of sugar and povidone iodine to enhance wound healing. Five years' experience. South Med J 74: 1329–1335

Koch G (1982) Evidence for declining caries prevalence in Sweden. J Dent Res 61(Sp Iss): 1340–1345

Koch OR, Porta EA, Hartroft WS (1968) A new experimental approach in the study of alcoholism. III. Role of alcohol versus sucrose or fat-derived calories in hepatic damage. Lab Invest 18: 379–386

Koh ET, Mueller J, Osilesi O, Knehans A, Reiser R (1985) Effects of fructose feeding on lipid parameters in obese and lean, diabetic and non diabetic Zucker rats. J Nutr 115: 1274–1284

Koivisto VA (1978) Fructose as a dietary sweetener in diabetes mellitus. Diabetes Care 1: 241–246

Koivisto VA, Karonen SL, Nikkila EA (1981) Carbohydrate ingestion before exercise: comparison of glucose, fructose, and sweet placebo. J Appl Physiol 51: 783–787

Koivisto VA, Harkonen M, Karonen SL, Groop PH, Elovaino R, Ferranini E, et al. (1985) Glycogen depletion during prolonged exercise. Influence of glucose, fructose or placebo. J Appl Physiol 58: 731–737

Koivistoinen P, Hyvonen L (1985) The use of sugar in foods. Int Dent J 35: 175–179

Kolata G (1982) Food affects human behavior. Science 218: 1209–1210

Kolber AR, LeFevre PG (1967) Evidence for carrier-mediated transport of monosaccharides in the Ehrlich ascites tumor cell. J Gen Physiol 50: 1907–1928

Kolinska F (1968) Physiological studies. Significance of sodium cation in transport of sugar by epithelial membranes. Cesk Fysiol 17: 123–130

Konig KG (1969) Caries activity induced by frequency-controlled feeding of diets containing sucrose or bread to Osborne–Mendel rats. Arch Oral Biol 14: 991–993

Konig KG (1982) Impact of decreasing caries prevalence: implications for dental research. J Dent Res 61(Sp Iss): 1378–1383

Konig KG, Grenby TH (1965) The effect of wheat grain fractions and sucrose mixtures on rat caries developing in two strains of rats maintained on different regimes and evaluated by two different methods. Arch Oral Biol 10: 143–153

Korsrud GO, Trick KD (1977) Sucrose, fructose, and glucose contents of Canadian breakfast cereals. Can Inst Food Sci Technol J 10: 134

Korsrud GO, Trick KD (1979) Sucrose, fructose, and glucose contents of infant cereals. J Can Diet Assoc 40: 56

Kossovitch N, Adra Ch (1945) Nonspecific agglutination by mono- and disaccharides. C R Soc Biol (Paris) 139: 1082–1083

Koster JF, Slee RG, Fernandez J (1975) On the biochemical basis of hereditary fructose intolerance. Biochem Biophys Res Commun 64: 289–294

Kotrba C, Code CF (1969) Gastric acid secretory responses to some purified foods and to additions of sucrose or olive oil. Am J Dig Dis 14: 1–8

Kover G, Harza T, Malysuz M, Szocs E (1968) Protective effect of sucrose and mannitol against vasoconstriction induced by angiotensin II in the kidney. Acta Physiol Acad Sci Hung 33: 19–26

Koulourides T (1968) Remineralization methods. Ann NY Acad Sci 153: 84–101

Koulourides T (1982) Implications of remineralization in the treatment of dental caries. In: Proceedings of the symposium on current topics in dental caries in commemoration of the decennial anniversary of Nihon University School of Dentistry at Matsudo. Nihon University, Matsudo, Japan

Koulourides T, Housch T (1983) Hardness testing and microradiography of enamel in relation to intraoral de- and remineralisation. In: Demineralisation and remineralisation of the teeth. IRL Press, Oxford

Koulourides T, Volker JF (1964) Changes of enamel microhardness in the human mouth. Ala J Med Sci 1: 435–437

Koulourides T, Phantumvanit P, Munksgaard EC, Housch T (1974) An intra-oral model used for studies of fluoride incorporation in enamel. J Oral Pathol 3: 185–196

Koulourides T, Keller SE, Manson-Hing L, Lilley V (1980) Enhancement of fluoride effectiveness by experimental cariogenic priming of human enamel. Caries Res 14: 32–39

Kozhukhar EN (1967) Interactions between sugar and lipids in human blood at varied heavy physical loads. Fiziol Zh Akad Nauk Ukr Rsr 13: 784–787

Kozlovsky AS, Moser PB, Reiser R, Anderson RA (1986) Effects of diets high in simple sugars on urinary chromium losses. Metabolism 35: 515–518

Kral J, Sanchez BC (1935) Effects of sugar and insulin on muscular work. Cas Lek Cesk 73: 1333–1335

Krasilnikoff PA, Gudman-Hoyer E, Moltke HH (1975) Diagnostic value of disaccharide tolerance tests in children. Acta Paediatr Scand 64: 693–698

Krasse B (1965) The effect of caries-inducing streptococci in hamsters fed diets with sucrose or glucose. Arch Oral Biol 10: 223–226

Krasse B (1982) Oral effect of other carbohydrates. Int Dent J 32: 24–32

Krause HP, Keup U, Puls W, Thomas G (1982) Pharmacological studies on acarbose. III. Anti-obesity effects. Int Congr Ser Excerpta Med 594: 156–160

Krause HP, Keup U, Thomas G, Puls W (1982) Reduction of carbohydrate-induced hypertriglyceridemia in (fa, fa) "Zucker" rats by the x-glucosidase inhibitor acarbose (BAY g 5421). Metabolism 31: 710–714

Krause W, Taeufel K, Ruttloff H, Maune R (1968) Enzymic cleavage and absorption of di- and higher oligosaccharides in the intestinal tract of animals and humans. Ernaehrungsforschung 13: 161–169

Kraybill HF (1976) Food chemicals and food additives. In: Newberne PM (ed) Trace substances and health. Marcel Dekker, New York, pp 245–318

Krehl WA, Cowgill GW (1955) Nutrient content of cane and beet sugar products. Food Res 20: 449–468

Kreitzman SN (1976) Nutrition in the process of dental caries. Dent Clin North Am 20: 491–505

Kreitzman SN, Klein RM (1976) Non-linear relationship between dietary sucrose and dental caries. International Association of Dental Research, Washington DC (Abstract of General Meeting No 454)

Kristofferson K, Birkhed D (1987) Effects of partial sugar restriction for 6 weeks on numbers of *Streptococcus mutans* in saliva and interdental plaque in man. Caries Res 21: 79–86

Kritchevsky D, Tepper SA (1969) Influence of dietary carbohydrate on lipid metabolism in rats. Med Exp 19: 329–341

Kritchevsky D, Grant WC, Fahrenbach MJ, Riccardi BA, McCandless RFJ (1958) Effect of dietary carbohydrate on the metabolism of cholesterol-4-$C^{14}$ in chickens. Arch Biochem Biophys 75: 142–147

Kritchevsky D, Tepper SA, Kitagawa M (1973) Experimental atherosclerosis in rabbits fed cholesterol-free diets. III. Comparison of fructose and lactose with other carbohydrates. Nutr Rep Int 7: 193–202

Kroll RG, Stone JH (1967) Nocturnal bottle-feeding as a contributory cause of rampant dental caries in the infant and young child. J Dent Child 30: 454–459

Krondl M, Coleman P, Wade J, Milner J (1983) A twin study examining the genetic influence on food selection. Hum Nutr Appl Nutr 37A: 189–198

Kruesi MJP (1986) Carbohydrate intake and childrens' behaviour. Food Technol 40: 150–151

Kruesi MJP, Rapoport JL (1986) Diet and human behavior: how much do they affect each other? Ann Rev Nutr 6: 113–130

Kruhoffer P (1946) The significance of diffusion and convection for the distribution of solutes in

the interstitial space. Acta Physiol Scand 11: 37–47
Ksiazyk J, Milla JP, Harries JT (1982) A comparison of the effect of glucose and sucrose on cholera toxin induced secretion in the rat jejunum in vivo. J Trop Pediatr 28: 8–10
Kubanyi A (1926) Ileus and lowering of blood sugar. J Am Med Assoc 86: 1667
Kubo T (1935) Effect of sugar alone and in combination with insulin injections on formation of immune bodies in rabbits poisoned with carbon tetrachloride. Mitt Med Gesellsch Tokyo 49: 312
Kuehnau J (1971) Recent studies on fructose and saccharose metabolism. Dtsch Zabnaerztl Z 26: 1028–1036
Kulasek G, Barej W, Leontowicz H, Krasicka B, Chomyszyn M, Zawitkowski J (1980) The effect of feeding sugar-beet silage and non-protein-N on rumen and blood metabolites in bulls. Br J Nutr 43: 229–234
Kummevow FA (1985) Optimum nutrition through better planning of world agriculture. World Rev Nutr Diet 45: 1–41
Kuo PT, Bassett DR (1965) Dietary sugar in the production of hyperglyceridemia. Ann Intern Med 62: 1199–1212
Kuo PT, Feng L, Cohen NN, Fitts WT Jr, Miller LD (1967) Dietary carbohydrates in hyperlipemia (hyperglyceridemia): hepatic and adipose tissue lipogenic activities. Am J Clin Nutr 20: 116–125
Kuriyama S (1917) The fate of sucrose parenterally administered. Am J Physiol 43: 343–350
Kushi LH, Lew RA, Stare FJ et al. (1985) Diet and 20-year mortality from coronary heart disease. The Ireland-Boston diet-heart study. N Engl J Med 213: 811–818
Kushigemachi K (1932) Effects of salts and sugar on growth of Trypanosoma. J Kumamoto Med Soc 8: 31–32
Kyzlink V, Curda D (1970) Effect of sucrose concentration and oxygen availability on L-ascorbic acid oxidation in a liquid medium. Z Lebensm Forschung 143: 263–273
Lagerlof F, Dawes C (1985) The effect of swallowing frequency on oral sugar clearance and pH changes by *Streptococcus mitior* in vivo after sucrose ingestion. J Dent Res 64: 1229–1232
Lagerlof F, Dawes C (1985) Effect of sucrose as a gustatory stimulus on the flow rates of parotid and whole saliva. Caries Res 19: 206–211
Laird DA (1930) Effect of sugar in recovering mental and motor control after brief periods of exercise: preliminary investigation. Med Res Rev 36: 383–386
Lake N, Harrill I (1972) Blood constituents and hepatic lipids in rats fed sucrose or starch with or without cyclamate metabolism. Metabolism 21: 533–539
Lakshmanan FL, Schuster EM, Adams M (1967) Effect of dietary carbohydrate on the serum protein components of two strains of rats. J Nutr 93: 117–125
Lakshmanan FL, Howe JC, Schuster EM, Barnes RE (1981) Response of two strains of rats to a high-protein diet containing sucrose or cornstarch. Proc Soc Exp Biol Med 167: 224–232
Lamb AR (1950) Metabolism, relative nutritional efficiency of sucrose and glucose in albino rat. J Nutr 41: 545–554
Lamprecht W, Heinz F (1958) Isolierung von Glycerin-Aldehydehydrogenase aus Rattenleber. Zur Biochemie des Fructosestoffwechsels. Z Naturforsch 13b: 464–465
Lancet (1985) Sugar and health. Lancet I: 456
Landers RE (1979) More on Karo syrup in infant formulas. Letters to the editor. J Pediatr 95: 488–489
Landsberg L, Young JB (1982) Effects of nutritional status on automatic nervous system function. Am J Clin Nutr 32: 1234–1240
Lang CM, Barthel CH (1972) Effects of simple and complex carbohydrates on serum lipids and atherosclerosis in nonhuman primates. Am J Clin Nutr 25: 470–475
Lang K (1971) Übersichten. Xylit, Stoffwechsel und klinische Verwendung. Klin Wochenschr 49: 233–245
Langseth L, Dowd J (1978) Glucose tolerance and hyperkinesia. Food Cosmet Toxicol 16: 129–133
Lasker M (1941) Essential fructosuria. Hum Biol 13: 51–63
Laszt L, dalla Torre L (1942) Relation between sugar resorption and phosphate metabolism: problems of rickets. Schweiz Med Wochenschr 72: 817–819
Latta H, Polt SS (1954) Sucrose inhibition of resorcinol hemolysis. Science 120: 271–273
Laube H, Fussganger RD, Goberna R, Pfeifer EF (1972) The effect of starch and sucrose-rich diets on glucose tolerance and insulin response. Klin Wochenschr 50: 239–242
Laube H, Wojcikowski C, Schatz H, Pfeifer EF (1978) The effect of a high maltose and sucrose feeding on glucose tolerance. Horm Metab Res 10: 192–195
Lauder NM, Valentine AD (1980) British Dietetic Association policy statement. Diet and dental caries. Report of a joint working party of The British Dietetic Association and The British

Paedodontic Society. J Hum Nutr 34: 158–160

Launiala K (1969) The effect of unabsorbed sucrose and mannitol on the small intestinal flow rate and mean transit time. Scand J Gastroenterol 3: 665–671

Launiala K, Perheentupa J, Visakorpi J, Hallman N (1964) Disaccharidases of intestinal mucosa in a patient with sucrose intolerance. Pediatrics 34: 615–620

Lavan JN, Neale FC, Posen S (1971) Urinary calculi: clinical, biochemical and radiological studies in 619 patients. Med J Aust 2: 1049–1061

Law TJ, Hull RN (1968) The stabilizing effect of sucrose upon respiratory syncitial virus (RS). Proc Soc Exp Biol Med 128: 515–518

Lawless H (1982) Paradoxical adaptation to taste mixtures. Physiol Behav 25: 149–152

Layrisse M, Martinez-Torres C, Renzi M (1976) Sugar as a vehicle for iron fortification: further studies. Am J Clin Nutr 29: 274–279

Layrisse M, Martinez-Torres C, Renzi M, Velez F, Gonzalez M (1976) Sugar as a vehicle for iron fortification. Am J Clin Nutr 29: 8–18

Lazicki-Lakshmanan F, Howe J, Schuster EM, Barnes RE (1981) Response of two strains of rats to a high-protein diet containing sucrose or cornstarch. Proc Soc Exp Biol Med 167: 224–232

Leach SA, Green RM (1980) Effect of xylitol-supplemented diets on the progression and regression of fissure caries in the albino rat. Caries Res 14: 16–23

Leach SA, Green RM (1981) Reversal of fissure caries in the albino rat by sweetening agents. Caries Res 15: 508–511

Leach SA, Connel R, Speechley JA, Green RM (1984) Reversal of dental caries by the sugar substitute licasin in vivo. International Association of Dental Research, Washington DC (Abstract of General Meeting No 1466)

Leader VR (1930) Some factors involved in the experimental production of pellagra in rats. I. Preliminary. Biochem J 24: 1172–1180

Lebenthal E, Heitlinger L, Lee PC et al. (1983) Corn syrup sugars: in vitro and in vivo digestibility and clinical tolerance in acute diarrhoea of infancy. J Pediatr 103: 29–34

Leclerc J, Miller ML (1982) Hepatic lipidosis in lactating rat fed a low protein and methionin supplemented diet: comparison of effects of fructose and glucose. C R Acad Sci (Paris) 205: 389–392

Lecoq R (1933) The role of B vitamins and the use of sugar in dove. The relative influence of hexoses and disaccharides as 66% of the supplement of carbohydrate nutrients. C R Hebd Seance Acad Sci 197: 1155–1157

Lecoq R, Chauchard P, Mazoue H (1948) A flour-wheat germ-sugar mixture which does not harm the neuromuscular equilibrium of the user. Ann Falsificat Fraude 41: 198–208

Lederer J, Chakroun R (1978) Exercise physique, metabolisme glucidique et secretion d'insuline chez le rat mis a un regime riche en graisse de boeuf et saccharose. Ann Endocrinol 39: 349–350

Lederer J, Pottier-Arnould AM, Pottier C, Niethals E (1976) Influence réciproque du saccharose et de diverses graisses sur le métabolisme glucidique et l'insulinemie du rat. Ann Endocrinol 37: 399–400

Lederer J, Niethals E, Pottier-Arnould AM, Delhaye-Pottier C (1979) Influence de l'alcool ethylique sur l'action lipidogène des régimes enrichis à la fois en graisse et sucre chez le rat. Rev Alcool 25: 137–143

Lederer J, Pottier-Arnould AM, Delhaye C, Neithals E (1979) Influence réciproque du saccharose et de diverses graisses sur le métabolisme glucidique et l'insulinemie du rat. Ann Endocrinol 40: 247–248

Lee CC, Herrmann RG (1959) Effect of vitamin D, sucrose, corn oil and endocrines on tissue cholesterol in rats. Circ Res 7: 354–359

Lee CC, Herrmann RG (1963) Sucrose diet and biliary cholate-excretion in rats: with note on procedure for cholate determination. Arch Int Pharmacodyn Ther 141: 591–594

Lee CC, Liau YH (1968) Studies with radioactive tracers. XII. Further investigations on the neutral extracts from bread baked with sucrose-$^{14}$C. Cereal Chem 45: 66–72

Lee HK (1966) Studies on the body growth and free amino acids in plasma and liver of rats fed on sucrose diet, adding methionine, threonine, and tryptophan. Daehan Hwahak Hwoejee 10: 109–113

Lee VA (1981) The nutritional significance of sucrose consumption 1970–1980. CRC Press, Boca Raton, Florida, 47 pp (CRC Critical Reviews in Food Science and Nutrition)

Lees RS, Frederickson DS (1965) Carbohydrate induction of hyperlipemia in normal man. Clin Res 13: 327

Leese HJ, Astley NR, Lambert D (1981) Glucose and fructose utilization by rat spermatozoa

within uterine lumen. J Reprod Fertil 61: 435–437
Lehner T (1980) Future possibilities for the prevention of caries and periodontal desease. Br Dent J 149: 318–325
Leibowitz SF, Weiss GF, Yee F, Tretter JB (1985) Noradrenergic innervation of the paraventricular nucleus: specific role in control of carbohydrate ingestion. Brain Res Bull 14: 561–567
Leiter EH, Coleman DL, Ingram DK, Reynolds MA (1983) Influence of dietary carbohydrate on the induction of diabetes in C57BL/KsJ-db/db diabetes mice. J Nutr 113: 184–195
Lemann J Jr, Piering WF, Lennon EJ (1969) Possible role of carbohydrate-induced calciuria in calcium oxalate kidney-stone formation. N Engl J Med 280: 232–237
Lemann J Jr, Lennon E, Piering WR, Prien EL Jr, Ricanati ES (1970) Evidence that glucose ingestion inhibits net renal tubular reabsorption of calcium and magnesium in man. J Lab Clin Med 75: 578–585
Lemieux RU, Huber G (1956) A chemical synthesis of sucrose. A conformational analysis of the reactions of 1,2-anhydro-*d*-D-glucopyranose triacetate. J Am Chem Soc 20: 4117–4119
Lenner RA (1976) Studies of glycemia and glucosuria in diabetics after breakfast meals of different composition. Am J Clin Nutr 29: 716–725
Lenner RA (1976) Specially designed sweeteners and food for diabetics: a real need? Am J Clin Nutr 29: 726–733
Leonard BE (1979) Pharmacological and biochemical aspects of hyperkinetic disorders. Neuropharmacology 18: 923–929
Leonard HB, Barry MI, Schuck C (1955) Effect of alloxan in rats fed corn starch, sucrose, and lactose diets. J Am Diet Assoc 31: 708–710
Leprohon-Greenwood CE, Anderson GH (1986) An overview of the mechanisms by which diet affects brain function. Food Tech 40: 132–138
Lester D, Greenberg LA (1952) Nutrition and the etiology of alcoholism. The effect of sucrose, saccharin and fat on the self-selection of ethyl alcohol by rats. Q J Stud Alcohol 13: 553–560
Leuthardt F, Testa E (1951) Die phosphorylierung der fructose in der leber. II. Mitteilung. Helv Chim Acta 34: 931–938
Levesque J, Dreyfus S, Levy F (1930) Eczema in infant in course of high sugar and low nitrogen diet. Bull Soc Pediatr Paris 28: 565–572
Levin AC (1978) Hyperkinesis. Mod Med 46: 82–94
Levine DJ (1935) Experimental investigation on the capability of various carbohydrates, to substitute for the fat in fat free dried milk. Z Vitaminforschung 4: 92–105
Levine R (1968) Constant sucrose intake in a genetically heterogenous population of mice. J Comp Physiol [A] 66: 456–459
Levine RS, Hill FJ (1978) Rampant caries and its management. Br Dent J 145: 210–212
Levinson RA, Englert E, Smith SE (1970) Intestinal absorption of sugars, water and sodium in alloxan diabetic rats. Diabetes 19: 683–687
Lev-Ran A, Anderson RW (1981) The diagnosis of hypoglycemia. Diabetes 30: 996–999
Li BW, Schuhmann PJ (1980) Gas liquid chromatographic analysis of sugars in ready to eat breakfast cereals. J Food Sci 45: 138–141
Li BW, Schuhmann PJ (1981) Gas chromatographic analysis of sugars in granola cereals. J Food Sci 46: 425–427
Li BW, Schuhmann PJ (1983) Sugar analysis of fruit juice content and method. J Food Sci 48: 633–635
Li BW, Schuhmann PJ, Holden JM (1983) Determination of sugar in yogurt by gas-liquid chromatography. J Agric Food Chem 31: 985–989
Li MK, Kavanagh JP, Prendiville V, Buxton A, Moss DG, Blacklock NJ (1986) Does sucrose damage kidneys? Br J Urol 58: 353–357
Lieberman HR, Wurtman RJ (1986) Foods and food constituents that affect the brain and human behavior. Food Tech 40: 139–141
Lieberman HR, Corkin S, Spring BJ, Growden JW, Wurtman RJ (1982) Mood, performance, and pain sensitivity: changes induced by food constituents. J Psychiatr Res 17: 135–145
Lieberman HR, Spring BJ, Garfield G (1986) The behavioural effects of food constituents: strategies used in studies of amino acids, proteins, carbohydrates and caffeine. Nutr Rev 44(Suppl): 61–70
Lin W, Anderson JW (1977) Effects of high sucrose or starch bran diets on glucose and lipid metabolism on normal and diabetic rats. J Nutr 107: 584–595
Lindeman RD, Ginn HE, Kalbfleisch J, Smith WO (1964) Effect of galactose on urinary electrolyte excretion in man. Proc Soc Exp Biol Med 115: 264–267
Lindeman RD, Adler S, Yiengst MJ, Beard ES (1967) Influence of various nutrients on urinary

divalent cation excretion. J Lab Clin Med 70: 236–245

Linke HAB (1980) Inhibition of dental caries in the inbred hamster by saccharin. Ann Dent 39: 71–74

Linneweh F, Schaumlöffel E, Graul EH, Bode HH (1966) On the residual activity of defective enzymes in hereditary monosaccharide and disaccharide malabsorption. Schweiz Med Wochenschr 96: 424–427 (in German)

Lisy V, Kováru H, Lodin Z (1971) In vitro effects of polyvinylpyrrolidone and sucrose on the acetylcholinesterase, succinate dehydrogenase and lactate dehydrogenase activities in the brain. Histochemie 26: 205–211

Little JA, Shanoff HM, Csima A (1967) Dietary carbohydrate and fat, serum lipoproteins, and human atherosclerosis. Am J Clin Nutr 20: 133–138

Little JA, Birchwood BL, Simmons DA et al. (1970) Interrelationship between the kinds of dietary carbohydrate and fat in hyperlipoproteinaemic patients. Atherosclerosis 11: 173–181

Littleton NW (1967) Studies of oral health in person nourished by stomach tube. I. Changes in the pH of plaque material after the addition of sucrose. J Am Dent Assoc 74: 119–123

Litwack G, Hankes LV, Elvehem CA (1952) Effect of factors other than choline on liver fat deposition. Proc Soc Exp Biol Med 1: 441–445

Liu G, Couston A, Hollenbeck C, Reaven G (1984) The effect of sucrose content in high and low carbohydrate diets on plasma glucose, insulin, and lipid responses in hypertriglyceridemic humans. J Clin Endocrinol Metab 59: 636–642

Lloyd ML, Olsen WA (1987) A study of the molecular pathology of sucrase–isomaltase deficiency. A defect in the intracellular processing of the enzyme. N Engl J Med 316: 438–442

Lochwood DH, Amatruda J (1983) Cellular alterations for insulin resistance in obesity and type II diabetes mellitus. Am J Med 75: 23–31

Lock S, Ford MA, Bagley R, Green LF (1980) The effect on plasma-lipids of the isoenergetic replacement of table sucrose by dried-glucose syrup (maize-syrup solids) in the normal diet of adult men – over a period of 1 year. Br J Nutr 43: 251–256

Loeb H, Eggermont E, Van Geffel R (1967) Clinical and biochemical study of a congenital sucrose intolerance case. Acta Paediatr Belg 21: 105–112

Loeper M, Lemaire A, Figares E, Garcia de Soria M (1935) Action on resistance of heart to various drugs. C R Soc Biol (Paris) 119: 367–368

Loiseleur J (1946) Evidence of antigenic properties of organic molecules of low molecular weight. C R Acad Sci 22: 461–462

Lombardo YB, Nusimovich B, Chicco A, Yommi MR, Gutman R (1981) Effect of glucagon (Gn) on triglyceride (TG) and glycogen (Gly) content of the perfused isolated heart of hyperlipaemic rats. Diabetes 21: 298

Lombardo YB, Chicco A, Mocchiutti N, de Rodi MA, Nusimovich B, Gutman R (1983) Effect of sucrose diet on insulin secretion in vivo and in vitro and on triglyceride storage and mobilisation of the heart of rats. Horm Metab Res 15: 69–76

Lo Monaco D, Nazari V, Romolotti A (1925) Milk secretion in productive cows after subcutaneous injections of saccharose. Arch Farmacol Sper 39: 182–191 (in Italian)

Lonsdale D, Shamberger RJ (1980) Red cell transketolase as an indicator of nutritional deficiency. Am J Clin Nutr 33: 205–211

Lopez H, Navia JM, Hoskins WA (1978) Caries evaluation of sucrose substituted snack foods in rats. International Association of Dental Research, Washington DC (Abstract of General Meeting No 150)

Lopez R, Schwartz JV, Cooperman JM (1980) Riboflavin deficiency in an adolescent population in New York City. Am J Clin Nutr 33: 1283–1286

Lottes MT, Henderson HZ (1979) The relation of sucrose consumption to dental caries in a sample of Indianapolis children. J Indiana Dent Assoc 58: 22–25

Low-Beer TS (1985) Nutrition and cholesterol gallstones. Proc Nutr Soc 44: 127–134

Lubin JM, Burns PE, Blot WJ, Ziegler RG, Lees AW, Fraumeni JF (1981) Dietary factors and breast cancer. Int J Cancer 28: 685–689

Luckey TD, Moore PR, Elvehjem CA, Hart EB (1946) Effect of diet on the response of chicks to folic acid. Proc Soc Exp Biol Med 62: 307–312

Luoma H, Luoma AR (1967) Influences of sucrose and buffer additives on plaque and phosphorus-32 of the enamel. Dent Res 46: 1392–1399

Luoma H, Nuuja T, Mummikoski P (1975) Changes in dental caries and calculus development in rats through additions of magnesium, orthophosphate and fluoride to high-sucrose diets. Arch Oral Biol 20: 227–230

Luostarinen V, Mäkinen KK, Mäkinen PL (1984) Effects on oral health of mouthrinses containing xylitol, sodium cyclamate and sucrose sweeteners in the absence of oral hygiene. V. Response of hamster cheek pouch microcirculation to dental plaque. Proc Finn Dent Soc 80: 35–39

Lutjens A, Verleur H, Plooij M (1975) Glucose and insulin levels on loading with different carbohydrates. Clin Chim Acta 62: 239–243

Lynch CJ, Guarino JJ, Deth RC, Steer ML (1984) Effect of sucrose feeding on x-adrenergic responses in rat liver. Am J Physiol 246: E344–E349

Macdonald I (1962) Some influence of dietary carbohydrate on liver and depot lipids. J Physiol (Lond) 162: 334–344

Macdonald I (1964) The influence of dietary carbohydrates on the lipid pattern in serum and in adipose tissue. Clin Sci 27: 23–30

Macdonald I (1964) Dietary carbohydrates and lipid metabolism. Proc Nutr Soc 23: 119–123

Macdonald I (1965) The lipid response of young women to dietary carbohydrates. Am J Clin Nutr 16: 458–463

Macdonald I (1966) Lipid response of post-menopausal women to dietary carbohydrate. Am J Clin Nutr 18: 86–90

Macdonald I (1966) Influence of fructose and glucose on serum lipid levels in men and pre- and postmenopausal women. Am J Clin Nutr 18: 369–372

Macdonald I (1966) Lipid response to dietary carbohydrates. In: Paoletti R, Kritchevsky D (eds) Advances in lipid research. Academic Press, New York, vol 4, pp 39–67

Macdonald I (1967) Interrelationship between the influence of dietary carbohydrates and fats on fasting serum lipids. Am J Clin Nutr 20: 345–351

Macdonald I (1967) Dietary carbohydrates in normolipemia. Am J Clin Nutr 20: 185–190

Macdonald I (1968) Ingested glucose and fructose in serum lipids in healthy men and after myocardial infarction. Am J Clin Nutr 21: 1366–1373

Macdonald I (1969) Sucrose – what else besides caries. Guy's Hosp Rep 118: 489–493

Macdonald I (1969) Effect on serum lipids of dietary sucrose and fructose. Acta Med Scand 542: 215–219

Macdonald I (1973) Diet and triglyceride metabolism. J Clin Pathol 5: 22–25

Macdonald I (1973) Effects of dietary carbohydrates on serum lipids. Prog Biochem Pharmacol 8: 216–241

Macdonald I (1975) Diet and human atherosclerosis – carbohydrates. Adv Exp Med Biol 60: 57–64

Macdonald I (1976) Sex differences in the metabolic response to dietary carbohydrates. In: Berdanier CD (ed) Carbohydrate metabolism. Hemisphere, Washington, pp 211–222

Macdonald I (1981) Sucrose. In: Howard AN (ed) Nutritional problems in modern society. John Libbey, London, pp 60–67

Macdonald I (1984) Differences in dietary induced thermogenesis following ingestion of various sugars. Ann Nutr Metab 28: 226–230

Macdonald I (1986) Dietary carbohydrate and energy balance. Prog Biochem Pharmacol 21: 181–191

Macdonald I, Braithwaite DM (1964) The influence of dietary carbohydrates on the lipid pattern in serum and in adipose tissue. Clin Sci 27: 23–30

Macdonald I, Keyser A (1976) Some effects in baboons of chronic glycerol ingestion. Proc Nutr Soc 35: 70A–71A

Macdonald I, Keyser A (1977) Some effects in baboons of chronic ingestion of glycerol with sucrose or glucose. Am J Clin Nutr 30: 1661–1669

Macdonald I, Keyser A (1977) Changes in blood metabolites following the acute ingestion of various amounts of glucose, fructose, sucrose and sorbitol. Proc Nutr Soc 36: 117A

Macdonald I, Pacy D (1976) Some immediate metabolic responses in man to fructose ingestion. Proc Nutr Soc 35: 69A–70A

Macdonald I, Roberts JB (1965) The incorporation of various $^{14}C$ dietary carbohydrates into serum and liver lipids. Metabolism 14: 991–999

Macdonald I, Roberts JB (1967) The serum lipid response of baboons to various carbohydrate meals. Metab Clin Exp 16: 572–579

Macdonald I, Turner LJ (1968) Serum-fructose levels after sucrose or its constituent monosaccharides. Lancet I: 841–843

Macdonald I, Turner LJ (1971) Serum glucose and fructose levels after sucrose meals in atherosclerosis. Nutr Metab 13: 168

Macdonald I, Coles BL, Brice J, Jordan MH (1970) The influence of frequency of sucrose intake on serum lipid, protein and carbohydrate levels. Br J Nutr 24: 413–423

Macdonald I, Keyser A, Pacy D (1978) Some effects in man of varying the load of glucose,

sucrose, fructose or sorbitol on various metabolites in blood. Am J Clin Nutr 31: 1305–1311
Macdonald I, Grenby TH, Fisher MA, Williams C (1981) Differences between sucrose and glucose diets in their effects on the rate of body weight change in rats. J Nutr 111: 1543–1547
Mackay DAM (1985) Factors associated with the acceptance of sugar and sugar substitutes by the public. Int Dent J 25: 201–209
Mackay LL, Mackay EM (1924) Convulsions resulting from fluid administration following sucrose injections and water abstinence. Proc Soc Exp Biol Med 21: 286
Madsen KO (1970) Organic compounds and dental caries. Adv Chem Ser 94: 55–92
Maenpaa PG, Raivio KO, Kekomaki MP (1968) Liver adenine nucleotides: fructose induced depletion and its effects on protein synthesis. Science 161: 1253–1254
Mahalanabis D (1986) Development of an improved formulation of oral rehydration salts (ORS) with antidiarrhoeal and nutritional properties a "super ORS". In: Holmgren J, Lindberg A, Mollby R (eds) Development of vaccines and drugs against diarrhoea. 11th Nobel Conference Stockholm 1985. Studentlitteratur, Lund, pp 240–256
Mahalanabis D, Merson MH, Barua D (1981) Oral rehydration therapy: recent advances. World Health Organization, Geneva, pp 245–249 (World Health Forum No. 2)
Mahler P (1925) The action of various sugars on the salt acid secretion of the stomach. Wien Arch Inn Med 10: 549–558
Mainguet PR, Vanderhoeden HL, Eggermont E (1973) Congenital maltase–sucrase and maltase–isomaltase deficiency in an adult. Digestion 8: 353–359
Maiwald HJ (1981) Correlation between the consumption of sugar, acid solubility of the tooth enamel and caries attack in children. J Int Assoc Dent Child 12: 3–7
Maiwald HJ, Greiling B, Callmeier S, Westphal S, Taeufel A (1984) The cariogenicity of refreshing beverages. Biol Abstr 79: 184 (Abstract No 19998)
Mäkinen KK (1974) Sugars and the formation of dental plaque. In: Sipple H, McNutt KW (ed) Sugars in nutrition. Academic Press, New York, pp 645–687
Mäkinen KK, Philosophy L (1972) The role of sucrose and other sugars in the development of dental caries: a review. Int Dent J 22: 363–386
Mäkinen KK, Hurttia H, Mäkinen PL, Paunio K (1984) Effects on oral health of mouthrinses containing xylitol, sodium cyclamate and sucrose sweeteners in the absence of oral hygiene. II. Relative composition of free amino acids in human crevicular fluid. Proc Finn Dent Soc 80: 13–19
Maller O, Hamilton CL (1968) Sucrose and caloric intake by normal and diabetic monkeys. J Comp Physiol [A] 66: 444–449
Malone MH, Gibson RD, Miya TS (1960) A pharmacologic study of the effects of various pharmaceutical vehicles on the action of orally administered phenobarbital. J Am Pharm Assoc Sci Ed 49: 529–534
Malyoth G (1934) Feeding of nurslings with sugars, the structure of which resembles sugars of the body. Z Kinder 56: 590–608
Manchester AC, Farrell KR (1981) Measurement and forecasting of food consumption by USDA. In: Committee on Food Consumption Patterns of the National Research Council. Assessing changing food consumption patterns. National Academy of Sciences, Washington DC, pp 51–71
Mandel ID (1979) Dental caries. Am Sci 67: 680–688
Mann I (1964) The biochemistry of semen and of the male reproductive tract. Methuen, London
Mann JI, Truswell AS (1972) Effects of isocaloric exchange of dietary sucrose and starch on fasting serum lipids, postprandial insulin secretion and alimentary lipaemia in human subjects. Br J Nutr 27: 395–405
Mann JI, Truswell AS, Pimstone BL (1971) The different effects of oral sucrose and glucose on alimentary lipaemia. Clin Sci 41: 123–129
Mann JI, Watermeyer GS, Manning EB, Randles J, Truswell AS (1973) Effects on serum lipids of different dietary fats associated with a high sucrose diet. Clin Sci 44: 601–604
Manso JM, Jover E, Mayor F, Velasco R, Romero H (1979) Effects of galactose, glucose and fructose on carbohydrate–lipid metabolism. J Med 10: 479–486
Mansson B, Holm AK, Ollinen P, Grahnen H (1979) Dental health in 13-year-old children in the north of Sweden – changes during a 10-year period. Swed Dent J 3: 193–203
Mardones J, Segoviariquelme N, Hederra A, Alcaino F (1955) Effect of some self-selection conditions on the voluntary alcohol intake of rats. Q J Stud Alcohol 16: 425–437
Marino S (1926) Influence of the spleen on carbohydrate metabolism. Prob Nutr 3: 1–21
Markelova VF, Zalesskaya YM (1977) Influence of rations with qualitatively different carbohydrates on the lipids metabolism. Vopr Pitan 1: 17–22

Marshall MW, Holdebrand HE (1963) Rat strain differences in response to three diets. J Nutr 79: 227–238
Marshall MW, Wormack MJ (1954) Influence of carbohydrates, nitrogen source and prior state of nutrition balance and liver composition in adult rat. J Nutr 52: 51–64
Marthaler TM (1978) Sugar and oral health: epidemiology in humans. In: Health and sugar substitutes. Proceedings of the ERGOB conference, Geneva. Karger, Basel, pp 27–34
Martin AF, Young FG (1967) Specific activity of carbon dioxide expired by rats after oral sucrose and other sugars. Nature (Lond) 215: 885–886
Martin-DuPan R, Mauron C, Glaeser B, Wurtman RJ (1982) Effects of various oral glucose doses on plasma neutral amino acid levels. Metabolism 31: 937–943
Martinetti R, Della Maggiore B, Turillazzi T (1937) Pharmacodynamic effect of some hypertonic sugar solutions (glucose, galactose, sucrose) introduced intravenously. Boll Soc Ital Biol Sper 12: 785–787
Martini GA, Brandes JW (1976) Increased consumption of refined carbohydrates in patients with Crohn's desease. Klin Wochenschr 54: 367–371
Masek J (1968) Effect of sugars on the organism's regulatory mechanisms. Ernaehrungsforschung 13: 259–273
Masironi R (1970) Dietary factors and coronary heart disease. Bull WHO 42: 103–114
Masserman JH (1935) Effects of the intravenous administration of hypertonic solutions of sucrose, with special reference to the cerebrospinal fluid pressure. Bull Johns Hopkins Hosp 57: 12–21
Masuko C (1940) Sugar element in human amniotic fluid. Jpn J Obstet Gynecol 23: 102–103
Mateos GG, Sell JL (1980) Influence of carbohydrate and supplemented fat source on the metabolizable energy of the diet. Poultry Sci 59: 2129–2135
Matsuoka M (1936) The working of sugar, starch, and yeast on the growth of albino rats. Sci Pap Inst Phys Chem Res 30: 662–670
Mattock MB, Sheorain VS, Subrahmanyam D (1978) Lecithin-cholesterol acyltransferase activity in carbohydrate-induced hypertriglyceridemia in mice. An immunofluorescent method for identification of isolated thyrotropic cells. Experientia 34: 304–305
Maunsbach AB, Madden SC, Latta H (1962) Light and electron microscopic changes in proximal tubules of rats after administration of glucose, mannitol, sucrose, or dextran. Lab Invest 11: 421–432
Mautner H (1927) Water retention in the liver after intravenous sugar injection. Arch Exp Pathol Pharmakol 126: 255–266
May CD (1984) Food sensitivity: facts and fancies. Nutr Rev 42: 72–78
Mayberry JF, Rhodes J (1984) Epidemiological aspects of Crohn's disease: a review of the literature. Gut 25: 886–899
Mayer J (1977) The bitter truth about sugar! Pa Dent J 44: 24–27
McCann AW (1926) Food value of sugars. S Afr Sugar J 10: 93
McCarron DA (1983) Calcium and magnesium in human hypertension. Ann Int Med 98: 800–805
McCay CM (1954) A simplified nutrition program for the latter half of life. J Am Geriatr Soc 2(7): 417–421
McCay CM, Lovelace F, Sperling G et al. (1952) Age changes in relation to the ingestion of milk, water, coffee, and sugar solutions. J Gerontol 7: 161–172
McClure FJ (1952) Dental caries in rats fed diet containing processed cereal foods and low content of refined sugar. Science 116: 229–231
McComb JA (1940) The use of hypertonic sucrose in anaphylactic-like shock in serum-producing horses. J Am Vet Med Assoc 97: 50
McConnell RN (1978) The international setting for sugar and sweeteners. Food and Agricultural Outlook Conference, Washington DC
McDonald JL, Stookey GK (1977) Animal study concerning the cariogenicity of dry breakfast cereals. J Dent Res 56: 1001–1006
McGandy RB, Hegsted DM, Myers ML, Stare FJ (1966) Dietary carbohydrate and serum cholesterol levels in man. Am J Clin Nutr 18: 237–242
McGandy RB, Hegsted DM, Stare FJ (1967) Dietary fats, carbohydrates, and atherosclerotic vascular disease. N Engl J Med 277: 186–192
McGandy RB, Hegsted DM, Stare FJ (1967) Dietary fats, carbohydrates, and atherosclerotic vascular disease (concluded). N Engl J Med 277: 242–247
McHugh WD, McEwen JD, Hitchin AD (1964) Dental disease and related factors in 13-year-old children in Dundee. Br Dent J 147: 246–253
McIntyre ML, Holden JM, Ahrens RA (1975) Blood pressure changes caused by supplementing the diets of young adult volunteers with different amounts of sucrose. Fed Proc 34: 911

McKechnie JK, Reid JVO, Joubert SM (1970) The effect of dietary sucrose on the performance of marathon runners. S Afr Med J 44: 728–731

McKendrick AJW, Roberts GS, Duguid R (1975) Dental caries and sugar intake. Lancet II: 1086–1087

McLennan H (1957) The diffusion of potassium, sodium, sucrose and insulin in the extracellular spaces of mammalian tissues. Biochim Biophys Acta 24: 1–8

McMartin A (1964) The contribution of sugar-cane to our food supplies. S Afr Med J 38: 659–661

Medalie JH, Papier CM, Goldbourt U, Herman JB (1975) Major factors in the development of diabetes mellitus in 10,000 men. Arch Intern Med 135: 811–817

Medical Research Council Working Party (1970) Dietary sugar intake in men with myocardial infarction. Lancet II: 1265–1271

Medvedeva NB (1940) Sugar component of protein. Med Zhur 10: 327–339

Mehnert H (1978) Begünstigen Sorbit und Fructose die diabetische Lento- und Neuropathie? Dtsch Med Wochenschr 103: 229–231

Mehnert H, Foerster H (1968) Evacuation mechanism of the stomach after oral administration of different sugars in man and rat. Diabetologia 4: 26–33

Meier KE, Freitag G (1955) Antibiotic properties of saccharides and bee honey. Z Hyg Infektionskrankh 141: 326–332

Mellberg JR, Larson RH (1978) Effect of cariogenic challenge on fluoride uptake by enamel of rats receiving fluoridated drinking water. Caries Res 12: 137–141

Menezes DM, Shaw JG, Anderson RJ (1972) The dental condition of 10–12-year-old children in Rangoon and Wolverhampton. Arch Oral Biol 17: 1187–1195

Menon PVG, Kurup PA (1974) Hypolipidaemic action of the polysaccharide from *Phaseolus mungo* (blackgram). Effect on glycosaminoglycans, lipids and lipoprotein lipase activity on normal rats. Atherosclerosis 19: 315–326

Menon PVG, Kurup PA (1976) Nature of the dietary carbohydrate and metabolism of glycosaminoglycans and glycoproteins in rats. J Nutr 106: 555–562

Merkens LS, Tepperman HM, Tepperman J (1980) Effects of short-term dietary glucose and fructose on rat serum triglyceride concentration. J Nutr 110: 982–988

Merle J, David J (1966) Ovarian function and ovulation in virgin and fertile *Drosophila melanogaster* females on an exclusive carbohydrate diet. C R Acad Sci Paris Ser D 264: 2028–2030

Mersereau WA, Hinchey EJ (1982) Prevention of phenylbutazone ulcer in the rat by glucose: role of a glycoprivic receptor system. Am J Physiol 242(Gastrointest Liver Physiol 5): G429–G432

Messer E (1985) A field test for estimating sweetness preferences to improve estimates of sucrose intakes in individuals. Food Nutr Bull 7: 58–59

Metais P (1967) Anomalies of sugar metabolism. Cah Nutr Diet 2: 23–31

Metcalfe DD (1984) Diagnostic procedures for immunologically-mediated food sensitivity. Nutr Rev 42: 92–97

Metze H (1978) Der fructosemetabolismus bei neugeborenen in abhangigkeit von der hohe des bilirubinspiegels. Klin Padiatr 191: 354–358

Meyer FA (1940) Rhythmic variations of blood sugar in course of day. Z Ges Exp Med 107: 569–582

Meyer H, Kröger H (1972) Influence of feeding sugar and starch to sows before labor on the birth weight of piglets. Dtsch Tieraerztl Wochenschr 79: 16–18

Meyer JR (1951) Sucrose importance for development of antineoplastic action of fungus culture fluid. Arq Inst Biol 20: 193–196

Meyerhof O (1926) The enzymatic lactic acid formation in muscle extracts. II. The cleaving of polysaccharides and hexosediphosphoric acid. Biochem Z 178: 462–490

Michael P (1940) Studies on golfers. J Am Med Assoc 115: 286–287

Michaelis OE (1982) The disaccharide effect: a mechanism for carbohydrate-induced lipogenesis. In: Reiser S (ed) Metabolic effects of utilized dietary carbohydrates. Marcel Dekker, New York, pp 55–70

Michaelis OE, Szepesi B (1973) Effects of various sugars on hepatic glucose-6-phosphate dehydrogenase, malic enzyme and total liver lipid of the rat. J Nutr 103: 697–705

Michaelis OE, Szepesi B (1974) The mechanism of a specific metabolic effect of sucrose in the rat. J Nutr 104: 1597–1609

Michaelis OE, Szepesi B (1977) Specificity of the disaccharide effect in the rat. Nutr Metab 21: 329–340

Michaelis OE, Nace CS, Szepesi B (1975) Demonstration of a specific metabolic effect of dietary disaccharides in the rat. J Nutr 105: 1186–1191

Michaelis OE, Scholfield DJ, Nace CS, Reiser S (1978) Demonstration of the disaccharide effect in nutritionally stressed rats. J Nutr 108: 919–925

Michaelis OE, Martin RE, Gardner LB, Ellwood KC (1981) Effect of dietary carbohydrates on systolic blood pressure of normotensive and hypertensive rats. Nutr Rep Int 23: 261–266

Michaelis OE, Ellwood KC, Hallfrisch J, Hansen CT (1983) Effect of dietary sucrose and genotype on metabolic parameters of a new strain of genetically obese rat: LA/N-corpulent. Nutr Res 3: 217–228

Michaelis OE, Ellwood KC, Judge JM, Schoene NW, Hansen CT (1984) Effect of dietary sucrose on the SHR/N-corpulent rat: a new model for insulin-independent diabetes. Am J Clin Nutr 39: 612–618

Michalek SM, McGhee JR, Shiota T, Devenyns D (1977) Low sucrose levels promote extensive *Streptococcus mutans*-induced dental caries. Infect Immun 16: 712–714

Midrio M, Bettini V (1967) In vitro mechanical responses of coronary artery and hepatic artery segments to changes in the ionic composition of the medium. II. Effects of progressive substitution of NaCl with sucrose, in the presence or absence of calcium. Boll Soc Ital Biol Sper 43: 1002–1005

Migler R, Cascarano J (1982) Glycolytic enzyme activities in liver, heart and gastrocnemius of Sprague-Dawley and Wistar rats: effect of high sucrose diet. Comp Biochem Physiol 73B: 635–639

Migler R, Cascarano J (1983) Effect of chronic hypoxia and high sucrose diet on body weights and glycolytic enzyme activities in liver, heart, and gastrocnemius of Sprague-Dawley and Wistar rats. Comp Biochem Physiol 75B: 277–285

Mikkelson TJ, Robinson JR (1968) Identification of sugars from rates of oxime formation. J Pharm Sci 57: 1180–1183

Milich R, Pelham WE (1986) Effects of sugar ingestion on the classroom and playground behavior of attention deficit disordered boys. J Consult Clin Psychol 54: 714–718

Miller LA (1964) Activation of intracellular catalase of mouse liver. Proc Soc Exp Biol Med 115: 25–29

Miller SA (1978) The wise use of chemicals in foods. Presented at the 176th American Chemical Society National Meeting, Miami Beach, Florida, 11 September. American Chemical Society, Washington DC

Miller SA (1978) Sugar, oral health and regulation: a strategy for prevention of oral disease. In: Foods, nutrition and dental health. American Dental Health Foundation, Pathotox, Park Forest South

Milne JL, Thompson IM (1965) The effect of sucrose and intravenous mannitol upon induced pyelonephritis in rats. J Urol 94: 647–648

Minah GE, Lovekin GB, Finney JP (1981) Sucrose-induced ecological response of experimental dental plaques from caries-free and caries-susceptible human volunteers. Infect Immun 34: 672–675

Minah GE, Solomon ES, Chu K (1985) The association between dietary sucrose consumption and microbial population shifts at six oral sites in man. Arch Oral Biol 30: 397–401

Minn SK, Harnchonboth K, Mandel EE (1978) Combined effect of dextrose and sodium chloride on red cell osmotic fragility. Am J Hematol 5: 33–42

Mita J (1926) Action of non-electrolytes on the heart. Arch Exp Pathol Pharmakol 112: 17–21

Mitchell HH, Hamilton TS, Beadles JR (1937) The comparative nutritive value of glucose, fructose, saccharose and lactose as supplements for complete nutrition. J Nutr 14: 435–452

Mitchell HS (1927) Comparative physiological values of five carbohydrates, based on growth and fecal analysis. Am J Physiol 79: 537–541

Mito H (1970) Experiment in nutrition and biochemical study of intestinal absorption of sucrose and lactose in infants, with special reference to newborn infants. Acta Paediatr Jpn 74: 873–882

Miyake S (1929) Resorption of various carbohydrates in the intestinal tract. Orient J Dis Infant 5: 4

Mock DM, Perman JA, Michael Thaler M, Morris RC (1983) Chronic fructose intoxication after infancy in children with hereditary fructose intolerance. N Engl J Med 309: 764–770

Mogenson GJ, Box BM, Philbrick DJ (1983) Nutrition and hypertension. Can J Physiol Pharmacol 61: 260–270

Moinuddin JF, Lee HWT (1958) Effects of feeding diets containing sucrose, cellobiose, or glucose on the dry weights of cleaned gastrointenstinal organs in the rat. Am J Physiol 192: 417–420

Molaparast-Shahidsaless F, Shrago E, Elson CE (1979) $\alpha$-Glycerophosphate and dihydroxyacetone phosphate metabolism in rats fed high-fat or high-sucrose diets. J Nutr 109: 1560–1569

Momchilov BB, Vitanov T (1970) An attempt to preserve the complement activity of guinea pig

serum by addition of protective substance and lyophilization. Sci Work Res Inst Epidemiol Microbiol 13: 235–238
Monauni J (1929) The question of potassium permeability in the muscular fiber boundary layer and its acceleration by cane sugar. Pflugers Arch Physiol 221: 800–806
Moncrieff A, Wilkinson RH (1954) Sucrosuria with mental defect and hiatus hernia. Acta Paediatr 43: 495–516
Mond R, Hoffmann F (1928) The membrane structure of red blood corpuscles. The relationship between permeability and molecular volume. Pflugers Arch Physiol 219: 467–480
Monson WJ, Dietrich LS, Elvehjem CA (1950) Studies on the effect of different carbohydrates on chick growth. Proc Soc Exp Biol Med 75: 256–259
Monson WJ, Harper AE, Benton DA, Elvehjem CA (1954) The effect of level of dietary protein on the growth of chicks fed purified diets containing sucrose or dextrin. J Nutr 53: 563–573
Monson WJ, Harper AE, Benton DA, Winje M, Elvehjem CA (1955) Effect of arginine and glycine on the growth of chicks receiving complete, purified diets. Poultry Sci 34: 186–190
Moog O, Eimer K (1925) The influence of hypertonic sodium chloride, calcium chloride and cane sugar solution on the imperceptible emission of perspiration. Munch Med Wochenschr 72: 1912–1914
Moore KK (1977) Polymeric food additives – functionality with safety. Food Prod Dev 11: 63, 80
Moore MC, Guzman MA, Schilling PE, Strong JP (1976) Dietary-atherosclerosis study on deceased persons. J Am Diet Assoc 68: 216–223
Moore MC, Guzman MA, Schilling PE, Strong JP (1977) Dietary-atherosclerosis study on deceased persons. J Am Diet Assoc 70: 602–606
Moore MJ (1980) Reactive hypoglycemia precipitated by sucrose and glucose. Va Med Mon 107: 785–788
Moore WJ (1983) Sugar and the antiquity of dental caries. J Dent Res 11: 189–213
Moore WM, Helleger AE, Battaglia FC (1966) In vitro permeability of different layers of the human placenta to carbohydrates and urea. Am J Obstet Gynecol 96: 951–955
Moran TH, McHugh PR (1981) Distinctions among three sugars in their effect on gastric emptying and satiety. Am J Physiol 241: R25–R30
Moran TJ (1953) Pulmonary oedema produced by intratracheal injection of milk, feeding mixtures and sugars. Am J Dis Child 86: 45–50
Morard JCI, Rutishauser E, Chatillon J (1956) Sucrose damage to the kidney. J Urol Med Chir 61: 612–624
More NS, Rao NA, Preuss HG (1986) Early sucrose-induced retinal vascular lesions in spontaneously hypertensive rats (SHR) and Wistar Kyoto rats (WKY). Ann Clin Lab Sci 16: 419–246
Moretti P, Muscolino G (1930) Action of carbohydrate on the toxicity of calcium cyanide. Arch Farmacol Sper 51: 135–140
Morgan AF, Cook BB, Davison HG (1938) The action of nutritive carbohydrates on vitamin B2 deficiency. J Nutr 15: 27–43
Morgan KJ, Zabik ME (1981) Amount and food sources of total sugar intake by children ages 5 to 12 years. Am J Clin Nutr 34: 404–413
Morgan KJ, Zabik ME, Leveille GA (1978) Food behavior of children: consumption patterns and nutrient composition of breakfasts, snacks, and total day. Michigan State University, Lansing, pp 2–12 (Mich State Univ Agric Exp St Bull No 374)
Morgan WTJ, Watkins WM (1953) Inhibition of hemagglutinins in plant seeds by human blood group substances and simple sugars. Br J Exp Pathol 34: 94–103
Morita J, Ueda K, Nanjo S, Komano, T (1985) Sequence specific damage of DNA induced by reducing sugars. Nucleic Acids Res 13: 449–458
Morita K (1959) Inhibition of hemolysis by sugars. II. Inhibitory effect of sugars on acid hemolysis. Shikoku Igaku Zasshi 14: 983–986
Morita M (1936) Action of certain types of sugar on functional capacity and fatigue of skeletal muscle. Nagasaki Igakkwai Zassi 14: 1481–1483
Mormann JE, Muhlemann HR (1981) Oral starch degradation and its influence on acid production in human dental plaque. Caries Res 15: 166–175
Morris JN, Marr JW, Clayton DG (1977) Diet and heart: a postscript. Br Med J ii: 1307–1314
Moscatelli EA, Fujimoto K, Gilfoil TC (1975) Effects of chronic consumption of ethanol and sucrose on rat whole brain 5-hydroxytryptamine. J Neurochem 25: 273–276
Moscatelli EA, Switzer MK, Blaker WD (1978) A model for chronic ethanol consumption in rats. Res Commun Psychol Psych Behav 3: 267–283
Moschini A (1924) The action of the stomach in secreting mono- and disaccharides for insulin

hypoglycemia. Boll Soc Med Chir 36: 393–399

Moschini A (1924) The action of oral supplements of mono- and disaccharide on insulin hypoglycemia. Arch Ital Biol 74: 126–130

Moser PB (1976) The influence of sucrose and starch diets and oral contraceptives on the serum lipid responses of young women. Diss Abstr Int [B] 37: 5605

Moser PB, Berdanier CD (1974) Effect of early sucrose feeding on the metabolic patterns of mature rats. J Nutr 104: 687–694

Moskowitz HR (1978) Psychological correlates of sugar consumption. In: Guggenheim B (ed) Health and sugar substitutes. Proceedings of the ERGOB conference on sugar substitutes, Geneva, 30 Oct–1 Nov 1978. Karger, Basel, pp 10–16

Moutier F, Camus L (1929) Comparative study of graphic representation of hyperglycemia after ingestion of dextrose or saccharose. Arch Mal App Digest 19: 210–213

Muhlemann HR (1980) Gute Zähne dank vollwertigem Zucker? Swiss Dent 1: 51–53

Mukhopadhyay A, Patel ST, O'Brien M et al. (1983) Hypolipidemic effects of clofibrate and selected chroman analogs in fasted rats. II. High sucrose-fed animals. Lipids 18: 59–67

Muller RE (1974) Effect of dietary carbohydrates on blood lipids. Chem Abstr 80: 322

Muller RE, Gortner RA Jr (1948) The influence of sugar content on pH on in vivo decalcification of rat molar teeth by acid beverages. Arch Biochem 20: 153–158

Müller-Wiefel DE, Steinmann B, Holm-Hadulla M, Wille I, Schärer K, Gitzelmann R (1983) Infusionsbedingtes Nieren- und Leberversagen bei undiagnostizierter hereditärer Fructose-Intoleranz. Dtsch Med Wochenschr 108: 985–989

Mundorff SA, Featherstone JDB, Eisenberg AD, Espeland M, Shields CP, Curzon MEJ (1985) Cariogenicity of foods: rat study. Paper presented at a Symposium of the International Association of Dental Research, Las Vegas, Nevada. International Association of Dental Research, Washington DC

Muralt A von (1932) The amount of double refraction of transverse striated muscles during contraction. Pflugers Arch Ges Physiol 230: 298–326

Murlin JR, Nasset ES, Murlin WR, Manly RS (1936) The rate of ketogenesis in human subjects on high-fat diets, as influenced by different sugars. J Nutr 12: 645–670

Murphy FD, Hershberg RA, Katz AM (1936) The effect of intravenous injections of sucrose solution (50%) on the cerebrospinal fluid pressure, the blood pressure and clinical course in cases of chronic hypertension. Am J Med Sci 192: 510–517

Muschel LH, Larsen LJ (1970) Effect of hypertonic sucrose upon the immune bactericidal reaction. Infect Immun 1: 51–55

Mutolo V, Abrignami F (1958) The effect of anionic detergents and trypsin on mitochondria of normal and tumor tissues. Br J Cancer 12: 285–289

Nagy EA (1980) Sucrose and dental disease. Br Dent J 149: 68

Naismith DJ (1971) Differences in the metabolism of dietary carbohydrates studied in the rat. Proc Nutr Soc 30: 259–265

Naismith DJ, Cursiter MC (1972) Is there a specific requirement for carbohydrate in the diet? Proc Nutr Soc 31: 94A–95A

Naismith DJ, Khan NA (1970) Differences in the throughput of triglycerides in the plasma of rats fed various carbohydrates. Proc Nutr Soc 29: 64A (Abstract)

Naismith DJ, Khan NA (1971) Lipoprotein lipase activity in the adipose tissue of rats fed sucrose or starch. Proc Nutr Soc 30: 12A (Abstract)

Naismith DJ, Rana IA (1971) Lecithin: cholesterol acyltranferase activity in the plasma of rats fed sucrose or starch. Nutr Metab 13: 172–177

Naismith DJ, Rana IA (1974) Sucrose and hyperlipidaemia. II. The relationship between the rates of digestion and absorption of different carbohydrates and their effects on enzymes of tissue lipogenesis. Nutr Metab 16: 285–294

Nakamura RM (1980) Sugar-induced heart lesions and vitamin K deficiency. Tox Res Proj Dir 5: 11

Nakano K (1958) Effects of carbohydrate on the digestibility and absorbability of cow-milk proteins. I. Digestion by proteolytic enzymes in vitro. Shikoku Igaku Zasshi 12: 505–509

Nakazima H (1940) Fever produced by injection of sugar solution in rabbits. Jpn J Med Sci IV Pharmacol 13: 70–71

Nassau E, Schaferstein S (1926) Influence on the absorption of sugar of other foodstuffs given with it. Z Kinderheilk 40: 659–670

Nasset ES, Heald FP, Calloway DH, Margen S, Schneeman P (1979) Amino acids in human blood plasma after single meals of meat, oil, sucrose and whiskey. J Nutr 109: 621–630

Nassonov D, Aisenberg E (1937) The influence of nonelectrolytes on the water content of living

and dead muscle. Ann Physiol Physiochim Biol 13: 1179–1212
National Caries Program (1983) National Institute of Health: annual report. National Institute of Dental Research, Bethesda, Maryland, pp 1–8
Navia JM (1977) Experimental dental caries. In: Animal models in dental research. University of Alabama, Birmingham, Alabama, pp 257–297
Navia JM, Lopez H (1983) Rat caries assay of reference foods and sugar-containing snacks. J Dent Res 62: 893–898
Navia JM, Lopez H, Fischer JS (1974) Caries promoting properties of sucrose substitutes in foods: mannitol, xylitol and sorbitol. J Dent Res 53(Sp Abstr Iss): 207
Neale G, Clark M, Levin B (1965) Intestinal sucrase deficiency presenting as sucrose intolerance in adult life. Br Med J ii: 1223–1225
Nechkovitch M (1924) Glucose and coloidal equilibrium of proteins. Arch Int Physiol 24: 1–6
Neill JM, Hehre EJ, Sugg JY, Jaffe E (1939) Serologic studies: reactions between solutions of reagent sucrose and type II antipneumococcus serum. J Exp Med 70: 427–442
Nestel PJ, Carroll KF, Havenstein N (1970) Plasma triglycerides response to carbohydrates, fats and caloric intake. Metabolism 19: 1–18
Neuberg C, Kobel M, Laser H (1930) Mechanism of enzymatic splitting of sugar in tumor and in embyronic tissue. Z Krebsforsch 32: 92–98
Neubert B (1952) Cerebrospinal fluid–blood sugar relationship in healthy child: early diagnosis of tuberculous meningitis. Kinderarztl Prax 20: 199–202
Neumann HH (1950) Decrease of dental caries during the war. Br Dent J 88: 56–60
Newberne PM, Rogers AE (1986) The role of nutrients in cancer causation. In: Hayashi Y, Nagao M, Sugimura T et al. (eds) Diet, nutrition and cancer. Japan Scientific Press, Tokyo; VNU Science Press, Utrecht
Newbrun E (1978) Criteria indicative of cariogenicity or non cariogenicity of foods and beverages. In: Guggenheim B (ed) Health and sugar substitutes. Proceedings of the ERGOB conference on sugar substitutes, Geneva, 30 Oct–1 Nov 1978. Karger, Basel, pp 253–258
Newbrun E (1982) Sucrose in the dynamics of the caries process. Int Dent J 32: 13–23
Newbrun E (1982) Sugar and dental caries. A review of human studies. Science 217: 418–428
Newbrun E, Frostell G (1978) Sugar restriction and substitution for caries prevention. Caries Res 12(Suppl 1): 65–73
Newton JM, Wardrip EK (1974) Sucrose in food systems. In: Inglett BE (ed) Symposium on sweeteners. AVI Publishing, Westport, Connecticut, pp 87–89
Nichols AB, Ravenscroft C, Lamphiear DE, Ostrander LD Jr (1976) Daily nutritional intake and serum lipid levels. The Tecumseh study. Am J Clin Nutr 29: 1384–1392
Nicholls DG, Locke RM (1984) Thermogenic mechanisms in brown fat. Physiol Rev 64: 1–64
Nickerson PA, Molteni A (1971) Survival of bilaterally adrenalectomized hamsters drinking saline sucrose. J Appl Physiol 31: 675–678
Nieddu G (1956) Carbohydrates and dental caries etiopathogenesis. IV. Experimental research on animals. Clin Odontoiatr 11: 3–9
Niessen KH, Brugmann G, Schmidt K (1975) Hereditörer Isomaltase-Saccharase-Mangel der Dünndarm-schleimhaut als Ursache fur ein Malabsorptionssyndrom. Klin Padiatr 187: 163–170
Nikkilä EA (1969) Control of plasma and liver triglyceride kinetics by carbohydrates metabolism and insulin. Adv Lipid Res 7: 63–134
Nikkilä EA (1974) Influence of dietary fructose and sucrose on serum triglycerides in hypertriglyceridemia and diabetes. In: Sipple HL, McNutt KW (eds) Sugars in nutrition. Academic Press, New York, pp 439–448
Nikkilä EA, Ojala K (1965) Induction of hyperglyceridemia by fructose in the rat. Life Sci 4: 937–943
Nikkila EA, Huttenen JK, Ehnholm C (1977) Postheparin plasma lipoprotein lipase and hepatic lipase in diabetes mellitus: relationship to plasma triglyceride metabolism. Diabetes 26: 11–21
Nilsson B, Holm AK, Sjostrom R (1982) Taste thresholds, preferences for sweet taste and dental caries in 15-year-old children. Swed Dent J 6: 21–27
Nishioka K, Katayama I, Sano S (1983) Urticaria induced by d-psicose. Lancet II: 1417–1418
Nishiyama Y (1939) Experimentelle Erzeugung der Sarkome bei Ratten durch Wiederholte Injektionen von Gludoselosung. Gann 32: 85–99
Nizel AE, Miller SA, Cleary TM, Quinn RS (1974) Relative cariogenicity of natural refined sugars. J Dent Res 53 (Sp Abstr Iss): 208
Noakes TD (1981) Sugar for energy in marathon running. S Afr Med J 60: 46
Nobecourt P (1926) Treatment with sucrose. Bull Soc Pediatr Paris 24: 352–356

Nobecourt P (1927) High tolerance for saccharose in cachectic infants. Arch Med Enfant 30: 313–338

Noguchi T, Inoue H, Tanaka T (1985) Transcriptional and posttranscriptional regulation of L-type pyruvate kinase in diabetic rat liver by insulin and dietary fructose. J Biol Chem 26: 14393–14397

Noller CR (1966) Chemistry of organic compounds, 3rd edn. Saunders, Philadelphia, p 421

Nordio S, Lamedica GM, Vignolo L (1961) A case of chronic connatal diarrhoea due to intolerance to saccharose and dextrin. Minerva Pediatr 13: 1766–1773 (in Italian)

Nordstrom C, Dahlqvist A, Josefsson L (1968) Quantitative determination of enzymes in different parts of the villi and crypts of rat small intestine. Comparison of alklaine phosphatase, disaccharidases, and dipeptidases. J Histochem Cytochem 15: 713–721

Norman DA, Morawski SG, Fordtran JS (1980) Influence of glucose, fructose and water movement on calcium absorption in the jejunum. Gastroenterology 78: 22–25

Nunheimer TD, Fabian FW (1940) Influence of organic acids, sugars, and sodium chloride upon strains of food poisoning staphylococci. Am J Public Health 30: 1040–1049

Nuttall RQ (1983) Diet and the diabetic patient. Diabetes Care 6: 197–207

Nuttall FQ, Gannon MC (1981) Sucrose and disease. Diabetes Care 4: 305–310

Nuttall FQ, Maryniuk MD, Kaufman M (1983) Individualized diets for diabetic patients. Ann Intern Med 99: 204–207

Obell AE (1974) Recent advances in mechanism of causation of diabetes mellitus in man and *Acomys cahirinus*. East Afr Med J 51: 425–428

Ocaranza F (1934) Saccharose administered intraperitoneally. Rev Mex Biol 14: 165–169

Öckerman PA, Lundborg H (1956) Conversion of fructose to glucose by human jejunum. Absence of galactose-to-glucose conversion. Biochim Biophys Acta 105: 34–42

Oda T (1928) The parenteral administration of cane and invert sugars. J Biochem 9: 383–405

Ogata M, Mochizuki Y (1956) The inhibitory action of sugar and polyatomic alcohol for the heat activation of complement. Acta Med Okayama 10: 82–88

Ogundipe OO, Bray A (1974) The influence of diet and fat cell size on glucose metabolism, lipogenesis, and lipolysis in the rat. Horm Metab Res 6: 351–356

Oka Y, Akanuma Y, Kasuga M, Kosaka K (1980) Effects of a high glucose diet on insulin binding and on insulin action in rat adipocytes. Diabetologia 19: 468–474

O'Keefe SJD, Marks V (1977) Lunchtime gin and tonic a cause of reactive hypoglycemia. Lancet I: 1286–1288

Okuchi I (1937) Histopathologic changes in parotid gland of rabbit fed on cane sugar. Mitt Med Akad Kioto 20: 498–499

Olbrich SE, Wayman O (1972) Effect of feeding raw sugar on growth, performance and rumen fluid parameters of fattening beef cattle. J Anim Sci 34: 820–825

Olcott HS (1938) Paralysis in the young of female rats deficient in vitamin D. J Nutr 15: 221–227

Olefsky JM, Saekow M (1978) The effects of dietary carbohydrate content on insulin binding and glucose metabolism by isolated rat adipocytes. Endocrinology 103: 2252–2263

Olsen I, Birkeland JM (1976) Initiation and aggravation of denture stomatitis by sucrose rinses. Scand J Dent Res 84: 94–97

Olsen ME, Faber OK, Binder C (1983) Hepatic extraction of insulin after carbohydrate hyperalimentation. Acta Endocrinol 102: 416–419

O'Mullane DM (1982) The changing patterns of dental caries in Irish schoolchildren between 1961 and 1981. J Dent Res 61 (Sp Iss): 1317–1320

Opie LH (1975) Dietary sucrose in relation to the development of ischaemic heart disease. Am Heart J 89: 674–675

Oppel WW (1929) Characteristics of the alimentary glycemia curve. IV. Saccharosemia. Biochem Z 205: 31–46

Oppel WW, Fedorow PS (1928) The characteristic features of the alimentary glycemic curve. I. The glycemic curve in children following the administration of saccharose. Z Ges Exp Med 63: 314–330

Ores OR, Abelin I (1954) Influence of hormones and diet on the creatine metabolism of the organs. Acta Endocrinol 15: 302–312

Ornoy A, Cohen AM (1980) Teratogenic effects of sucrose diet in diabetic and nondiabetic rats. Isr J Med Sci 16: 789–791

Osborn TWB, Noriskin JN, Staz J (1937) Comparison of crude and refined sugar and cereals in their ability to produce in vitro decalcification of teeth. J Dent Res 16: 165–171

Oshima M (1938) The effect of carbohydrate metabolism on the growth and radiosensitivity of malignant tumours. Jpn J Obstetr Gynecol 21: 188–197

Ostrander LD Jr, Lamphiear DE, Block WD, Johnson BC, Eptein FH (1974) Biochemical precursors of atherosclerosis. Studies in apparently healthy men in a general population. Tecumseh, Michigan. Arch Intern Med 134: 224–230

Ostrom CA, Koulourides T, Hickman F, Phantumvanit P (1977) Combined effects of sucrose and flouride on experimental caries and on the associated microbial plaque. J Dent Res 56: 212–221

Otter WD, Van Boxtel ABThG (1971) Relation between the glycogen content of the liver and liver weight, and its meaning for enzymology. Experientia 27: 1271–1272

Otto PL, Sulzbacher SI, Worthington-Roberts BS (1982) Sucrose-induced behavior changes of persons with Prader-Willi syndrome. Am J Ment Defic 86: 335–341

Outhouse J, Smith J, Merritt L, White FR (1937) Growth promotion and bone calcification action of more carbohydrates. J Nutr 14: 579–595

Outhouse J, Smith J, Twomey I (1938) Relative effects of certain saccharides and of vitamin D on mineral metabolism of rats. J Nutr 15: 257–268

Ove P, Coetzee ML, Morris HP (1971) DNA synthesis and the effect of sucrose in nuclei of host liver and Morris hepatomas. Cancer Res 31: 1389–1395

Oxley JA, Bruckdorfer KR, Edwards G, Yudkin J (1979) The effect of dietary carbohydrates and other diet constituents on nephrocalcinosis in the rat. Proc Nutr Soc 38: 85A

Pacheva G (1968) Carbohydrates in the food of contemporary man. Khranit Prom 17(6): 29–30

Paffrath H, Siering-Kaulla H (1929) Digestibility as shown by sugar tolerance tests on health nursing. Kinder 48: 76–85

Page E, Gingras R (1946) Etude sur les régimes alimentaires durant l'inanition chez le rat. Lav Med 11: 969–975

Page L, Friend B (1978) The changing United States diet. Bioscience 28: 192–197

Palm JD (1975) Benefits of dietary fructose in alleviating the human stress response. In: Jeanes A, Hodge J (eds) Physiological effect of food carbohydrates. American Chemical Society, Washington DC, pp 54–72

Palmer AK, Bottomley AM (1977) Effect of xylitol during a modified teratology study in rats (unpublished report from the Huntingdon Research Center submitted to the World Health Organization by Hoffman La Roche and Co, Ltd, Basel. In: Joint FAO/WHO Expert Committee on Food Additives (ed) Summary of toxicological data of certain food additives and contaminants. World Health Organization, Geneva, p 32 (WHO Food Additives Series No 13, 1978)

Palmer AK, Bottomly AM, Wight DGD, Cherry CP (1978) Effect of xylitol on reproductive functions of multiple generations in the rat (unpublished report from the Huntingdon Research Center submitted to the World Health Organization by Hoffman La Roche and Co, Ltd, Basel. In: Joint FAO/WHO Expert Committee on Food Additives (ed) Summary of toxicological data of certain food additives and contaminants. World Health Organization, Geneva, pp 31–32 (WHO Food Additives Series No 13, 1978)

Palumbo PJ, Briones ER, Nelson RA, Kottke BA (1977) Sucrose sensitivity of patients with coronary artery disease. Am J Clin Nutr 30: 394–401

Panikarovskii VV, Girgor'yan AS, Zhizhina NA, Prokhonchukov AA, Sazonova VI (1968) State of some internal organs of Wistar rats kept for long periods on a sucrose–casein caries-producing diet. Stomatologiya 47(4): 23–27

Panikarovskil VV, Girgor'yan AS, Sazonova VI (1970) State of the adrenal cortex in rats kept on a sucrose-casein cariogenic diet. Stomatologiya 49(1): 9–14

Panksepp J, Meeker R (1977) Effects of insulin and hypothalmic lesions on glucose preference in rats. In: Katsuki Y (ed) Food intake chemical senses international symposium. University Park Press, Baltimore, Maryland, pp 343–356

Paolino VJ (1982) Anti-plaque activities of cocoa. Presented at the third annual conference on foods, nutrition and dental health. American Dental Association, Chicago, Illinois

Papachristodoulou D, Heath H, Kang SS (1976) The development of retinopathy in sucrose fed and streptozotocin diabetic rats. Diabetologia 12: 367–374

Pariza MW, Foster EM (1983) Determining the safety of enzymes used in food processing. J Food Proc 46: 453–468

Parker KJ (1974) Sucrose as an industrial raw material. Sucrerie Belg 93: 15–27

Parker KJ (1978) Alternatives to sugar. Nature (Lond) 271: 493–495

Parks RE, Ben-Gershom E, Lardy HA (1957) Liver fructokinase. J Biol Chem 227: 231–242

Parmentier R, Corvilain J (1956) Nephrotic lesions provoked by sucrose. Rev Belg Path Med Exp 25: 210–218

Parr DR, Harris EJ (1975) Effects of sucrose and dextran on the toxicity of lead to mitochondria in the presence of inorganic phosphate in vitro. Biochem Soc Trans 3: 951–953

Parsons EI, McCollum EV, Frobisher M Jr (1946) The effects of dietary carbohydrates on the dental flora of the rat. Am J Hyg 44: 249–256

Pascual CG, Singh R, Juliano BO (1978) Free sugars of rice grain. Carbohydr Res 62: 381–385

Pastorino D (1952) Capillary-venous glycemic curves after single glucose, or double glucose, galactose, lactose, or sucrose administration, or atropine, pilocarpine, adrenaline or insulin treatment (in children). Riv Pediatr Siciliana 8: 465–476

Patel JC, Metha AB, Dhirawani MK, Juthani VJ, Aiyer L (1969) High carbohydrate diet in the treatment of diabetes mellitus. Diabetologia 5: 243–247

Paterson JH (1942) Effect of intravenous infections of sucrose on cerebrospinal fluid pressure as measured by lumbar puncture. Proc R Soc Med 35: 530–534

Paton JHP (1932) National consumption of carbohydrates in relation to disease. Edinburgh Med J 39: 556–570

Paul O, Lepper MH, Phelan WH et al. (1963) A longitudinal study of coronary heart disease. Circulation 28: 20–31

Paul O, MacMillan A, McKean H, Park H (1968) Sucrose intake and coronary heart disease. Lancet II: 1049–1051

Paunio K, Hurttia H, Tenovuo J, Makinen KK, Tiekso J (1984) Effects on oral health of mouth rinses containing xylitol, sodium cyclamate and sucrose sweeteners in the absence of oral hygiene. I. Clinical findings and analysis of gingival exudate. Proc Finn Dent Soc 80: 3–12

Pearce EIF, Gallagher IHC (1979) The behaviour of sucrose and xylitol in an intra-oral caries test. NZ Dent J 75: 8–14

Pedersen ME, Schotz MC (1980) Rapid changes in rat heart lipoprotein lipase activity after feeding carbohydrates. J Nutr 110: 481–487

Pedoe HT, Rose G (1979) Atherosclerosis as related to diet. Int Rev Biochem 27: 245–280

Pennington SN, Barakat HA, Kohm GL, Smith CP (1978) Effect of sucrose on lipogenesis of rats chronically treated with ethanol. Experientia 34: 564–565

Penny WJ, Mayberry JF, Aggett PJ, Gilber JO, Newcombe RG, Rhodes J (1983) Relationship between trace elements, sugar consumption and taste in Crohn's disease. Gut 24: 288–292

Perazzo JC, Monserrat AJ (1978) Effects of repeated injections of sucrose on the kidney: role of the route of administration. IRCS medical science. Cell and membrane biology. Kidneys and urinary system. Metabolism and nutrition. Pathology 6: 41

Permutt MA (1980) Is it really hypoglycemia? If so, what should you do? Med Times 103: 35–43

Perroni L, Ensoli G, Nunziata A et al. (1980) Comparative study of the effects of fructose-1,6-diphosphate, fructose and physiological saline on the blood levels of glucose and adenosine-triphosphate after an oral glucose load. Pharmacol Res Commun 12: 147–153

Persky V, McDonald A, Molteni A, Liu K (1986) The effect of dietary sucrose on blood pressure in spontaneously hypertensive rats. Nutr Res 6: 1111–1115

Peskin AR (1948) Increased sugar tolerance as a factor in the production of a sympton complex simulating peptic ulcer, neurocirculatory asthenia, and psychoneurosis. Am J Digest Dis 15: 92–101

Petersen DB, Lambert J, Gerring S et al. (1986) Sucrose in the diet of diabetic patients: just another carbohydrate? Diabetologia 29: 216–220

Peterson RE, O'Toole JJ, Kirkendall WM, Kempthorne O (1959) The variability of extracellular fluid space (sucrose) in man during a 24 hour period. J Clin Invest 38: 1644–1658

Petrescu AD (1978) Fructoza in pediatrie. Pediatrics 27: 203–216

Petrescu AD (1979) Glucoza in oftalmologie. Oftalmologia 23: 5–8

Petrescu AD (1979) Fructoza in oftalmologie. Oftalmologia 23: 81–83

Petrovich YuA, Kolosovskii VM (1968) The secretion of iodine, carbonate, and phosphate in the saliva of rats kept on a normal and cariogenic diet. Stomatologiya 47(5): 5–8

Pfisterer HG (1953) Intolerance for milk and sugar following gastric resection. Arch Klin Chir 275: 528–543

Philipson H (1983) Dietary fibre in the diabetic diet. Acta Med Scand 671 (Suppl): 91–93

Phillips MI, Davies DR (1985) The mechanism of guanosine triphosphate depletion in the liver after a fructose load. The role of fructokinase. Biochem J 228: 667–671

Phillips RW, Westmoreland N, Panepinto L, Case GL (1982) Dietary effects on metabolism of Yucatan miniature swine selected for low and high glucose utilization. J Nutr 112: 104–111

Piatkowski B, Steger H (1966) Nutritional studies of calves. V. Effect of sugar-containing food and pelletized rations on the metabolism and growth of early-weaned calves. Arch Tierernahr 17: 209–216

Pictet A, Vogel H (1928) Synthese du saccharose. Helv Chim Acta 11: 436–442

Pierce NF, Hirschhorn N (1977) Oral fluid: a simple weapon against dehydration in diarrhoea.

How it works and how to use it. World Health Organization, Geneva, pp 87–93 (WHO Chronicle No 31)

Pittet DL, Chappius PL, Acheson K, De-Techtermann F, Jequier E (1976) Thermic effect of glucose in obese subjects studied by direct and indirect calorimetry. Br J Nutr 35: 282–292

Platt L, Ball KP, Bridgen WW et al. (1970) Dietary sugar intake in men with myocardial infarction. Report to the Medical Research Council by its working party on the relationship between dietary sugar intake and arterial disease. Lancet II: 1265–1271

Pliner P (1982) The effects of mere exposure on liking for edible substances. Appetite 3: 283–290

Podolsky E (1948) The discovery of sugar as a medicine. Sugar 43(6): 33–35

Pollak A, Coradello H, Leban J et al. (1983) Inhibition of alkaline phosphatase activity by glucose. Clin Chim Acta 133: 15–24

Pollak OJ (1961) Cholesterolemia of rabbits fed carbohydrate diets. J Am Geriatr Soc 9: 349–358

Polonovski M (1940) Effects of ingestion of sugar on resistance to anoxemia and to acapnia at high altitudes. Bull Acad Natl Med (Paris) 123: 688–694

Ponz F, Larralde J (1952) Relation between selective absorption of sugars and enzyme inhibitors. Rev Esp Fisiol 8: 71–82

Porcher C, Auger L, Brigando (1928) Utilization of various sugars carried in the organisms parenteral passage ways. C R Soc Biol (Paris) 98: 51–52

Porikos KP, Booth G, van Itallie TB (1977) Effect of covert nutritive dilution on the spontaneous food intake of obese individuals. Am J Clin Nutr 30: 1638–1644

Porikos KP, Hesser MF, van Itallie TB (1982) Caloric regulation in normal-weight men maintained on a palatable diet of conventional foods. Physiol Behav 29: 293–300

Porte D Jr, Brierman EL, Bagdade JD (1966) Substitution of dietary starch for dextrose in hyperlipemic subjects. Proc Soc Exp Biol Med 123: 814–816

Portha B, Giroix MH, Picon L (1982) Effect of diet on glucose tolerance and insulin response in chemically diabetic rats. Metabolism 21: 1194–1199

Portmann OW, Mann CV, Wysocki AP (1955) Bile acid excretion by the rat: nutritional effects. Arch Biochem Biophys 59: 224–232

Portman OW, Lawry EY, Bruno D (1956) Effect of dietary carbohydrate on experimentally induced hypercholesteremia and hyperbetalipoproteinemia in rats. Proc Soc Exp Biol Med 91: 321–323

Poryadkov LF (1970) Influence of diets enriched with different sugars on the excretion of a lipid bile complex. Vopr Pitan 29: 71–72

Poulsom R (1986) Morphological changes in organs after sucrose or fructose feeding. Prog Biochem Pharmacol 21: 104–134

Powell HC, Costello ML, Myers RR (1981) Endoneurial fluid pressure in experimental models of diabetic neuropathy. J Neuropathol Exp Neurol 40: 613–624

Powell HC, Ivor LP, Costello ML, Wolf PL (1982) Elevated haemoglobin A in streptozocin diabetic rats and in rats on sucrose and galactose enriched diets. Clin Biochem 15: 133–137

Powers HWS Jr (1978) Sucrose consumption in our society. Am J Clin Nutr 31: 1301

Powers MA, Crapo PA (1982) The fructose story. Diabet Educ 7: 22–25

Poyser RH, West GB (1968) Structural requirements of sugars as antagonists of the vascular response to dextran in rat skin. Br J Pharmacol Chemother 32: 219–226

Poznanska H (1964) Blood fructose and inorganic phosphorus after oral sucrose load as an index of fructose metabolism. I. Indirect determination of liver fructokinase activity. Pol Hyg Lek 19: 1675–1677

Poznanska H (1967) Blood fructose and inorganic phosphorus after oral saccharose load as index of fructose metabolism. Disturbance of fructose metabolism in patients after viral hepatitis. Pol Med J 6: 576–580

Prather ES (1979) Influence of estrogen content of oral contraceptives and consumption of sucrose on blood parameters. Tox Res Proj Dir 4: 1–14

Preston E, Allen M, Haas N (1983) A modified method for measurement of radiotracer permeation across the rat blood–brain barrier: the problem of correcting brain uptake for intravascular tracer. J Neurosci Meth 9: 45–55

Preuss HG, Fournier RD (1982) Effects of sucrose ingestion on blood pressure. Life Sci 30: 879–886

Preuss MB, Preuss HG (1980) The effects of sucrose and sodium on blood pressures in various substrains of Wistar rats. Lab Invest 43: 101–107

Pribylova H, Melichar V, Sabata V (1977) The effect of prenatal fructose infusions upon metabolic condition of the newborn. Biol Neonate 32: 108–112

Price RG, Kang SS, Bruckdorfer KR, Worcester NA, Yudkin J (1977) Renal damage in rats

caused by long-term feeding of sucrose rich diets. Kidney Int 12: 447–448
Prinsloo JG, Wittmann W, Pretorius PJ, Kruger H, Fellingham SA (1969) Effects of different sugars on diarrhoea of acute kwashiorko. Arch Dis Child 44: 593–599
Prinz RJ, Roberts WA, Hantman E (1980) Dietary correlates of hyperactive behaviour in children. J Consult Clin Psychol 48: 760–769
Procaccio P, Castellana G, Cerini G, Rizzo A (1977) l'Iperuricemia indotta dal fruttoso. I. Ipotesi di inquadramento chimico-clinico. Boll Soc Ital Biol Sper 53: 1980–1985
Procaccio P, Dolce E, Cerini G, Castellana G, Pirelli A (1977) l'Iperuricemia indotta dal fruttoso. II. Escrezione cetecolaminica a seguito di infusione rapida di fruttoso. Boll Soc Ital Biol Sper 53: 1986–1987
Prokhonchukov AA, Zhizhina AN (1967) Effect of a sucrose–casein cariogenic diet on the $P^{32}$ phosphate uptake by proteins of the mineralized tissues of rats. Stomatologiya 46(3): 21–23
Pron L (1929) Diluted sugar produces good effects on stomach. Rev Gén Clin Thér 43: 710 (in French)
Pruett EDR (1970) Glucose and insulin during prolonged work stress in men living on different diets. J Appl Physiol 28: 199–208
PSCC (1986) Simple and complex carbohydrates. Proceedings of a conference on diet and health: scientific concepts and principles. Am J Clin Nutr 45: 1039–1040
Puls W, Keup U, Krause HP, Thomas G, Hoffmeister F (1977) A new approach to the treatment of diabetes, obesity, and hyperlipoproteinaemia. Naturwissenschaften 64: 536–537
Quatraro A, Minei A, Donzella C, Caretta F, Consoli G, Giugliano G (1985) Sugar and wound healing. Lancet II: 664
Quednau HO (1956) Production of experimental cataract by feeding of sugar. Arch Ophthalmol 158: 64–70
Quevedo TD (1947) Cane sugar in nutrition of man. Med Matanzas 6: 42–46
Quevedo TD (1948) Cane sugar and dextrose as foods of man: comparison of nutritive value after ingestion by mouth. Salud Beileza 3: 15–17
Quereshi RU, Akinyanju PA, Yudkin J (1970) The effect of an "atherogenic" diet containing starch or sucrose upon carcass composition and plasma lipids in the rat. Nutr Metab 12: 347–357
Rabasse L (1932) Role of sugared milk soup in nutrition of children. Ann Hyg 10: 273–276
Rabinowitch IM (1945) Short period blood sugar time curves following ingestion of sucrose. J Nutr 29: 99–105
Randoin L, Lecoq R (1929) Biological testing of milk altered by the removal of fat and addition of sugar. C R Soc Biol (Paris) 102: 528–531
Randoin L, Lecoq R (1931) The changing value of vitamin B content in carbohydrates. C R Soc Biol (Paris) 108: 1041–1043
Randolph TG, Rollins JP (1950) Allergic reactions from ingestion of intravenous injection of cane sugar (sucrose). J Lab Clin Med 36: 242–248
Randolph TG, Rollins JP (1950) Beet sensitivity: allergic reactions from ingestion of beet sugar (sucrose) and monosodium glutamate of beet origin. J Lab Clin Med 36: 407–415
Rangaraj N, Kalant H (1982) Effect of chronic ethanol treatment on temperature dependence and on norepinephrine sensitization of rat brain ($Na^+$ + $K^+$)-adenosine triphosphatase. J Pharmacol Exp Ther 223: 536–539
Rao PN, Gordon C, Davies D, Blacklock NJ (1982) Are stone formers maladapted to refined carbohydrates? Br J Urology 54: 575–577
Rapoport JL (1982/83) Effects of dietary substances in children. J Psychiatr Res 17: 187–191
Rapoport JL, Ferguson JL (1981) Biological validation of the hyperkinetic syndrome. Dev Med Child Neurol 23: 667–682
Rapoport S, Brodsky WA, West CD, Mackler B (1948) Urinary flow excretion of solutes, and osmotic work during diuresis of solute loading in hydropenia in man. Science 108: 630–632
Rapp DJ (1978) Does diet affect hyperactivity? J Learn Disabil 11: 56–62, 383–389
Raschewskaja A (1934) Addition of sugar to drinking water of workmen in overheated rooms. Arch Gewerbepath Gewerbehyg 5: 303–310
Rath R (1977) Lipogenesis and obesity in energy storage. Prog Food Nutr Sci 2: 307–321
Rath RJ, Masek V, Kujalova, Slabochova Z (1974) Effect of a high sugar intake on some metabolic and regulatory indicators in young men. Nahrung 18: 343–353
Raul F, Von der Decken A (1983) Modulation of RNA polymerase activities in the intestines of adult rats by dietary sucrose. J Nutr 113: 2134–2140
Raul F, Von der Decken A (1984) Dietary sugar promotes gene activation in intestinal cell chromatin of adult rats. Experientia 40: 364–365

Raul F, Van der Decken A (1985) Changes in chromatin structure and transcription activity by starvation and dietary sucrose in mature and immature intestinal epithelial cells of the rat. Cell Mol Biol 31: 299–304

Raul F, Simon PM, Kedinger M, Grenier JF, Haffen K (1980) Effect of sucrose refeeding on disaccharidase and aminopeptidase activities of intestinal villus and crypt cells of adult rats. Evidence for a sucrose-dependent induction of sucrose in the crypt cells. Biochim Biophys Acta 630: 1–9

Raul F, Pousse A, Grenier JF, Haffen K (1982) Is stimulation of intestinal sucrase activity by dietary sucrose dependent on RNA synthesis in adult rats? Gastroenterol Clin Biol 6: 424–429

Ravich WJ, Bayless TM, Thomas M (1983) Fructose: incomplete intestinal absorption in humans. Gastroenterology 84: 26–29

Rawat AK (1977) Effects of fructose and other substances on acetaldehyde and ethanol metabolism in the rat. Res Commun Chem Pathol Pharmacol 18: 665–676

Rawat AK (1977) Effects of fructose and other substances on ethanol and acetaldehyde metabolism in man. Res Commun Chem Pathol Pharmacol 16: 281–290

Ray RM Jr, Young F (1978) Effect of dietary sucrose and cholesterol on atherosclerosis, live malate dehydrogenase and plasma lipoprotein lipase levels in pigeons. J Nutr 108: 944–953

Rayssiguier Y, Gueux E, Weiser D (1981) Effect of magnesium deficiency on lipid metabolism in rats fed a high carbohydrate diet. J Nutr 111: 1876–1883

Reaven GM (1979) Effects of differences in amount and kind of dietary carbohydrate on plasma glucose and insulin responses in man. Am J Clin Nutr 32: 2568–2578

Reaven GM (1982) Effects of fructose on lipid metabolism. Am J Clin Nutr 35: 627

Reaven GM, Risser TR, Chen Y-DI, Reaven EP (1979) Characterization of a model of dietary-induced hypertriglyceridemia in young, nonobese rats. J Lipid Res 20: 371–378

Reaven GM, Zavaroni I (1982) Effect of acarbose treatment on sucrose induced hypertriglyceridaemia in the rat. Int Congr Ser Excerpta Med 594: 55–62

Rebello T, Hodges RE, Smith JL (1983) Short-term effects of various sugars on antinatriuresis and blood pressure changes in normotensive young men. Am J Clin Nutr 38: 84–94

Reboud AM, Dubost S, Reboud JP (1984) Sucrose modifies ribosomal stability and conformation. Biochemistry 66: 251–255

Reddy BS, Pleasants JR, Wostmann BS (1968) Effect of dietary carbohydrates on intestinal dissaccharidases in germ free and conventional rats. J Nutr 95: 413–419

Reding R, Slosse A (1929) Alkalosis and disturbances of sugar metabolism in cancerous and precancerous states. Bull Assoc France Etude Cancer 18: 122–151

Reed DJ, Woodbury DM (1960) Kinetics of $C^{14}$-sucrose distribution in cerebral cortex, cerebrospinal fluid, and plasma of rats. Fed Proc 19: 80

Regan JJ Jr, Doorneweerd DD, Gilboe DP, Nuttall FQ (1980) Influence of fructose on the glycogen synthase and phosphorylase systems in rat liver. Metabolism 29: 965–969

Regan TJ, Weisse AB, Oldewurtel HA, Hellems HK (1964) The hyperosmolic effects of ethanol and sucrose on the left ventricle (dog). J Clin Invest 43: 1289–1290

Reichen J, Le M (1983) Taurocholate, but not taurodehydrocholate, increases biliary permeability to sucrose. Am J Physiol 245: G651–G655

Reintam EA (1960) The volatile fatty acid and sugar content in the blood of farm animals. Dokl Vses Akad Sel'skokhoz Nauk Im V I Lenima 6: 36–40

Reiser S (1982) Health implications of food carbohydrates: heart disease and diabetes. In: Lineback DR, Inglett GE (eds) Food carbohydrates. AVI Publishing, New York, pp 170–205

Reiser S (1985) Effect of dietary sugars on metabolic risk factors associated with heart disease. Nutr Health 3: 203–216

Reiser S, Hallfrisch J (1977) Insulin sensitivity and adipose tissue weight of rats fed starch or sucrose diets *ad libitum* or in meals. J Nutr 107: 147–155

Reiser S, Kelsay JL (1981) Metabolic effects of dietary sucrose and fiber on humans. In: Beecher GR (ed) Human nutrition research. Allanheld, Osmun, Totowa, New Jersey, pp 143–163

Reiser S, Szepesi B (1978) SCOGS report on the health aspects of sucrose consumption. Am J Clin Nutr 31: 9–11

Reiser S, Michaelis OE, Putney J, Hallfrisch J (1975) Effect of sucrose feeding on the intestinal transport of sugars in two strains of rats. J Nutr 105: 894–905

Reiser S, Hallfrisch J, Putney J, Lev F (1976) Enhancement of intestinal sugar transport by rats fed sucrose as compared to starch. Nutr Metab 20: 461–470

Reiser S, Hallfrisch J, Michaelis OE IV, Lazar FL, Martin RE, Prather ES (1978) Isocaloric exchange of dietary starch and sucrose in humans. I. Effects on levels of fasting blood lipids. Am J Clin Nutr 32: 1659–1669

Reiser S, Handler HB, Gardner LB, Hallfrisch JG, Michaelis OE, Prather ES (1979) Isocaloric exchange of dietary starch and sucrose in humans. II. Effects on fasting blood insulin, glucose and glucagon and on insulin and glucose response to a sucrose load. Am J Clin Nutr 32: 2206–2216

Reiser S, Michaelis OE, Cataland S, O'Dorisio TM (1980) Effect of an isocaloric exchange of dietary starch and sucrose in humans on the gastric inhibitory polypeptide response to a sucrose load. Am J Clin Nutr 33: 1907–1911

Reiser S, Bickard MC, Hallfrisch J, Michaelis OE, Prather ES (1981) Blood lipids and their distribution in lipoproteins in hyperinsulinemic subjects fed three different levels of sucrose. J Nutr 11: 1045–1057

Reiser S, Bohn E, Hallfrisch J, Michaelis OE, Keeney M, Prather ES (1981) Serum insulin and glucose in hyperinsulinaemic subjects fed three different levels of sucrose. Am J Clin Nutr 34: 2348–2358

Reiser S, Ferretti FJ, Fields M, Smith JC (1983) Role of dietary fructose in the enhancement of mortality and biochemical changes associated with copper deficiency in rats. Am J Clin Nutr 38: 214–222

Reiser S, Hallfrisch J, Lyon R, Michaelis OE IV (1983) Effect of chronic hyperinsulinism on metabolic parameters and histopathology in rats fed sucrose or starch. J Nutr 113: 1073–1080

Reiser S, Smith JC, Mertz W et al. (1985) Indices of copper status in humans consuming a typical American diet containing either fructose or starch. Am J Clin Nutr 42: 242–251

Reiser S, Hallfrisch J, Fields M et al. (1986) Effects of sugars on indices of glucose tolerance in humans. Am J Clin Nutr 43: 151–159

Remington JW (1940) Fluid and electrolyte shifts in the normal and adrenalectomized rat after the intraperitoneal injection of isotonic sugar solutions. Endocrinology 26: 631–640

Renkin EM (1955) Effects of blood flow on diffusion kinetics in isolated, perfused hindlegs of cats. A double circulation hypothesis. Am J Physiol 183: 125–136

Renold AE, Thorn GW (1955) Clinical usefulness of fructose. Am J Med 19: 163–168

Renold AE, Berger M, Jeanrenaud B (1978) Obesity and diabetes. In: Guggenheim B (ed) Health and sugar substitutes. Proceedings of the ERGOB conference on sugar substitutes, Geneva, 30 Oct–1 Nov 1978. Karger, Basel, pp 17–21

Retief DH, Cleaton-Jones PE, Walker ARP (1975) Dental caries and sugar intake in South African pupils of 16 to 17 years in four ethnic groups. Br Dent J 138: 463–469

Rey J, Frezal J, Jos J, Bauche P, Lamy M (1963) Diarrhoea due to impaired hydrolysis of sucrose, maltose and isomaltose in the gut. Arch Fr Pediatr 20: 381–400

Reyes FGR, Toledo MCF (1982) Maillard browning reaction of sucrose–glycine model system. Effect of citrate buffer on the color formation and on the hydrolysis of sucrose. Ciênc Technol Aliment (Brazil) 2: 134–142

Reyes FGR, Wrolstad RE, Cornwell CJ (1982) Comparison of enzymic gas-liquid chromatographic and high-performance liquid chromatographic methods for determining sugars and organic acids in strawberries at three stages of maturity. J Assoc Off Anal Chem 65: 126–131

Reyes FGR, Poocharoen B, Wrolstad RE (1982) Maillard browning of sugar-glycine mode system: changes in sugar concentration, colour and appearance. J Food Sci 47: 1376–1377

Riby JE, Kretchmer N (1984) Effect of dietary sucrose on synthesis and degradation of intestinal sucrose. Am J Physiol 246: G757–G763

Richardson AE (1965) Clinical (and pharmacotherapeutical) aspects of cerebral edema. Proc R Soc Med 58: 604–606

Richardson BD, Cleaton-Jones P, McInnes PM, Rantsho JM, Pieters L (1978) Total sucrose intake and dental caries in black and in white South African children of 1–6 years. Part II: Dental caries and sucrose intake. J Dent Assoc S Afr 33: 539–544

Richardson BD, Rantsho JM, Pieters L, Cleaton-Jones P (1978) Total sucrose intake and dental caries in black and white South African children of 1–6 years. J Dent Assoc S Afr 33: 533–537

Richardson DP, Scrimshaw NS, Young VR (1980) The effect of dietary sucrose on protein utilization in healthy young men. Am J Clin Nutr 33: 264–272

Richardson HL, Kennedy JC, West ES (1943) Diuretic and other effects of intravenous sorbitol and sucrose. Removal of edema fluid by the combined action of these substances and salyrgan. Northwest Med 42: 80–83

Richardson JF (1971) The dietary sugar intake of businessmen; the results of a cross-sectional survey. Postgrad Med J 47 (Suppl): 423–426

Richardson JF (1972) The sugar intake of businessmen and its inverse relationship with relative weight. Br J Nutr 27: 449–460

Richmond ML, Brandao SCC, Gray JI, Markakis P, Stine CM (1981) Analysis of simple sugars and sorbitol in fruit by high performance liquid chromatography. J Agric Food Chem 29: 4–7

Richter AW, Granath K, Ostling G (1976) Anaphylactoid reactions in connection with infusion of invert sugar solutions are due to macromolecular contaminants. Int Arch Allergy Appl Immunol 50: 606–612

Richter CP, Rice KK (1945) A comparison of the nutritive value of dextrose and sucrose and of the effects produced on their utilization by thiamine hydrochloride. Am J Physiol 143: 336–343

Richter V (1985) Sucrose induced hyperlipoproteinaemia and carnitine. Nahrung 29: 626–627

Riemann JF, Kolb S (1984) Low sugar and fibre-rich diet in Crohn's disease. Fortschr Med 102: 67–70

Rifkind BM, Lawson DH, Gale M (1966) Effect of short-term sucrose restriction on serum lipid levels. Lancet II: 1379

Rifkind BM, Goor RS, Levy RI (1979) Current status of the role of dietary treatment in the prevention and management of coronary heart disease. Med Clin North Am 63: 911–925

Rigdon RH, Cardwell ES (1942) Renal lesions following the intravenous injection of hypertonic sucrose. A chemical and experimental study. Arch Intern Med 69: 670–690

Riñón L, Yanguas E, Ortiz M, Lluch M, Ponz F (1986) Effect of luminal $Na^+$ on the kinetics of intestinal absorption of sugars in vivo. Rev Esp Fisiol 42: 265–270

Roberts IF, Roberts GJ (1979) Relation between medicines sweetened with sucrose and dental disease. Br Med J ii: 14–16

Robrish SA, Krichevsky MI (1972) Acid production from glucose and sucrose by growing cultures of caries-conducive streptococci. J Dent Res 51: 734–739

Rodin J (1977) Implications of responsiveness to sweet taste for obesity: In: Weiffenback J (ed) Taste and development: the genesis of sweet preference (DHEW Publ No 77–1068)

Roe DA (1983) Diet, nutrition and cancer: from basic research to policy implications. Proceedings of a workshop held at Cornell University. Alan R Liss, New York

Roe FJC (1979) Mineral deposition in the renal pelvis of rats: a brief review. In: Joint FAO/WHO Expert Committee on Food Additives (ed) Twenty-sixth report. World Health Organization, Geneva, p 11 (WHO Technical Report Series No 683, 1982)

Roe FJC (1984) Perspectives in carbohydrate toxicology with special references to carcinogenicity. Swed Dent J 8: 99–111

Roe FJC, Levy LS, Carter RL (1970) Feeding studies on sodium cyclamate, saccharin and sucrose for carcinogenic and tumor promoting activities. Food Cosmet Toxicol 8: 135–145

Rogel AM, Vohra P (1983) Hypocholesterolemia and growth-depression in chicks fed guar gum and konjac mannan. J Nutr 113: 873–879

Rogers BJ, Yanagimachi R (1975) Retardation of guinea pig sperm acrosome reaction by glucose: the possible importance of pyruvate and lactate metabolism in capacitation and the acrosome reaction. Biol Reprod 13: 568–575

Rohr HP, Zollinger HU (1966) The so-called sucrose nephrosis studied by electronmicroscopic autoradiography with sacchrose-$C^{14}$. Virchows Arch Pathol Anat Physiol Klin Med 341: 115–122

Rolla G (1971) Sucrose, dental plaque and caries. Tidsskr Nor Laegeforen 91: 697–700

Rolla G, Scheie AA, Ciardi JE (1985) Role of sucrose in plaque formation. Scand J Dent Res 93: 105–111

Romsos DR, Leveille GA (1974) Effect of meal frequency and diet composition on glucose tolerance in the rat. J Nutr 104: 1503–1512

Rosen S (1969) Comparison of sucrose and glucose in the causation of dental caries in gnotobiotic rats. Arch Oral Biol 14: 445–450

Rosen S, Weisenstein PR (1965) The effect of sugar solutions on pH of dental plaques from caries-susceptible and caries-free individuals. J Dent Res 44: 845–849

Rosen S, Min DB, Harper DS, Harper WJ, Beck WX, Beck FM (1984) Effect of cheese, with and without sucrose, on dental caries and recovery of *Streptococcus mutans* in rats. J Dent Res 63: 894–896

Rosenbaum H (1933) Sucrose of nonmedullated mammalian nerves. Biochem Z 257: 307–309

Rosenbrough RW, Rosenbrough RH, Gels E, Henderson K (1979) Effect of supplemental glucose or sucrose on liver and carcass glycogen metabolism of young chicks. Poultry Sci 58: 1524–1528

Rosenkranz A, Rantlitschko M (1965) Sucrose and isomaltose intolerance in an infant. Wien Klin Wochenschr 77: 103–106

Rosenmann E, Teitelbaum A, Cohen AM (1971) Nephropathy in sucrose-fed rats. Electron and light microscopic studies. Diabetes 20: 803–810

Rosenmann, E. Palti Z, Teitelbaum A, Cohen AM (1974) Testicular degeneration in genetically selected sucrose fed diabetic rats. Metabolism 23: 343–348

Rosensweig NS, Herman RH (1968) Control of jejunal sucrase and maltase activity by dietary

sucrose or fructose in man: a model for the study of enzyme regulation in man. J Clin Invest 47: 2253–2262
Rosensweig NS, Herman RH (1969) Time response of jejunal sucrase and maltase activity to a high sucrose diet in normal man. Gastroenterology 56: 500–505
Rosensweig NS, Herman RH (1970) Dose response of jejunal sucrase and maltase activities to isocaloric high and low carbohydrate diets in man. Am J Clin Nutr 23: 1373–1377
Rosensweig NS, Stifel FB, Herman RH (1968) Dietary regulation of the glycolytic enzymes. II. Adaptive changes in human jejunum. Biochim Biophys Acta 170: 228–234
Rosenthal O (1930) Fermentation of various types of sugar by liver of rats: dependence on nutritional condition of experimental animal, and on seasonal factors. Biochem Z 227: 354–384
Ross AC, Minick CR, Zilversmit DB (1978) Equal atherosclerosis in rabbits fed cholesterol free low fat diet or cholesterol supplemented diet. Atherosclerosis 29: 301–315
Rowe, HG, Anderson RH, Wanninger LA (1974) Effects of ready to eat breakfast cereals on dental caries experience in adolescent children. A three year study. J Dent Res 53: 33–36
Rowland N, Stricker EM (1979) Differential effects of glucose and fructose infusions on insulin induced feeding in rats. Physiol Behav 22: 387–389
Roy AC, Sen KC (1929) Hemolysis in cane sugar solution and the behaviour of normal serum in the presence of chemically hemolytic substances. J Indian Chem Soc 6: 171–180
Rozenfel'd EL, Lukomskaya IS, Gorodetskii VK, Zarubina NA, Zaretskii MM (1964) Sucrose synthesis in human beings. Vopr Med Khim 10: 554–556
Rubner M (1931) History of the development of energy utilization in vertebrates. Sber Preuss Akad Wiss Phys-Math Kl 17: 272–316 (in German)
Rugg-Gunn AJ (1983) Diet and dental caries. In: Murrary JJ (ed) The prevention of dental disease. Oxford University Press, Oxford, pp 3–82
Rugg-Gunn AJ, Edgar WM (1984) Sugar and dental caries. A review of the evidence. Community Dent Health 1: 85–92
Rugg-Gunn AJ, Edgar WM, Geddes DAM, Jenkins GN (1975) The effect of different meal patterns on plaque pH in human subjects. Br Dent J 139: 351–356
Rugg-Gunn AJ, Hackett AF, Appleton DR, Jenkins GN, Eastoe JE (1984) Relationship between dietary habits and caries increment assessed over two years in 405 English adolescent school children. Arch Oral Biol 29: 983–992
Rugg-Gunn AJ, Roberts GJ, Wright WG (1985) Effect of human milk on plaque pH in situ and enamel dissolution in vitro compared with bovine milk, lactose, and sucrose. Caries Res 19: 327–334
Rugg-Gunn AJ, Hackett AF, Appleton DR (1986) Cariogenicity of starch in a two-year longitudinal study of 405 Northumbrian school children. J Den Res 65: 511 (Abstract No 210)
Rummel W, Wilbrandt W (1951) Apparent inhibition of osmotic hemolysis by sugar. Arch Ges Physiol 253: 194–204
Rumsey JM, Rapoport JL (1983) Assessing behavioural and cognitive effects of diet. In: Wurtman RJ, Wurtman JJ (eds) Pediatric populations. In: Nutrition and the brain, Raven Press, New York, vol 6, pp 101–161
Runner MN (1959) Inheritance of susceptibility to congenital deformity. Metabolic clues provided by experiments with teratogenic agents. Pediatrics 23: 245–251
Rupe BD, Mayer J (1967) Endogenous glucose release stimulated by oral sucrose administration in rats. Experientia 23: 1009–1010
Rylnikov YP (1974) Changes in the lipid composition of the blood and certain tissues under sustained administration of qualitatively different carbohydrates. Kardoliology 14: 100–106
Sakal T (1941) Sugar assimilation and sugar excretion threshold in pituitary diseases without endocrine disturbances: 10 cases of tumours. Tokyo Igakkwai Zassi 55: 234
Salel AF, Riggs K, Mason DT, Amsterdam EA, Zelis R (1974) The importance of type IV hypolipoproteinemia as a predisposing factor in coronary artery disease. Am J Med 57: 897–903
Salerno GL, Pontis HG (1978) Studies on sucrose phosphate synthetase. The inhibitory action of sucrose. FEBS Lett 86: 263
Saller CF, Chiodo LA (1980) Glucose suppresses basal firing and haloperidol-induced increases in the firing rate of central dopaminergic neurons. Science 210: 1269–1271
Salma GP, Goicolea J-JI, Elgarbly F et al. (1984) Sucrose taken during mixed meal has no additional hyperglycaemic action over isocaloric amounts of starch in well-controlled diabetics. Lancet II: 122–125
Salter AJ, Grenby TH (1969) Dental caries in a black-hooded rat strain in a nutritional experiment. Caries Res 3: 201–204
Salter AJ, Yudkin (1978) Dental caries and between-meal snacks. Br Med J i: 577

Samson HH (1978) Sucrositis. S Afr Med J 54: 590–591

Sanada I (1968) Effect of dietary sucrose on vascular lesions in the rabbit. Jpn Circ J 32: 1613–1620

Sanchez A, Reeser JL, Lau HS et al. (1973) Role of sugars in human neutrophilic phagocytosis. Am J Clin Nutr 26: 1180–1184

Sanchez JJ, Gonzales NS, Pontis HG (1971) Fructokinase from rat liver. I. Purification and properties. Biochim Biophys Acta 227: 67–78

Sato K, Ueda K, Kurokasa T (1930) The fate of infused sugar with special consideration of sugar absorbability by tissue cells. Tohoku J Exp Med 14: 335–340

Sato S, Ohta T, Nagao M, Tsuji K, Kosuge T (1979) Reduction in mutagenicity of cigarette smoke condensate by added sugars. Mutat Res 60: 155–161

Scavotto SP (1982) Coping with less caries in the private practice of dentistry. J Dent Res 61 (Sp Iss): 1364–1368

Schachtele CF (1982) Changing perspectives on the role of diet in dental caries formation. Nutr News 45: 13–14

Schachtele CF, Hagen EW, Anderson DL (1971) Temperature shift analysis of bacteriophage 29 gene expression in *Bacillus amyloliquefaciens*. J Virol 8: 352–354

Schachtele CF, Loken AE, Knudson DJ (1972) Preferential utilization of the glucosyl moiety of sucrose by a cariogenic strain of *Streptococcus mutans*. Infect Immun 5: 531–536

Schachtele CF, Loken AE, Schmitt MK (1972) Use of specifically labelled sucrose for comparison of extracellular glucan and fructan metabolism by oral streptococci. Infect Immun 5: 263–266

Schachtele CF, De Sain CV, Hawley LA, Anderson DL (1972) Transcription during the development of bacteriophage 29 production of host and 29 specific ribonucleic acid. J Virol 10: 1170–1178

Schachtele CF, De Sain CV, Anderson DL (1973) Transcription during the development of bacteriophage 29 definition of "early" and "late" 29 ribonucleic acid. J Virol 11: 9–16

Schachtele CF, Reilly BE, De Sain CV, Whittington MO, Anderson DKL (1973) Selective replication of bacteriophage 29 deoxyribonucleic acid in 6-(*p*-hydroxyphenylazo-uracil)-treated *Bacillus subtilis*. J Virol 11: 153–155

Schachtele CF, Germaine GR, Harlander SK (1975) Production of elevated levels of dextransucrase by a mutant of *Streptococcus mutans*. Infect Immun 12: 934–937

Schachtele CF, Staat RH, Harlander SK (1975) Dextranases for oral bacteria: inhibition of water insoluble glucan production and adherence to smooth surfaces by *Streptococcus mutans*. Infect Immun 12: 309–317

Schachtele CF, Harlander SK, Germaine GR (1976) *Streptococcus mutans* dextransucrase: availability of disaggregated enzyme after growth in a chemically defined medium. Infect Immun 13: 1522–1524

Schachtele CF, Harlander SK, Fuller DW, Zollinger PK, Woon-Lam S Leung (1977) Bacterial interference with sucrose dependent adhesion of oral streptococci. In: Microbial aspects of dental caries. IRL Press, Arlington, Virginia, pp 401–412

Schaefer DE (1959) The distribution of insulin and sucrose in the perfused heart. Diss Abstr Int [B] 20: 2359

Schaefer DE, Johnson JA (1961) Permeability of mammalian heart capillaries to sucrose and insulin. Am J Physiol 206: 985–991

Schaefer O (1977) Changing dietary patterns in the Canadian north: health, social and economic consequences. J Can Diet Assoc 38: 17–25

Schaferstein SJ (1930) Beet sugar as valuable as dextrinmaltose mixture for infant feeding. Monatsschr Kinderheilkd 46: 385–402

Schapira F (1986) L'intolérance héréditaire au fructose. Cah Nutr Diet 21: 72–75

Schapira F, Schapira G, Dreyfus JC (1961–62) La lesion enzymatique de la fructosurie benigne. Enzym Biol Clin 1: 170–175

Scharff TG, Badr M, Doyle RJ, Zaleski C (1981) Antagonism between salicylic acid and glucose in *Saccharomyces cerevisiae*. Toxicol Appl Pharmacol 60: 485–491

Scheinin A (1976) Caries control through the use of sugar substitutes. Int Dent J 26: 4–13

Scheinin A (1978) Sugar substitutes in relation to the incidence of clinical and experimental caries. Pharmacol Ther Dent 3: 95–100

Scheinin A, Mäkinen KK (1971) The effect of various sugars on the formation and chemical composition of dental plaque. Int Dent J 21: 302–321

Scheinin A, Odont L, Odont D (1974) Dietary sugars and sugar substitutes. Int Dent J 23: 427–431

Scheinin A, Mäkinen KK, Ylitalo K (1975) Turku sugar studies. V. Final report on the effect of sucrose, fructose and xylitol diets on the caries incidence in man. Acta Odontol Scand 33: 67–104

Scheinin A, Mäkinen KK, Tammisalo E, Rekola M (1975) Turku sugar studies. XVIII. Incidence of dental caries in relation to 1 year consumption of xylitol chewing gum. Acta Odontol Scand 33: 269–278

Schemmel RA, Teague RJ, Bray GA (1982) Obesity in Osborne-Mendel and S5B/P1 rats: effects of sucrose solutions, concentration, and treatment with estradiol or insulin. J Physiol (Lond) 243: R347–R353

Scherf D, Weissberg J (1940) The ECG after intravenous injection of hypertonic sucrose and saline solutions. Cardiologia 4: 260–266

Scherp HW (1971) Dental caries: prospects for prevention. Science 173: 1199–1205

Schill H (1965) Electron microscopic and enzymic histochemical investigations in experimental sucrose nephrosis. Zentralbl Allg Pathol Pathol Anat 107: 389–405

Schiweck H (1970) Is sugar harmful for the human organism? Zucker 23: 247–257

Schlesinger M, Cohen A, Hurvitz D (1966) Inhibition of carbohydrates of the cytotoxicity of heterologous sera for mouse thymus cells. Isr J Med Sci 2: 616–619

Schmaehl D, Habs M (1984) Investigations on the carcinogenicity of the artificial sweeteners sodium cyclamate and sodium saccharin in rats in a two-generation experiment. Arzneim-Forsch Drug Res 34: 604–606

Schmidt EG, Eastland JS, Burns JH (1935) Comparative study of glucose and sucrose tolerance tests. J Lab Clin Med 21: 13–25

Schobel C (1968) Comparative examinations with four different blood sugar determination methods and blood sugar values of clinically healthy dogs. Kleintier Praxis 13: 171–172

Schoenle E, Zapf J, Foresch ER (1979) Transport and metabolism of fructose in fat cells of normal and hypophysectomized rats. Am J Physiol 237: E325–E330

Schouten H, Giterson A (1963) Blood, urine, and liquor sugar. Clin Chim Acta 8: 802–803

Schramek J (1932) Application and effect of hypertonic sugar solutions. Gyogyaszat 72: 474–475

Schrek R (1945) Studies in vitro on the physiology of cells. Histologic reactions of living tissues to hypotonic solutions. Am J Pathol 21: 1101–1111

Schrier AM (1963) Sucrose concentration and response rates of monkeys. Psychol Rep 12: 666

Schrier AM (1965) Response rates of monkeys (*Macaca mulatta*) under varying conditions of sucrose reinforcement. J Comp Physiol [A] 59(3): 378–384

Schroeder H (1938) Sugar and teeth. Ernahrg 3: 199–203

Schubert J, Watson JA (1967) Hydroxyalkyl peroxides and the toxicity of irradiated sucrose. Int J Radiat Biol 13: 485–489

Schubert J, Watson JA, White R (1967) Hydroxyalkyl peroxides and the toxicity of irradiated sucrose. Int J Radiat Biol 13: 485–489

Schubert J, Watson JA (1969) Organic peroxides and the antibacterial action of irradiated sucrose as affected by catalase. Radiat Res 37: 531–538

Schuetze U (1941) Effect of various types of sugar on altitude tolerance of aviation crews. Luftfahrtmedizin 5: 97–102

Schultz AL, Grande F (1968) Effects of starch and sucrose on the serum lipids of dogs before and after thyroidectomy. J Nutr 94: 71–73

Schultz FW, Knott EM, Gedgoud JL, Loewenstamm I (1938) Comparative values of carbohydrates in infant feeding. J Pediatr 12: 716–724

Schulz RW, Lawrence DH (1958) Learning of a discrimination by satiated and deprived rats with sucrose as incentive. Am J Psychol 71: 563–567

Schuzdziarra V, Dangel G, Klier M, Henrichs I, Pfeiffer EF (1981) Effect of solid and liquid carbohydrates upon postprandial pancreatic endocrine function. J Clin Endocrinol Metab 53: 16–20

Schusdziarra V, Rewes B, Lenz N, Maier V, Pfeifer EF (1983) Carbohydrates modulate opiate receptor mediated mechanisms during postprandial endocrine function. Regul Pept 7: 243–252

Schutte U, Schutte E, Reim M (1977) Fructose levels of aphotic rabbit eyes. Trans Ophthalmol Soc UK 97: 714–715

Schwarz LS, Pokrowskaja GN (1935) The glycogen in blood by diabetes mellitus. Arch Sci Biol 38: 789–793

Schwartz D, Patois E, Beaumont J-L (1967) Les triglycérides sanguins dans un groupe professionnel. J Atheroscler Res 7: 537–548

Schwartz HG, Elman R (1938) Effects of sorbitol and sucrose on cerebrospinal fluid pressure and urine output. Proc Soc Exp Biol Med 39: 506–508

Schwartz RS, Ravrissin E, Massari M, O'Connell M, Robbins DC (1985) The thermic effect of carbohydrate versus fat feeding in man. Metabolism 34: 285–292

SCOGS/FASEB (1976) Evaluation of the health aspects of sucrose as a food ingredient. Life Sciences Research Office (LSRO)/Federation of American Societies for Experimental Biology (FASEB), US Food and Drug Administration (Prepared for FDA under contract No FDA 223-75-2004)

Scott DW, Gorry GA, Gotto AM (1981) Diet and coronary heart disease: the statistical analysis of risk. Circulation 63: 516–518

Scott PP, Greaves JP, Scott MG (1964) Nutritional blindness in the cat. Exp Eye Res 3: 357–364

Scragg RKR, McMichael AJ, Baghurst PA (1984) Diet, alcohol and relative weight in gall stone disease: A case-control study. Br Med J 288: 1113–1119

Sebastian KL, Zacharias NT, Philip LB, Augusti KT (1979) The hypolipidemic effect of onion (*Allium cepa* Linn) in sucrose fed rabbits. Indian J Physiol Pharmacol 23: 27–30

Secher-Hansen E (1970) Studies on the subcutaneous absorption in mice. VII. Absorption of $^{3}H_2O$ and $^{14}C$-sucrose fron non-buffered solutions at different pH values. Acta Pharmacol (Kobenhavn) 29: 97–101

Secher-Hansen E, Langgàrd H, Schou J (1968) Studies on the subcutaneous absorption in mice. IV. Absorption of carbohydrates with different molecular weights from connective tissue. Acta Pharmacol (Kobenhavn) 26: 9–14

Seely, S, Horrobin DF (1983) Diet and breast cancer: the possible connection with sugar consumption. Med Hypotheses 11: 319–327

Seham M, Seham G (1929) The relation between malnutrition and nervousness. Am J Dis Child 37: 1–38

Seidenstadt RM, Eaton KE (1978) Adrenal and ovarian regulation of salt and sucrose consumption. Physiol Behav 21: 313–316

Select Committee on Nutrition and Human Needs (1977) Dietary goals for the United States, 2nd edn. US Government Printing Office, Washington DC

Semenza G (1967) Digestion and absorption of sugars in the intestinal tract. Caries Res 1: 187–207

Sen RN (1964) Neurogenic influence on absorption of sugar in rat intestine. J Exp Med Sci 7: 101–103

Senti FR, Schaefer WC (1972) Corn, its importance in food, feed and industrial uses. Cereal Sci Today 17: 352–356

Serbource-Goguel N, Porque D, Feger J, Durand G, Appel M, Agneray J (1985) Utilisation du fructose par l'hépatocyte isolé de rat nourri Zucker obèse fa/fa. I. Orientation métabolique du fructose en l'absence de modulateur. Ann Nutr Metab 29: 119–128

Seronde J Jr (1969) Influence of dietary carbohydrate on duodenal lesions in the mouse. J Nutr 99: 191–195

Sestoft L (1979) Fructose and the dietary therapy of diabetes mellitus. Diabetes 17: 1–3

Sestoft L, Fleron P (1974) Determination of the kinetic constants of fructose transport and phosphorylation in the perfused rat liver. Biochim Biophys Acta 345: 27–38

Sestoft L, Folke M, Kristensen LO, Gammeltoft S, Bartels P (1979) The effect of fructose on potassium ion movements across the cell membrane in liver. Biochem Soc Trans 7: 204–206

Shafer WG (1949) The caries-producing capacity of starch, glucose, and sucrose diets in the syrian hamster. Science 110: 143–144

Shafrir E (1985) Effect of sucrose and fructose on carbohydrate and lipid metabolism and the resulting consequences. Regulation of carbohydrate metabolism, vol 2, chap 5. CRC Press, Boca Raton, Florida, pp 95–140

Shafrir E, Orevi M (1984) Response of hepatic fructokinase to long-term sucrose diets and diabetes in spiny mice, albino mice and rats. Comp Biochem Physiol 78B: 493–498

Shallenberger RS (1974) Occurrence of various sugars in foods. In: Sipple HL, McNutt KW (eds) Sugars in nutrition. Academic Press, New York, pp 67–80

Shallenberger RS, Birch GG (1975) Sugar chemistry. AVI Publishing, Westport, Connecticut, pp 46–88

Shannon IL (1974) Sucrose and glucose in dry breakfast cereals. J Dent Child 41: 347–359

Shannon IL (1978) Concentration of sugar in commercial baby foods. J Dent Child 45: 19–22

Shannon IL, McCartney JC (1981) Presweetened dry breakfast cereals: potential for dental danger. J Dent Child 48: 215–218

Shannon IL, Wescott WB (1975) Sucrose and glucose concentrations of frequently ingested foods. J Acad Gen Dent 23: 37–43

Shannon IL, Edmonds EJ, Madsen KO (1979) Honey: sugar content and cariogenicity. J Dent Child 46: 29–33

Shapland CD (1926) Blood-sugar in the normal and diabetic. Lancet II: 589–594

Share L (1960) Volumes of distribution of hemoglobin, of ($^{14}$C) carboxypolyglucose and of ($^{14}$C) sucrose in pellets of rat-liver mitochondria. Biochim Biophys Acta 39: 154–155

Share L, Hansrote RW (1960) Permeability of rat liver microsomes to sucrose and carboxypolyglucose in vitro. J Biophys Biochem Cytol 7: 239–242

Sharief NN, Macdonald I (1982) Differences in dietary induced thermogenesis with various carbohydrates in normal and overweight men. Am J Clin Nutr 35: 267–272

Sharief NN, Macdonald I (1982) Different effects of various carbohydrates on the metabolic rate in rats. Ann Nutr Metab 26: 66–72

Sheehan PM (1981) Blood glucose and plasma lipids of Zucker fatty and lean rats fed diets containing cornstarch and sucrose. Diss Abstr Int [B] 42: 2312B

Sheehan PM, Reynolds LR, Thye FW, Ritchey SJ (1984) Blood glucose and plasma lipids of Zucker rats fed diets containing cornstarch or sucrose. Nutr Rept Int 29: 1337–1344

Sheiham A (1979) The epidemiology of dental caries and periodontal disease. J Clin Periodontol 6: 7–15

Sheiham A (1983) Nutrition: the changing scene. Lancet I: 282–284

Sheiham A (1983) Sugars and dental decay. Lancet I: 282–284

Sheiham A (1984) Changing trends in dental caries. Int J Epidemiol 13: 142–147

Sheorain VS, Mattock MB, Subrahmanyam D (1980) Mechanism of carbohydrate induced hypertriglyceridemia: plasma lipid metabolism in mice. Metabolism 29: 924–929

Shern RJ, Groh R, Kennedy J, Monell-Torrens E, Stiles HM (1984) Measurement of pH levels in plaque on rat molars. International Association of Dental Research, Washington DC (Abstract of General Meeting No 1468)

Shiff TS, Roheim PS, Eder HA (1971) Effects of high sucrose diets and 4-aminopyrazolopyrimidine on serum lipids and lipoproteins in the rat. J Lipid Res 12: 596–603

Shigeaki S, Tomoko O, Minako N, Kuniro T, Takuo K (1979) Reduction in mutagenicity of cigarette smoke condensate by added sugars. Mutat Res 60: 155–161

Shigeoka T, Izawa O, Kitazawa K, Yamauchi F (1984) Studies on the metabolic fate of sucrose esters in rats. Food Chem Toxicol 22: 409–414

Shinomiya M, Nakagaki H, Sakakibara Y (1982) Determination of sucrose contents in available foods and drinks. II. Determination by sugar analyzer. Aichi Gakuin Daigaku Shigakkai Shi 20: 171–180

Shoemaker WC, Yanof HM, Turk LN III, Wilson TH (1963) Glucose and fructose absorption in the unanaesthetized dog. Gastroenterology 44: 654–662

Shore V, Shore B (1960) Effect of mercuric chloride on some kidney enzymes in chow-fed and sucrose-fed rats. Am J Physiol 198: 187–190

Shrestha BM, Kreutler PA (1983) A comparative rat caries study on cariogenicity of foods using the intubation and gel methods. J Dent Res 62 (Sp Iss): 685 (Abstract No 318)

Shrestha BM, Kreutler PA, Stein MJ, Zarcone AM (1984) Corn bran as a potential cariostatic food additive. International Association of Dental Research, Washington DC (Abstract of General Meeting No 1470)

Shyu KW, Hsu MY (1980) The cariogenicity of xylitol, mannitol, sorbitol and sucrose. Proc Natl Sci Counc Repub China [B] 4: 21–26

Siddons RC, Smith RH, Henschel MJ, Hill WB, Porter JWG (1969) Carbohydrate utilization in the preruminant calf. Br J Nutr 23: 333–341

Sidi AD, Ashley FP (1984) Influence of frequent sugar intakes on experimental gingivitis. J Periodontal 55: 419–423

Siebert C, Traenckner K, Lang K (1954) Behavior of kidney mitochondria after intravenous sucrose injection in rats. Naturwissenschaften 41: 460

Siedler AJ, Rice MS, Maloney PA, Lushbough CH, Schweigert BS (1962) The influence of varying levels of dietary protein, carbohydrate, and fats in the nutrition of the rat. J Nutr 77: 149–154

Sillero MAG, Sillero A, Sols A (1969) Enzymes involved in fructose metabolism in liver and the glyceraldehyde metabolic crossroads. Eur J Biochem 10: 345–350

Silva Abiaka M de (1969) Dietary sucrose and cardiovascular disease. A review. J Nutr Diet 6: 343–355

Silverstone LM (1977) Preventive dentistry. Diet and dental caries, Part I. Dent Update 4: 271–274

Simic BS, Stefanovic S, Simic-Penezic S et al. (1970) Nitrogen balance in patients with chronic renal insufficiency receiving a protein-poor diet with too little or too much saccharose. Srp Arh Celok Le 98: 1432–1443 (in Czech)

Simko V (1980) Increase in serum lipids of feeding sucrose: The role of fructose and glucose. Am J Clin Nutr 33: 2217

Simko V, Kelley RE (1979) Physical exercise modifies the effect of high cholesterol–sucrose feeding in the rat. Eur J Appl Physiol 40: 145–153

Simmons WK, de Mello AV (1975) Blindness in the nine states of Northeast Brazil. Am J Clin Nutr 28: 202

Simon I (1938) Studies on the osmotic pressure of organs. IV. The variations of osmotic pressure of organs after intravenous injection of hypertonic solutions of hexamethylene tetramine, urea, glucose, sucrose and uroselectan. Arch Farmacol Sper 65: 153–176

Simon I (1938) Osmotic pressure of the organs. XI. Variations in osmotic pressure of the blood and of organs at various times after the intravenous injection of hypertonic solutions of hexamethylene-tetramine, urea, glucose and sucrose. Arch Farmacol Sper 66: 210–234

Simon J (1978) Les sirops de glucose issus du mais. Ann Nutr Aliment 32: 681–688

Simpson HCR, Lousley S, Geekie M et al. (1981) A high carbohydrate leguminous fibre diet improves all aspects of diabetic control. Lancet I: 1–5

Sims EAH, Danforth JRE, Horton ES, Bray GA, Glennon JA, Salans LB (1973) Endocrine and metabolic effects of experimental obesity in man. Recent Prog Horm Res 29: 457–496

Sinclair B, Cameron DA, Goldsworthy (1950) Some observations on dental conditions in Papua-New Guinea, 1947, with special reference to dental caries. Dent J Aust 22: 58–93, 120–157

Singh VN, Brin M (1981) Fructose ingestion does not affect serum lipids in human subjects. Am J Clin Nutr 34: 1630–1632

Sjostrom L, William-Olsson T (1981) The effects of a new glycoside hydrolase inhibitor on glucose and insulin levels during sucrose loads in obese subjects. Curr Ther Res Clin Exp 30: 351–366

Sjostrom L, William-Olsson T (1982) Effects of acarbose on glucose and insulin levels during sucrose loading in obese subjects. Int Congr Ser Excerpta Med 594: 73–85

Skinner A, Woods A (1984) An investigation of the effects of maltose and sucrose in the diet on the microbiology of dental plaque in man. Arch Oral Biol 29: 323–326

Skinner A, Connolly, P, Naylor MN (1982) Influence of the replacement of dietary sucrose by maltose in solid and in solution on rat caries. Caries Res 16: 443–452

Skinner A, Connolly P, Naylor MN (1982) The influence of the replacement of dietary sucrose by maltose on the formation and biochemistry of human dental plaque. Arch Oral Biol 27: 603–608

Slama G, Haardt MJ, Jean-Joseph P et al. (1984) Sucrose taken during mixed meal has no additional hyperlgycaemic action over isocaloric amounts of starch in well controlled diabetes. Lancet II: 122–124

Sleder J, Chen Y-Di, Cully MD, Reaven GM (1980) Hyperinsulinemia in fructose induced hypertriglyceridemia in the rat. Metabolism 29: 303–305

Slein MW, Cori GT, Cori CF (1950) Comparative study of hexokinase from yeast and animal tissue. J Biol Chem 186: 763–780

Smith LH Jr, Ettinger RH, Seligson D (1953) A comparison of the metabolism of fructose and glucose in hepatic disease and diabetes mellitus. J Clin Invest 32: 273–282

Smith M, Kinney GC (1956) Sugar as a reward for hungry and non-hungry rats. J Exp Psychol 51: 348–352

Smith M, Duffy M (1957) Consumption of sucrose and saccharin by hungry and satiated rats. J Comp Physiol [A] 50: 65–69

Smith-Barbaro PA, Quinn MR, Fisher H, Hegsted DM (1980) Pressor effects of fat and salt in rats. Proc Soc Exp Biol Med 165: 283–290

Sognnaes RF (1949) Further analysis of wartime caries observations. An opportunity and a responsibility. Br Dent J 87: 291–294

Sols, A, Crane RK (1954) Substrate specificity of brain hexokinase. J Biol Chem 210: 581–595

Solyst JT, Michaelis OE IV, Reiser S, Ellwood KC, Prather ES (1980) Effect of dietary sucrose in humans on blood uric acid, phosphorus, fructose and lactic acid responses to a sucrose load. Nutr Metab 24: 182–188

Sommariva D, Scotti L, Fasoli F (1978) Low-fat diet versus low-carbohydrate diet in the treatment of type IV hyperlipoproteinaemia. Atherosclerosis 29: 43–51

Soterakis J, Iber FL (1975) Increased rate of alcohol removal from blood with oral fructose and sucrose. Am J Clin Nutr 28: 254–257

Soukupova K, Prusova F (1970) Nutrition and the ischemic heart disease mortality rates in 33 countries. Nutr Metab 12: 240–244

Southgate DAT, Paul AA, Dean AC, Christie AA, Fric C (1978) Free sugars in foods. J Hum Nutr 32: 335–347

Sparrow MP, Mayrhofer G, Simmonds WJ (1967) Uptake and increased binding by smooth muscle in half isotonic sucrose and its relationship to contractility. Aust J Exp Biol Med Sci 45: 469–484

Spealman CR (1940) The action of osmotically active substances on the heart rate. Am J Physiol 129: 293–296

Spektor YaZ (1942) Comparative data on liver and muscle glycogen storage in rats fed different sugars. Med Exp 8: 28–35

Sreebny LM (1982) Sugar and human dental caries. World Rev Nutr Diet 40: 19–65

Sreebny LM (1982) The sugar-caries axis. Int Dent J 32: 1–12

Sreebny LM, Chatterjee R, Kleinberg I (1985) Clearance of glucose and sucrose from the saliva of human subjects. Arch Oral Biol 30: 269–274

Srinivasan SR, Radhakrishnamurthy B, Webber LS, Dalferes ER Jr, Kokatnur MG, Berenson GS (1978) Synergistic effects of dietary carbohydrate and cholesterol on serum lipids and lipoproteins in squirrel and spider monkeys. Am J Clin Nutr 31: 603–613

Srinivasan SR, Clevidence BA, Pargaonkar PS, Radhakrishnamurthy B, Berenson GS (1979) Varied effects of dietary sucrose and cholesterol on serum lipids, lipoproteins and apolipoproteins in rhesus monkeys. Atherosclerosis 33: 301–314

Srinivasan SR, Berenson GS, Radhakrishnamurthy B, Dalferes ER Jr, Underwood D, Foster TA (1980) Effects of dietary sodium and sucrose on the induction of hypertension in spider monkeys. Am J Clin Nutr 33: 561–569

Srinivasan SR, Radhakrishnamurthy B, Berenson GS (1982) Dietary sucrose and cardiovascular disease risk factors in nonhuman primate models. In: Reiser S (ed) Metabolic effects of utilizable carbohydrates. Marcel Dekker, New York, pp 119–139

Staat RH, Schachtele CF (1974) Evaluation of dextranase production by the cariogenic bacterium *Streptococcus mutans*. Infect Immun 9: 467–469

Staat RH, Schachtele CF (1975) Characterization of a dextranase produced by an oral strain of *Actinomyces israelii*. Infect Immun 12: 556–563

Staat RH, Schachtele CF (1976) Analysis of the dextranase activity produced by an oral strain of *Bacteroides ochraceus*. J Dent Res 55: 1103–1110

Staat RH, Gawronski TH, Schachtele CF (1973) Detection and preliminary studies on dextranase producing microorganisms from human dental plaque. Infect Immun. 8: 1009–1016

Stacpoole P, Swift L. Greene H, Slonim A, Burr I (1980) Homozygous familial hypercholesterolemia: cholesterol reduction in a patient by oral vivonex administration. Clin Res 28: 473A

Stare FJ (1967) Dietary fats and carbohydrates, blood lipids, and coronary heart disease. Am J Clin Nutr 20: 149–151

Stare FJ (1975) Sugar in the diet of man. World Rev Nutr Diet 22: 237–326

Stare FJ (1975) Role of sugar in modern nutrition. World Rev Nutr Diet 22: 239–247

Staub HW, Thiessen R Jr (1968) Dietary carbohydrate and serum cholesterol in rats. J Nutr 95: 633–638

St Clair RW, Bullock BC, Lehner NDM, Clarkson TB, Lofland HB Jr (1971) Long-term effect of dietary sucrose and starch on serum lipids and atherosclerosis in miniature swine. Exp Mol Pathol 15: 21–33

Steel JM, Mitchell D, Prescott RL (1983) Comparison of the glycaemic effect of fructose, sucrose and starch containing mid-morning snacks in insulin dependent diabetics. Hum Nutr Appl Nutr 37A: 3–8

Steel R (1951) Disappearance from mouse tissues of carbon from assimilated $C^{14}$-labelled sucrose. Fed Proc 10: 251–252

Steinmann B, Gitzelmann (1981) The diagnosis of hereditary fructose intolerance. Helv Paediatr Acta 36: 297–300

Stephan RM (1944) Intra-oral hydrogen-ion concentrations associated with dental caries activity. J Dent Res 23: 257–266

Stephan RM (1966) Effects of different types of human foods on dental health in experimental animals. J Dent Res 45: 1551–1561

Stepp W, Schroeder H (1936) Beriberi in man produced by eating excessive amounts of sugar. Munch Med Wochenschr 83: 763–764

Sterkin E, Wengerowa FM (1934) Studies on fructose and lactose metabolism. Biochem Z 272: 246–258

Stevens HA, Ohlson MA (1966) Estimated intake of simple and complex carbohydrates. J Am Diet Assoc 48: 294–296

Stewart CC (1932) Pyuria and pyelitus: the influence of exsiccosis on their experimental production. Am J Dis Child 43: 632–644

Stewart WH, Hoppert CA, Hunt HR (1953) Incidence of dental caries in caries-susceptible and caries-resistant albino rats (*Rattus norvegicus*) when fed diets containing granulated and powdered sucrose. J Dent Res 32: 210–214

Stifel FB, Herman RH, Rosensweig NS (1968) Dietary regulation of galactose-metabolizing enzymes: adaptive changes in rat jejunum. Science 162: 692–693

Stifel FB, Rosensweig NS, Zakim D, Herman RH (1968) Dietary regulation of glycolytic enzymes. I. Adaptive changes in rat jejunum. Biochim Biophys Acta 170: 221–227

Stookey GK (1978) In vitro methods of studying the cariogenic potential of comestibles. In: Foods, nutrition and dental health. American Dental Health Foundation, Pathotox, Park Forest South, Illinois, pp 170–180

Story JA (1982) Dietary carbohydrate and atherosclerosis. Fed Proc 41: 2797–2800

Stout RW (1977) The relationship of abnormal circulating insulin levels to atherosclerosis. Atherosclerosis 27: 1–13 (Review)

Stout RW (1979) Diabetes and atherosclerosis – the role of insulin. Diabetes 16: 141–150

Stralfors A (1966) Inhibition of hamster caries by substances of brown sugar. Arch Oral Biol 11: 617–626

Stralfors A, Thilander H, Bergenholtz A (1967) Caries and periodontal disease in hamsters fed cereal foods varying in sugar content and hardness. Arch Oral Biol 12: 1361–1365

Strasser T (1972) Atherosclerosis and coronary heart disease: the contribution of epidemiology. World Health Organization, Geneva, pp 7–11 (WHO Chronicle No 26)

Strautmane IK (1966) Comparative biological evaluation of glucose and sucrose given intravenously (dog). Izv Akad Nauk Latv SSR 8: 229

Stroder U, Neumann H (1953) Behaviour of blood sugar, leukocytes and differential blood picture during change from horizontal posture to active upright position; studies on normal persons and patients with vegetative disorders with and without orthostatic disturbances. Verh Dtsch Ges Inn Med Kong 59: 156–159

Strohm JG, Osgood SB (1936) Intravenous sucrose as diuretic. Northwest Med 35: 89

Sugawa-Katayama Y, Morita N (1975) Effects of a high fructose diet on lipogenic enzyme activities in some organs of rats fed *ad libitum*. J Nutr 105: 1377–1383

Sugawa-Katayama Y, Morita N (1977) Effect of a high fructose diet on lipogenic enzyme activities of a meal-fed rats. J Nutr 107: 534–538

Sumi T (1960) Experimental studies on the congenital hydrocephalus due to excessive sugar. J Osaka City Med Center 9: 351–359

Sundin B, Birkhed D, Granath L (1983) Is there not a strong relationship nowadays between caries and consumption of sweets? Swed Dent J 7: 103–108

Surtshin A (1957) Protective effect of a sucrose diet in mercuric chloride poisoning. Am J Physiol 190: 271–277

Surtshin A, Yagi K (1958) Distribution in renal cell fractions of sulfhydryl groups in rats on normal and sucrose diets and its relation to renal mercury distribution after mercuric chloride injection. Am J Physiol 192: 405–409

Suzuki M, Hashiba N, Kajuu T (1982) Influence of timing of sucrose meal feeding and physical activity on plasma triacylglycerol levels in rat. J Nutr Sci Vitaminol (Tokyo) 29: 295–310

Suzuki M, Ide K, Saituh S (1983) Diurnal changes in liver and skeletal muscle of rats in relation to the feeding time of sucrose. J Nutr Sci Vitaminol (Tokyo) 29: 545–552

Sved AF (1983) Precursor control of the function of monoaminergic neurons. In: Wurtman RJ, Wurtman JJ (eds) Nutrition and the brain. Raven Press, New York, vol 6, pp 223–275

Swan DC, Davidson P, Albrink MJ (1966) Effect of simple and complex carbohydrates on plasma nonesterified fatty acids, plasma sugar and plasma insulin during oral carbohydrate tolerance tests. Lancet I: 60–63

Swann AC (1984) (Na+, K+)-ATPase stimulation by sucrose feeding: prevention by 6-hydroxydopamine. Life Sci 34: 353–357

Sweeney JA (1983) A spoonful of sugar. J Irish Dent Assoc 29: 42–47

Szanto S, Yudkin J (1969) The effect of dietary sucrose on blood lipids, serum insulin, platelet adhesiveness and body weight in human volunteers. Postgrad Med J 45: 602–607

Szanto S, Yudkin J (1970) Dietary sucrose and platelet behaviour. Nature (Lond) 225: 467–468

Szepesi B (1982) Evidence that sucrose alters food efficiency in the rat: interaction of dietary sucrose and fat. Nutr Rep Int 25: 31–47

Takasaki Y (1979) Protective effect on mono- and disaccharides on glutamate-induced brain damage in mice. Toxicol Lett 4: 205–210

Takekoshi T (1970) On the role of dietary sucrose in the development of vascular disorders. Jpn Circ J 34: 959–969

Takemoto TI (1975) Sex differences in the response of different kinds of dietary carbohydrates in rats. Tohoku J Exp Med 115: 213–222

Taketa ST, Alexander CG, Feigen GA (1960) Effect of nitrate ion on the distribution of sucrose

in the rat ventricle. Nature (Lond) 187: 419–420
Takizawa N (1940) Experimental production of sarcoma in mouse by injection of dextrose, levulose and galactose; histogenesis of fibroplastic sarcoma. Gann 34: 1–5
Talbot JM, Fischer KD (1978) The need for special foods and sugar substitutes by individuals with diabetes mellitus. Diabetes Care 1: 231–240
Talbott JH, Coombs FS, Consolazio WV, Pecora LJ (1938) Respiratory exchange during high carbohydrate ingestion. Am J Physiol 124: 246–253
Tamir I, Epstein D, Heldenberg D, Leuton O, Werbin B (1972) Serum lipids during short-term high glucose and high sucrose feeding in infants. Pediatrics 50: 84–91
Tanaka H, Suzuki N, Arima M (1982) Hypoglycemia in the fetal alcohol syndrome in rat. Brain Dev 4: 97–103
Tange U (1938) The vitamin B2 complex. IV. The action of carbohydrate on a vitamin B2 decificent condition. Sci Papers Inst Phys Chem Res 35: 64–72
Tange U, Kaneko (1938) Vitamin B2 complex. II. Nutritive value of sugars. Sci Papers Inst Phys Chem Res 35: 47–55
Tansey MJB, Opie LH, Kennelly BM (1979) The effects of oral sucrose and of estimated infarct size on plasma free fatty acids, plasma glucose and serum insulin in the early stages of acute myocardial infarction. Eur J Clin Invest 9: 81–88
Tanzer JM, McCabe RM (1968) Selection of plaque-forming streptococci by serial passage of wires through sucrose containing broth. Arch Oral Biol 13: 139–143
Tarjan I, Linden LA (1982) Isotope studies on the permeability of the dental enamel to sucrose and xylitol. J Int Assoc Dent Child 13: 53–56
Tarjan R, Kramer M, Szole-Szotyori K, Hafiez AA (1967) Effect of nutritional factors on fat metabolism. Z Ernahrungswiss 8: 32–41
Taruis, Nonaka K, Saito Y (1970) Hormonal regulation on the membrane permeability of sugar. Folia Endocrinol Jpn 46: 639–644
Taskinen MR (1986) Effects of dietary carbohydrates in metabolic disturbances in man. Prog Biochem Pharmacol 21: 160–180
Tateda H (1965) Sugar receptor and alpha-amino acid in the rat. Olfaction taste. In: Proceedings of the 2nd international symposium, Tokyo, pp 383–397
Taterka H (1929) Sugar metabolism and household water. II. Insulin and sugar carrying capacity by cardiacal diabetics. Klin Wochenschr 8: 1763–1764
Taub C (1973) Basic rules of the food allergy diet. In: Miller JB (ed) Food allergy. Thomas CC, Springfield, Illinois, pp 26–59
Taskinen MR (1986) Effects of dietary carbohydrates on liver content and on serum lipids in relation to age and strain of rats. J Nutr 91: 275–282
Taylor SA, Price RG, Kang SS, Yudkin J (1980) Modification of the glomerular basement membrane in sucrose-fed and streptozotocin-diabetic rats. Diabetes 19: 364–372
Tehrani A, Brudevold F, Attarzadeh F, Van Houte J, Russo J (1983) Enamel demineralization by mouth rinses containing different concentrations of sucrose. J Dent Res 62: 1216–1217
Tenovuo J, Makinen KK, Paunio K (1984) Effects on oral health of mouthrinses containing xylitol, sodium cyclamate and sucrose sweeteners in the absence of oral hygiene. IV. Analysis of whole saliva. Proc Finn Dent Soc 80: 28–34
Terzioglu M (1949) The effect of glucose injections on the regeneration of blood in the rabbit. Arch Int Pharmacodyn Ther 80: 276–300
Terzioglu M, Eroz K (1949) The fate of subcutaneously injected sucrose in the rabbit. Arch Int Pharmacodyn 80: 269–275
Thenen SW, Mayer J (1976) Effects of fructose and other dietary carbohydrates on plasma glucose, insulin, and lipids in genetically obese (ob/ob) mice. Proc Soc Exp Biol Med 153: 464–467
Thiebaud D, Jacot E, Schmitz H, Spengler M, Felber JP (1984) Comparative study of isomalt and sucrose by means of continuous indirect calorimetry. Metabolism 33: 808–813
Thirion J, Thibaut-Vercruyssen R, Ronveaux-Dupal MF, Wattiaux R (1983) Experimental sucrose overloading of rat liver lysosomes: effect of pre-treatment with invertase. Eur J Cell Biol 31: 107–113
Thomas G, Keup U, Krause HP, Puls W (1982) Pharmacological studies on acarbose. II. Anti-hyperlipaemic effects. Int Congr Ser Excerpta Med 594: 151–155
Thomlinson RH (1980) Kitchen remedy for necrotic malignant breast ulcers. Lancet II: 707
Thompson AM, Hintz LK, Fleming J (1962) Time studies on distribution of T-1824 and sucrose in the rat liver. In: 46th annual meeting, Atlantic City, New Jersey, April 1962. Fed Proc 21: 122
Thompson CM, Funk K, Schemmel R, Mickelsen O (1974) Dental caries: possible sugar substitutes

for sucrose. Ecol Food Nutr 3: 231–236

Thompson RG, Hayford JT (1982) Influence of dietary carbohydrate on 24 hour concentrations of glucose, lipids, and glucoregulatory hormones. In: Reiser S (ed) Metabolic effects of utilizable dietary carbohydrates. Marcel Dekker, New York, pp 175–207

Thompson RG, Hayford JT, Danney MM (1978) Glucose and insulin responses to diet. Effect of variations in source and amount of carbohydrate. Diabetes 27: 1010–1026

Thompson RG, Hayford JT, Hendris JA (1979) Triglyceride concentrations: the disaccharide effect. Science 206: 838–839

Thornber JM, Eckhert CD (1984) Protection against sucrose-induced retinal capillary damage in the Wistar rat. J Nutr 114: 1070–1075

Thornton JR, Emmett PM, Heaton KW (1979) Diet and Crohn's disease: characteristics of the pre-illness diet. Br Med J ii: 762–764

Thornton JR, Emmett PM, Heaton KW (1983) Diet and gallstone: effects of refined and unrefined carbohydrate diets on bile cholesterol saturation and bile acid metabolism. Gut 24: 2–6

Thornton JR, Emmett PM, Heaton KW (1985) Smoking, sugar and inflammatory bowel disease. Br Med J 290: 1786–1787

Thornton SN (1986) Drinking in the pigeon (*Columba livia*) in response to water deprivation and the influence of intracerebroventricular infusions of NaCl or sucrose solutions. Physiol Behav 38: 719–724

Timofeeva NM, Iezuitova NN, Chernyakhovskaya MYu, De Laey P, Ugolev AM (1967) Reaction between tributyrin, dipeptide, and sucrose in mucous membrane of the small intestine. Dokl Akad Nauk SSSR 176: 1451–1454

Tobey TA, Mondon CE, Zavaroni E, Reaven GM (1982) Mechanism of insulin resistance in fructose fed rats. Metabolism 31: 608–612

Tomassi G, Lintas C (1985) Attuali orientamenti in tema di livelli di assunzione raccomandati di carboidrati con particolore riguarto al saccarosio. Riv Soc Ital Sci Alim 14: 299–306

Tomatis L (1986) Diet, nutrition and cancer: concluding remarks and future perspectives. In: Hayashi Y, Nagao M, Sugumura T et al. (ed) Diet, nutrition and cancer. Japan Scientific Societies Press, Tokyo; VNU Science Press BV, Utrecht

Topping DL, Mayes PA (1976) Comparative effects of fructose and glucose on the lipid and carbohydrate metabolism of perfused rat liver. Br J Nutr 36: 113–126

Toverud G (1949) Dental caries in Norwegian children during and after the last world war. A preliminary report. Proc R Soc Med 42: 249–258

Toverud G (1957) The influence of war and postwar conditions on the teeth of Norwegian school children. Milbank Memorial Fund Q 35: 371–454

Trautner K (1983) Zucker-plaque-Karies. Swiss Dent J 4: 25–26, 31

Triner L, Mraz M, Chmelarova M (1963) The effect of glucose and glucose together with insulin on the resistance of fasted rats to trauma in the Noble-Collip drum. Physiol Bohemoslov 12: 136–144

Trouillet JL, Chastre J, Fagon JY, Pierre J, Domart Y, Gilbert C (1985) Use of granulated sugar in treatment of open mediastinitus after cardiac surgery. Lancet II: 180–184

Trounce JQ, Walker-Smith JA (1985) Sugar intolerance complicating acute gastroenteritis. Arch Dis Child 60: 986–990

Trowell HC (1978) The development of the concept of dietary fiber in human nutrition. In: Symposium on role of dietary fiber in health. Am J Clin Nutr 31: S3–S11

Trugnan G, Thomas-Benhamou G, Cardot P, Rayssiguier Y, Bereziat G (1985) Short term essential fatty acid deficiency in rats. Influence of dietary carbohydrates. Lipids 20: 862–868

Trump BF, Janigan DT (1962) The pathogenesis of cytologic vacuolization in sucrose nephrosis. An electron microscopic and histochemical study. Lab Invest 11: 395–411

Truswell AS (1978) Diet and plasma lipids – a reappraisal. Am J Clin Nutr 31: 977–989

Truswell AS (1985) The ABC of nutrition: nutritional advice for other chronic diseases. Br Med J 291: 197–200

Truswell AS, Mann JI, Campbell GD (1971) Serum-lipids in sugar-cane cutters. Lancet I: 602

Tsurumi K (1955) About the influence of experimental glucosuria upon the development of teeth. Report 1. What is the influence of experimental glucosuria upon the growth rate of teeth? Jpn J Constitutional Med 20: 111–119

Tsurumi K (1956) About the influence of experimental glycosuria upon the development of teeth. Report 6. Pathological observations about the teeth. Jpn J Constitutional Med 20: 278–281

Turner JL, Bierman EL, Brunzell JD, Chait A (1979) Effect of dietary fructose on triglyceride transport and glucoregulatory hormones in hypertriglyceridemic men. Am J Clin Nutr 32: 1043–1050

Ueyama T (1969) Digestion and absorption of sucrose in full term and premature newborn infants. Bull Kobe Med Coll 29(4): 58–66
Ulshen MH, Grand RJ (1979) Site of substrate stimulation of jejunal sucrase in the rat. J Clin Invest 64: 1097–1102
Ungari C (1942) Sucrose, behaviour of glycemia, saccharosemia and saccharosuria following ingestion of sucrose in infants with acute digestive disorders; 12 cases. Clin Pediatr 24: 539–560
Urbach E, Willheim R (1932) Rare, formerly disregarded, nutritive allergens (sodium chloride, organic acids, sugar). Klin Wochenschr 11: 1012–1014
Urbanowicz M, Chalcarz W, Jelone B, Czanocinska J (1985) Plasma lipid and protein responses in rats fed diets containing pectin and varying levels of sucrose and starch. Nutr Rep Int 32: 649–658
US/USSR Steering Committee for Problem Area I (1984) The pathogenesis of atherosclerosis. Nutrient intake and its association with high-density lipoprotein and low-density lipoprotein cholesterol in selected US and USSR subpopulations. Am J Clin Nutr 39: 942–952
Utter O (1927) Saccharose tolerance and saccharose elimination in urine of children. Finska Lak Sallsk Handl 69: 613–623
Vaaler S, Hanssen KF, Aagenaes O (1980) Sucrose and sorbitol as sweeteners in the diet of insulin-dependent diabetics. Acta Med Scand 207: 371–373
Vallerand AL, Lupien J, Bukowiecki LJ (1986) Synergistic improvement of glucose tolerance by sucrose feeding and exercise training. Am J Physiol 250: E607–E614
Van den Berghe G (1978) Metabolic effects of fructose in the liver. Curr Top Cell Regul 13: 97–135
Van den Berghe G (1978) Safety of nutritive sweeteners: fructose and sorbitol. Health and sugar substitutes. In: Proceedings of the ERGOB conference, Geneva. Karger, Basel, pp 92–97
Van den Berghe G (1986) Fructose: metabolism and short-term effects on carbohydrate and purine metabolic pathways. Prog Biochem Pharmacol 21: 1–32
Van den Berghe G, Hers HG (1978) Dangers of intravenous fructose and sorbitol. Acta Pediatr Belg 31: 115–123
Vanecek R (1983) Stanoveni obsahu fruktózy v cisté glukóze. Cesk Farm 32: 127–129
Van Handel E, Zilversmit CB (1959) Comparisons of artherogenesis in rabbits fed liquid oil, hydrogenated oil, wheat germ and sucrose. J Nutr 69: 202–208
Van Heyningen R (1966) Sugar cataracts in animals and man (rat, diabetes, lens, galactosemia). Proc 3rd Annual Symposium Brit Small Animal Veter Assoc 111: 120
Van Houte J (1978) Carbohydrates, sugar substitutes and oral bacterial colonization. Health and sugar substitutes. In: Proceedings of the ERGOB conference, Geneva. Karger, Basel, pp 199–204
Van Houte J (1980) Bacterial specificity in the etiology of dental caries. Int Dent J 30: 305–326
Varela RM, Teixeira SG, Batista M (1972) Hypovitaminosis A in the sugar cane zone of southern Pernambuco State, Northwest Brazil. Am J Clin Nutr 25: 800–804
Varley CK (1984) Diet and the behaviour of children with attention deficit disorder. J Am Acad Child Psychol 23: 182–185
Vassel B (1939) A comparative study of the effects of glucose, sucrose, fructose and invert sugar on growth, fat and glycogen formation in the white rat. University of Michigan, 2 pp (Thesis)
Vassel B (1943) A comparative study of the effects of glucose, sucrose, fructose and invert sugar on growth, fat and glycogen formation in the white rat. University Microfilms, Ann Arbor, Michigan, 140 pp (No 572)
Vasselli JR, Haraczkiewicz E (1982) Intestinal glucosidase inhibition: effects upon food intake and the development of obesity in 2 rat models. Int Cong Ser Excepta Med 594: 161–166
Vasselli JR, Cleary MP, Van Itallie TB (1984) Obesity. In: Present knowledge in nutrition, 5th edn. The Nutrition Foundation, Washington DC, pp 35–36
Vaughan OW, Filer LF (1960) The enhancing action of certain carbohydrates on the intestinal absorption of calcium in the rat. J Nutr 71: 10–14
Velazques M, Ly J, Preston TR (1969) Digestible and metabolizable energy values for pigs of diets based on high-test molasses or final molasses and sugar. J Anim Sci 29: 578–580
Verzar F, von Kuthy A (1930) The exhaustion of insulin formation by carbohydrate overload. Pflugers Arch Physiol 225: 606–612
Victor J (1933) The effects of sugar and electrolyte solutions on the metabolism and irritability of heart muscle. Am J Physiol 103: 620–630
Videla E, Blanco AM, Galli ME, Fernandez-Collazo E (1981) Human seminal biochemistry: fructose, ascorbic acid, citric acid, acid phosphatase and their relationship with sperm count. Andrology 13: 212–214

Vijayagopalan P, Kurup PA (1970) Effect of dietary starches on the serum, aorta, and hepatic lipid levels in cholesterol-fed rats. Atherosclerosis 11: 252–264

Vijayagopalan P, Srinivasan SR, Radhakrishnamurthy B, Berenson GS (1980) Decreased secretion of triacyglycerol by exogenous cholesterol in high sucrose-fed rabbits. Biochem Med 24: 49–59

Vinay P, Bourbeau D, Duranceau A et al. (1981) Metabolic effects of total parenteral nutrition with or without fructose, with special reference to the liver adenine nucleotide content. Clin Invest Med 4: 87–96

Vinogradova NA, Doroshenko NV, Nikol'skii NN, Troshin AS (1968) Regulation of sugar transport in muscle fibers. Biofizika 13: 365–368

Vinuela E, Salas M, Sols A (1963) Glucokinase and hexokinase in liver in relation to glycogen synthesis. J Biol Chem 238: 1175–1177

Virkkunen M, Huttunen MO (1982) Evidence for abnormal glucose tolerance test among violent offenders. Neuropsychobiology 8: 30–34

Viswanatha T, Gander JE, Liener IE (1954) Inter-relation of fat, carbohydrate and vitamin E in the diet of the growing rat. J Nutr 52: 613–626

Vlitos AJ (1977) Sweets for starters. Chem Br 13: 340–345

Voegtlin C, Dunn ER, Thompson JW (1925) Antagonistic action of sugar, amino acid and alcohol on insulin poisoning. Am J Physiol 71: 574–582

Volker JF, Pinkerton DM (1947) Acid production in saliva-carbohydrate mixtures. J Dent Res 26: 229–232

Volpe JJ, Vagelos PR (1974) Regulation of mammalian fatty-acid synthetase. The roles of carbohydrate and insulin. Proc Natl Acad Sci USA 71: 889–893

Von der Decken A, Astrom S, Arrhenius EK (1981) Increased transcription activity of rat liver chromatin after protein restriction and limited digestion of nuclei with micrococcus nuclease. J Nutr 111: 1249–1257

Von Der Fehr FR (1982) Evidence of decreasing caries prevalence in Norway. J Dent Res 61 (Sp Iss): 1331–1335

von Gordon L (1936) Causes of dental caries: possible relationship to sugar consumption. Veroffentl Geb Med Verwalt 46: 175–286

von Pressentin R (1953) Question of substitution of other sugars for dextrose; experimental studies on heart and intestine of guinea pig. Arztl Forsch 7: 1/389–1/390

Vorontsov DS (1964) Protoplasmatic membrane of muscle fibers as an active cell apparatus. Fiziol Zh Akad Nauk Ukr Rsr 10: 439–449

Vrána A, Fabry P (1983) Metabolic effects of high sucrose or fructose intake. World Rev Nutr Diet 42: 56–101

Vrána A, Kazdová L (1986) Effects of dietary sucrose or fructose on carbohydrate and lipid metabolism. Prog Biochem Pharmacol 21: 59–73

Vrána A, Slabochova Z, Kazdová L, Fabry P (1971) Insulin sensitivity of adipose tissue and serum insulin concentration in rats fed sucrose or starch diets. Nutr Rep Int 3: 31–37

Vrána A, Fabry P, Kazdová L (1973) Effect of dietary fructose on fatty acid synthesis in adipose tissue and on triglyceride concentration in blood in the rat. Nutr Metab 15: 305–313

Vrána A, Fabry P, Kazdová L (1974) Lipoprotein lipase activity in heart, diaphragm and adipose tissue in rats fed various carbohydrates. Nutr Metab 17: 282–288

Vrána A, Fabry P, Slabochova Z, Kazdová L (1974) Effect of dietary fructose on free fatty acid release from adipose tissue and serum free fatty acid concentration in the rat. Nutr Metab 17: 74–83

Vrána A, Slabochova Z, Fabry P, Kazdová L (1974) Influence of diet with a high starch or sucrose content on glucose tolerance, serum insulin level and insulin sensitivity in rats. Physiol Bohemoslov 23: 305–310

Vrána A, Fabry P, Kazdová L (1976) Effects of dietary fructose on serum triglyceride concentrations in the rat. Nutr Rep Int 14: 593–596

Vrána A, Fabry P, Kazdová L (1977) Diet induced adaptation of intestinal fructose absorption in the rat. Physiol Bohemoslov 26: 225–234

Vrána A, Fabry P, Kazdová L (1978) Liver glycogen synthesis and glucose tolerance in rats adapted to diets with a high proportion of fructose or glucose. Nutr Metab 22: 262–268

Vrána A, Fabry P, Poledne R, Kazdová L (1978) Palmitate and glucose oxidation by diaphragm of rats with fructose-induced hypertriglyceridemia. Metabolism 27: 885–888

Vrana A, Fabry P, Kazdová L, Zvolankova K (1978) Effect of the type and proportion of dietary carbohydrate on serum glucose levels and liver and muscle glycogen synthesis in the rat. Nutr Metab 22: 313–320

Vrána A, Fabry P, Kazdová L, Slabochova Z, Poledne R (1978) Metabolic effects of a high

sucrose or fructose intake. Baroda J Nutr 5: 67–79
Vrána A, Fabry P, Kazdová L, Poledne R, Slabochova Z (1982) Sucrose induced hypertriglyceridaemia: Its mechanism and metabolic effects. Czech Med 5: 9–15
Waddel CA, Desai ID (1981) The use of laboratory animals in nutrition research. World Rev Nutr Diet 36: 206–222
Wahlquist ML, Wilmhurst EG, Murton CR, Richardson EN (1978) The effect of chain length on glucose absorption and the related metabolic response. Am J Clin Nutr 31: 1998–2001
Wales A (1975) The role of the combination of sucrose and milk products in diabetes mellitus and ischaemic heart disease. Med Hypotheses 1: 191
Wales A (1976) The role of the combination of sucrose and milk in diabetes mellitus. Am J Clin Nutr 29: 689–690
Wales A (1977) The role of the combination of sucrose and milk in diabetes mellitus. Am J Clin Nutr 30: 300
Wales A (1978) The role of the combination of sucrose and milk in diabetes mellitus. Am J Clin Nutr 31: 559–560
Walgren MC, Young JB, Kaufnan LN, Llandsberg L (1987) The effects of various carbohydrates on sympathetic activity in heart and interscapular brown adipose tissue of the rat. Metabolism 36: 585–594
Walker ARP (1969) Sugar intake and coronary heart disease. Lancet II: 1071
Walker ARP (1971) Sugar intake and coronary heart disease. Atherosclerosis 14: 137–152
Walker ARP (1975) Sugar and dental caries. S Afr Med J 49: 1458
Walker ARP (1975) Are high compared with low consumers of sugar more prone to obesity, diabetes and coronary heart disease? S Afr J Sci 71: 201–205
Walker ARP (1975) Sucrose, hypertension, and heart disease. Am J Clin Nutr 28: 195–196
Walker ARP (1977) Sugar intake and diabetes mellitus. S Afr Med J 49: 842–851
Walker ARP (1977) Controversy over carbohydrate intake in diabetes mellitus. S Afr J Sci 73: 74–77
Walker ARP (1979) The relative risks of saccharin and sucrose ingestions. Am J Clin Nutr 32: 727–728
Walker ARP (1984) How practicable are meaningful reductions in dental caries by dietary means? Nutr Abstr Rev 54: 211–217
Walker ARP, Burkitt DP (1976) Colonic cancer–hypotheses of causation, dietary prophylaxis, and future research. Digest Dis 21: 910–917
Walker ARP, Cleaton-Jones PE (1978) Is sugar good for you? S Afr Med J 54: 589–590
Walker ARP, Dison E, Duvenhage A, Walker BF, Friedlander E, Aucamp V (1981) Dental caries in South African black and white high school pupils in relation to sugar intake and snack habits. Community Dent Oral Epidemiol 9: 37–43
Walker H (1925) Influence of various substances on the diastatic action of saliva. Biochem J 19: 221–225
Walker RA, Ahrens RA (1975) Sucrose, hypertension, and heart disease. Am J Clin Nutr 28: 195–202
Wall JR, Pyke DA, Oakley WG (1973) Effect of carbohydrate restriction in obese diabetics: relationship of control to weight loss. Br Med J i: 577–578
Wander RC, Berdanier CD (1986) Effects of type of dietary fat and carbohydrate on gluconeogenesis in isolated hepatocytes from BHE rats. J Nutr 116: 1156–1164
Wang K-M (1964) Morphological and histochemical changes in the kidney of the sucrose-fed rats. Acta Histochem 18: 95–105
Wang YM, Van Eys J (1981) Nutritional significance of fructose and sugar alcohols. Ann Rev Nutr 1: 437–475
Warnick CT, Lazarus HM (1982) Effect of a protein-free diet and fasting on RNA polymerase activity in mice. J Nutr 112: 293–298
Waterman RA, Romsos DR, Tsai AC, Miller ER, Leveille GA (1975) Effects of dietary carbohydrate source on growth, plasma metabolites and lipogenesis in rats, pigs and chicks. Proc Soc Exp Biol Med 150: 220–225
Watkins PJ (1982) ABC of diabetes–hypoglycaemia. Br Med J 285: 278–279
Wattiaux R, Wattiaux-De Coninck S, Rutgeerts MJ, Tulkens P (1964) Influence of the injection of a sucrose solution on the properties of rat-liver lysosomes. Nature (Lond) 203: 757–758
Wei YSH (1982) Diet and dental caries. In: Stewart RE, Barber TK, Wei S, Troutman KC (eds) Pediatric dentistry, CV Mosby, St Louis, pp 576–589
Wei YSH (1982) Fluoride supplementation. In: Stewart RE: Barber TK, Wei S, Troutman KC (eds) Pediatric dentistry. CV Mosby, St Louis, pp 737–746

Weick BG, Ritter S, McCarty R (1983) Plasma catecholamines in fasted and sucrose supplemented rats. Physiol Behav 30: 247–252

Weinstein L, Colucci VM (1972) Effects of magnitude of sucrose and saccharine on body weight. J Psychol 80: 157–160

Weismann K, Hagdrup HK (1981) Hair changes due to zinc deficiency in the case of sucrose malabsorption. Acta Derm Venereol (Stockh) 61: 444–447

Weiss RL, Trithart AH (1960) Between-meal eating habits and dental caries experience in preschool children. Am J Public Health Nations Health 50: 1097–1104

Welch K, Sadler K (1966) Permeability of the choroid plexus of the rabbit to several solutes. Am J Physiol 210: 652–660

Wells WW, Quan-Ma R, Cook CR, Anderson SC (1962) Lactose diets and cholesterol metabolism. II. Effect of dietary cholesterol, succinylsufathiazole, and mode of feeding on atherogenesis in the rabbit. J Nutr 76: 41–47

Werner D, Emmett PM, Heaton KW (1984) Effects of dietary sucrose on factors influencing cholesterol gallstone formation. Gut 25: 269–274

Weser E, Sleisenger MH, Dickstein M, Bartley FH (1967) Metabolism of circulating disaccharides in man and the rat. J Clin Invest 46: 499–505

Whang R, Welt LG (1963) Observations in experimental magnesium depletion. J Clin Invest 42: 305–313

White CE, Head HH, Bazer FW (1984) Response of plasma glucose, fructose and insulin to dietary glucose and fructose in the lactating sow. J Nutr 114: 361–368

White CE, Head HH, Bachman KC, Bazer FW (1981) Yield and composition of milk and weight gain of nursing pigs from sows fed diets containing fructose or dextrose. J Anim Sci 59: 141–150

White GE (1973) The use of continuous culture and an enamel dissolution assay in studying the relationship between *Streptococcus mutans* and dental caries. PhD thesis. Massachusetts Institute of Technology, Cambridge, Mass. pp 139

White LW, Landau BR (1965) Sugar transport and fructose metabolism in human intestine in vitro. J Clin Invest 44: 1200–1213

Whitnah CH, Bogart R (1936) Reproductive capacity of female rats as affected by kind of carbohydrates in the ration. J Agric Res 53: 527–532

WHO/FAO (1970) Thirteenth report of the Joint FAO/WHO Expert Committee on Food Additives. World Health Organization, Geneva (WHO Technical Report Series No 445)

WHO/FAO (1977) Summary of toxicological data of certain food additives. World Health Organization, Geneva (WHO Food Additives Series No 12)

WHO/FAO (1978) Summary of toxicological data of certain food additives and contaminants. World Health Organization, Geneva (WHO Food Additives Series No 13)

WHO/FAO (1982) Twenty-sixth report of the Joint FAO/WHO Expert Committee on Food Additives. World Health Organization, Geneva (WHO Technical Report Series No 683)

WHO/FAO (1987) Toxicological evaluation of certain food additives and contaminants. Cambridge University Press, Cambridge (WHO Food Additives Series No 20)

Widdas WF (1968) Membrane transport of sugars. Carbohyd Metab Its Disord 1: 1–23

Wiese HF, Bennett, MJ, Coon E, Yamanaka W (1965) Lipid metabolism of puppies as affected by kind and amount of fat and of dietary carbohydrate. J Nutr 86: 271–280

Wiesner F, Gerner R, Batholmes A, Volk M (1979) Untersuchungen zum Übertritt von Glukose und Fruktose in das Fruchtwasser und den fetalen Kreislauf. Arch Gynecol 228: 216–217

Wilhelmsen L, Wedel H, Tibblin G (1973) Multivariate analysis of risk factors for coronary heart disease. Circulation 48: 950–958

Wille LE, Vellar OD, Hermansen L (1973) Changes in serum pre-betalipoprotein following the feeding of sucrose. A preliminary trial in physically active and inactive young male students. J Oslo City Hosp 23: 141–156

Williams C (1985) Human metabolic response to exercise. BNF Nutrition Bulletin 10: 20–27

Williams CA, Macdonald I (1982) Metabolic effects produced in baboons associated with the ingestion of diets based on lactose hydrolysate. Ann Nutr Metab 26: 374–383

Wilmer HA (1944) The mechanism of sucrose damage to the kidney tubules. Am J Physiol 141: 431–438

Wilson TH, Vincent TN (1955) Absorption of sugars in vitro by the intestine of the golden hamster. J Biol Chem 216: 851–866

Winand J, Christophe J (1976) Effects of high fructose and high sucrose diets on normal and obese hyperglycemic Bar Harbor mice. Arch Int Physiol Biochem 84: 209–210

Winand J, Christophe J (1977) Effects of high sucrose and high fructose diets on non-obese and

congenitally obese hyperglycemic Bar Harbor mice. Bibl Nutr Dieta 25: 83–91
Winholt AS (1970) Sucrose content and plaque formation in extracts from various food products. Odontol Revy 21: 301–307
Winitz M, Birnbaum SM, Greenstein JP (1957) Quantitative nutritional studies with water-soluble, chemically defined diets. IV. Influence of various carbohydrates on growth, with special reference to D-glucosamine. Arch Biochem Biophys 72(2): 437–447
Winitz M, Seedman DA, Graff J (1970) Studies in metabolic nutrition employing chemically defined diets. I. Extended feeding of normal human adult males. Am J Clin Nutr 23: 525–545
Winter GB, Hamilton MC, James PMC (1966) Role of the comforter as an aetiological factor in rampant caries of the deciduous condition. Arch Dis Child 41: 207–212
Wise A, Suzangar M, Messripour M, Mohammadi J (1978) Urinary excretion of aflatoxin M1 after administration of aflatoxin B1 in sucrose- or starch-rich diets. Br J Nutr 40: 397–401
Wislick L (1971) Diabetes mellitus and cancer. Br Med J 349
Witztum JL, Scholfeld G (1978) Carbohydrate diet-induced changes in very low density lipoprotein composition and structure. Diabetes 27: 1215–1229
Wogan GN, Newberne PM (1967) Dose-response characteristics of aflatoxin B-1 carcinogenesis in the rat. Cancer Res 27: 2370–2376
Wold JK, Heen T (1978) Polysaccharide contaminants in ultra pure grade sucrose with relation to anaphylactoid reactions in the clinical use of invert sugar solutions II. Studies on the chemical structure of two native dextrans. Acta Pharm Suec 15: 51–58
Wolf HP, Leuthardt F (1953) Ueber die glycerin-dehydrogenase der leber. Helv Chim Acta 36: 1463–1467
Wolfe BM, Ahuja SP, Marliss EB (1975) Effects of intravenously administered fructose and glucose on splanchnic amino acid and carbohydrate metabolism in hypertriglyceridemic men. J Clin Invest 56: 970–977
Wolfe BM, Ahuja SP (1977) Effects of intravenously administered fructose and glucose on splanchnic secretion of plasma triglycerides in hypertriglyceridemic men. Metabolism 26: 963–978
Wolraich M, Milich R, Stumbo P, Schultz F (1985) Effects of sucrose ingestion on the behaviour of hyperactive boys. J Pediatr 106: 675–682
Womack M, Marshall MW, Parks AB (1953) Some factors affecting nitrogen balance in the adult rat. J Nutr 51: 117–130
Wood DS, Stern MP, Silvers A, Reaven GM, Groeben J (1972) Prevalence of plasma lipoprotein abnormalities in a free-living population of the Central Valley, California. Circulation 45: 114–126
Wood FC Jr, Bierman EL (1972) New concepts in diabetic dietetics. Nutr Today 7: 4–10
Wood SD, Reid JT (1975) The influence of dietary fat metabolism and body fat composition in meat feeding and nibbling rats. Br J Nutr 34: 15–24
Woods HF, Eggleston LV, Krebs HA (1970) The cause of hepatic accumulation of fructose-1-phosphate on fructose loading. Biochem J 119: 501–510
Woods SC, Lotter EC, McKay LD, Porte D Jr (1979) Chronic intercerebroventricular infusion of insulin reduces food intake and body weight of baboons. Nature (Lond) 282: 503–505
Wooley DW, Wooley SC, Daunkam RB (1972) Calories and sweet taste. Affects on sucrose preference in the obese and non-obese. Physiol Behav 9: 765–768
Worcester NA, Bruckdorfer KR, Yudkin J (1975) The effect of dietary sucrose and dietary cholesterol on hyperlipidaemia and atherosclerosis in White Leghorn cockerels (Gallus *domesticus*). Proc Nutr Soc 34: 81A–82A
Worcester NA, Bruckdorfer KR, Yudkin J (1975) The effect of dietary sucrose and different dietary fats on hyperlipidaemia and atherosclerosis in White Leghorn cockerels (*Gallus domesticus*) Proc Nutr Soc 34: 82A–83A
World Health Organization (1986) Oral rehydration therapy for treatment of diarrhoea in the home. World Health Organization, Geneva (WHO/CDD/SER/86.9)
World Health Organization/Pan American Health Organization (1983) Oral rehydration therapy: an annotated bibliography. Pan American Health Organization, Washington DC
World Health Organization/United Nations Children's Fund (1985) The management of diarrhoea and use of oral rehydration therapy. World Health Organization, Geneva
Woteki CE, Welsh SO, Raper N, Marston RW (1982) Recent trends and levels of dietary sugars and caloric sweeteners. In: Reiser S (ed) Metabolic effects of utilizable dietary carbohydrates. Marcel Dekker, New York, pp 1–27
Wright DW, Hansen RI, Mondon CE, Reaven GM (1983) Sucrose-induced insulin resistance in the rat: Modulation by exercise and diet. Am J Clin Nutr 38: 879–883

Wunderly Ch (1947) Hemolytic systems with glutathione as a component. Helv Physiol Pharmacol Acta 5: 91–104

Wurtman JJ (1984) The involvement of brain serotonin in excessive carbohydrate snacking by obese carbohydrate cravers. J Am Diet Assoc 94: 1004–1007

Wurtman JJ, Wurtman RJ (1979) Drugs that enhance central serotoninergic transmission diminish elective carbohydrate consumption by rats. Life Sci 24: 895–904

Wurtman JJ, Wurtman RJ (1979) Sucrose consumption early in life fails to modify the appetite of adult rats for sweet foods. Science 205: 321–322

Wurtman JJ, Wurtman RJ (1982) Studies on the appetite for carbohydrate in rats and humans. In: Lieberman HR, Wurtman RJ (eds) Research strategies for assessing the behavioural effects of food and nutrients. Proceedings of a conference at Massachusetts Institute of Technology. Pergamon Press, Oxford, pp 247–267

Wurtman RJ, Hefti F, Melamed E (1980) Precursor control of neurotransmitter synthesis. Pharmacol Rev 32: 315–335

Wykcham-Martin J (1981) Hidden sugar. Dent Health (Lond) 20: 14–15

Wynn W, Haldi J (1954) Dental caries in the albino rat on high-sucrose diets containing different amounts of aluminum. J Nutr 54: 285–290

Wynn W, Haldi J, Shaw JH, Sognnaes RF (1953) Difference in caries-producing effects of 2 purified diets containing same amount of sugar. J Nutr 50: 267–274

Wynn W, Haldi J, Bentley KD, Law ML (1959) Further studies on the difference in cariogenicity of two diets comparable in sucrose content. J Nutr 67: 569–580

Wynn W, Haldi J, Law ML (1960) Dental caries in the albino rat fed high sucrose diets of relatively high and low pyridoxine content. J Nutr 70: 69–71

Wynn W, Haldi J, Law ML (1962) Cariogenicity of acidic and basic high sucrose diets fed albino rats. J Dent Res 41: 861–865

Wysocki E (1965) Nutrition and dental caries. VII. Susceptibility to dental caries of four breeds of rats kept on diets with high sugar content. Rocz Panstw Zakl Hig 16: 429–435

Wysokinski A, Wysokinska MS (1978) Metabolism and importance of fructose in human nutrition. Wiad Lek 31: 1607–1612

Yager J, Young RT (1974) Non-hypoglycemia is an epidemic condition. N Engl J Med 291: 907–908

Yagi K, White HL (1958) Comparison of ammonium sulfate fractionation of proteins and of protein-bound mercury in kidney soluble fraction of chow-fed and sucrose-fed rats. Am J Physiol 194: 547–552

Yamada K (1932) Fate of sucrose parenterally injected into rabbits. Sei-i-Kwai Med J 51: 2–5

Yamada K (1932) Effect of sucrose parenterally applied upon amount of some nitrogenous substances in urine. Sei-i-Kwai Med J 51: 6–7

Yamada K (1932) Effect of sucrose parenterally applied upon blood constituents in rabbits. Sei-i-Kwai Med J 51: 11

Yamada K, Goda T, Bustamante S, Koldovsky O (1983) Different effects of starvation on activity of sucrose in rat jejuno-ileum. Am J Physiol 244: 449–455

Yamagata S, Arai I, Unoura K, Aratani T, Miura K (1954) Experimental studies on function of reticuloendothelial system; relation between function of Kupffer's stellate cells and liver function with special consideration of effect of sugars. Tohoku J Exp Med 59: 265–273

Yamagiwa K (1935) On the effect of a non-electrolyte upon the size of the action potential of nerve. J Physiol (Lond) 84: 83–89

Yamaguchi S (1924) The secretion of saliva from the mouth. II. The glycogen with special consideration of the separation of sugar and glycogen. Beitr Pathol Anat Allerg Pathol 73: 123–141

Yamane T (1932) Pathologic changes in bones provoked by ingestion; experiments on rabbits. Arch Jpn Chir 9: 1028–1029

Yamane T (1932) Osseous changes produced by saccharose in adult rabbits. Mitt Med Akad Kyoto 6: 2541–2542

Yambe M (1936) Blood platelets in diabetics and their fluctuations following ingestion of cane sugar. Mitt Med Akad Kyoto 17: 983–984

Yano K, Rhoads GG, Kagan A, Tillotson J (1978) Dietary intake and risk of coronary heart disease in Japanese men living in Hawaii. Am J Clin Nutr 31: 1270–1279

Yasui T (1982) Basic study on experimental rat caries. Koku Eisei Gakkai Zasshi 31: 25–38

Yasui T, Morita J, Takahashi K (1966) Inhibition of some neutral organic solutes of the denaturation of myosin A-adenosinetriphosphatase. J Biochem 60: 303–316

Yeager JF (1929) Studies on the acceleration and inhibition of haemolysis. I. The inhibition of

saponin and taurocholate haemolysis by sucrose. Q J Exp Physiol 19: 219–235
Yeh KY, Du FW, Holt PR (1986) Endogenous corticosterone rather than dietary sucrose as a modulator for intestinal sucrase activity in artificially reared rat pups. J Nutr 116(7): 1334–1342
Yonger J, Weill J, Kaplan M (1963) Intolerance to saccharose. Clinical and biological study in 5 cases. Arch Fr Pediatr 20: 253–263
Young J, Landsberg L (1977) Stimulation of the sympathetic nervous system during sucrose feeding. Nature (Lond) 269: 615–617
Young J, Landsberg L (1977) Suppression of sympathetic nervous system during fasting. Science 196: 1473–1475
Young JB, Landsberg L (1981) Effects of oral sucrose on blood pressure in the spontaneously hypertensive rat. Metabolism 30: 421–424
Young JB, Mullen D, Landsberg L (1978) Caloric restriction lowers blood pressure in the spontaneously hypertensive rat. Metabolism 27: 1711–1714
Yousufzai SYK, Siddiqi M (1977) Serum and liver lipid responses to 3-hydroxy-3-methylglutaric acid in rats on different carbohydrate diets. Lipids 12: 262–266
Yudkin J (1957) Diet and coronary thrombosis. Hypothesis and fact. Lancet II: 155–162
Yudkin J (1963) Dietary carbohydrate and ischaemic heart disease. Am Heart J 66: 835–836
Yudkin J (1964) Dietary fat and dietary sugar in relation to ischemic heart disease and diabetes. Lancet II: 2–5
Yudkin J (1964) Patterns and trends in carbohydrate consumption and their relation to disease. Proc Nutr Soc 23: 149–162
Yudkin J (1968) Dietary sugar and serum-cholesterol. Lancet I: 917
Yudkin J (1968) Sugar and coronary thrombosis. Postgrad Med 44: 67–70
Yudkin J (1970) Sucrose, insulin, and coronary heart disease. Am Heart J 80: 844–846
Yudkin J (1971) Sugar consumption and myocardial infarction. Lancet I: 296–297
Yudkin J (1972) Sugar and disease. Nature (Lond) 239: 197–199
Yudkin J (1972) Sucrose and cardiovascular disease. Proc Nutr Soc 31: 331–337
Yudkin J (1978) Dietary factors in arteriosclerosis: sucrose. Lipids 13: 370–372
Yudkin J (1979) The avoidance of sucrose by thiamine-deficient rats. Int J Vitam Nutr Res 49: 127-135
Yudkin J (1979) Sugar and diabetes mellitus. Br Med J i: 820
Yudkin J (1981) Increased serum lipids and other effects of dietary sucrose. Am J Clin Nutr 34: 1453–1454
Yudkin J, Krauss R (1967) Dietary starch, dietary sucrose and hepatic pyruvate kinase in rat. Nature (Lond) 215: 75
Yudkin J, Morland J (1967) Sugar intake and myocardial infarction. Am J Clin Nutr 20: 503–506
Yudkin J, Roddy J (1964) Levels of dietary sucrose in patients with occlusive atherosclerotic disease. Lancet II: 6–8
Yudkin J, Roddy J (1966) Assessment of sugar intake: validity of the questionnaire method. Br J Nutr 20: 807–811
Yudkin J, Szanto S (1970) The relationship between sucrose intake, plasma insulin and platelet adhesiveness in men with and without occlusive atherosclerosis. Proc Nutr Soc 29 (Suppl): 2A–3A
Yudkin J, Szanto S (1971) Hyperinsulinism and atherogenesis. Br Med J i: 349
Yudkin J, Szanto S (1972) Increased levels of plasma insulin and 11 hydroxycorticosteriod induced by sucrose, and their reduction by phenformin. Horm Metab Res 4: 417–420
Yudkin J, Szanto S, Kakkar VV (1970) The relationship between sucrose intake, plasma insulin and platelet adhesiveness in men with and without occlusive atherosclerosis. Proc Nutr Soc 29: 2A–3A
Yudkin J, Kang SS, Bruckdorfer KR (1980) Effects of high dietary sugar. Br Med J ii: 1396
Yue DK, Hanwell MA, Satchell PM, Turtle JR (1982) The effect of aldose reductase inhibition on motor nerve conduction velocity in diabetic rats. Diabetes 31: 789–794
Zabik ME, Schemmel R (1980) Influence of diet on hexachlorobenzene acccumulation in Osborne–Mendel rats. J Environ Pathol Toxicol 4: 97–103
Zabik ME, Morgan KJ (1984) Elderly food intake patterns: amount and food sources of total sugar intake by US elderly aged 62 years and older. J Can Diet Assoc 45: 336–343
Zacharias NT, Sebastian KL, Philip B, Augusti KT (1980) Hypoglycemic and hypolipidaemic effects of garlic in sucrose fed rabbits. Indian J Physiol Pharmacol 24: 151–154
Zahavi I, Dinari G, Rosenbach Y, Nitzan M (1981) Metabolic effects of short-term fructose administration in infants. Curr Ther Res 29: 757–760

Zakim D (1972) The effect of fructose on hepatic synthesis of fatty acids. Acta Med Scand 542 (Suppl): 205–214

Zakim D, Herman RH, Gordon WC Jr (1969) The conversion of glucose and fructose to fatty acids in human liver. Biochem Med 2: 427–437

Zarattini A (1941) Sulla produzione sperimentale del sarcoma nei ratti mediante somministrazione parenterale di glucosio. Tumori 14: 77–84

Zavaroni I, Reaven GM (1981) Inhibition of carbohydrate-induced hypertriglyceridemia by a disaccharidase inhibitor. Metabolism 30: 417–420

Zavaroni I, Chen Y-DI, Reaven GM (1982) Studies of the mechanism of fructose-induced hypertriglyceridemia in the rat. Metabolism 31: 1077–1083

Zavaroni I, Sander S, Scott S, Reaven GM (1980) Effect of fructose feeding on insulin secretion and insulin action in the rat. Metabolism 29: 970–973

Zellner DA, Rozin P, Aron M, Kulish C (1983) Conditioned enhancement of human liking for flavor by pairing with sweetness. Learn Motiv 14: 338–350

Zengo AN, Mandel ID (1972) Short communication: sucrose tasting and dental caries in man. Arch Oral Biol 17: 605–607

Ziegler E (1976) Zuckerkonsum und pränatale Akzeleration. II. Betrachtungen über Atiologie und Pathophysiologie der säkularen pränatalen Akzeleration. Helv Paediatr Acta 31: 365–373

Ziegler I, Jasinski B (1968) Investigations on the influence on the growth of rats by adding sucrose to the diet. Pathol Microbiol 31: 202–208

Zimmerman M (1976) Utilisation des sucres issus du mais dans les boissons sucrées en France. Sirops de glucose, sirops de glucose riches en fructose–dextrose, etc. Bios 7: 7–8

Zingg W (1951) Experimental sucrose storage in the mitochondria of kidney tubules. Schweiz Z Pathol Bakteriol 14: 1–16

Zita AC, McDonald RE, Andrews AL (1959) Dietary habits and the dental caries experience in 200 children. J Dent Res 38: 860–865

Zorzi DL (1930) Carbohydrates on blood-vessels and on dilated epithelium. Arch Farmacol Sper Aff 50: 167–218, 239–250

Zuckerman AJ, Macdonald I (1964) The role of dietary carbohydrates and infection on liver lipid and collagen. Br J Exp Pathol 45: 589–594

Zummo C (1934) Sur le métabolisme de l'azote (série II). Action d'un sel alcalinisant (citrate de sodium) sur le métabolisme azoté endogène au cours d'une alimentation exclusivement lipidique. Arch Int Physiol 40: 129–139

Zuniga MA, Koulourides T (1969) Experimental plaque activity in three areas of the human alveolar–dental arch. Alabama J Med Sci 6: 442–446

Zuniga MA, Lopez H, Sandham HJ, Bradley EL, Koulourides T (1973) Calcium and phosphorus contents of dental plaque and micro-hardness changes of sample enamel in the human mouth. Alabama J Med Sci 10: 3–10

# Selected Books


Aykroyd WR (1967) The story of sugar. Quadrangle Books, Chicago, Illinois

Aykroyd WR (1967) Sweet male factor: sugar, slavery and human society. Heinemann, London

Barnes AC (1964) The sugar cane. Leonard Hill, London

BNF/TFSS (1987) Sugars and syrups. The report of the British Nutrition Foundation's Task Force. The British Nutrition Foundation, London

Deer N (1949) The history of sugar vol 1. Chapman and Hall, London

Deer N (1950) The history of sugar vol 2. Chapman and Hall, London

FAO/WHO (1980) Carbohydrates in human nutrition. Food and Agriculture Organization of the United Nations, Rome, p 27 (Food and Nutrition Paper 15)

Garrow JS (1981) Treat obesity seriously: a clinical manual. Churchill Livingstone, Edinburgh

Greenwald P, Ershow AG, Novelli WD, Benton CM (1985) Cancer, diet and nutrition. A comprehensive source book, 1st edn. Marquis WHO's WHO, Chicago, Illinois

Hayashi Y, Sugimura T, Takayama S et al. (eds) (1986) Diet, nutrition and cancer. Japan Scientific Societies Press, Tokyo

Hugill JAC (ed) (1949) Sugar. Cosmo Publications, London

International Sugar Research Foundation (1970) Sucrose chemicals. International Sugar Research Foundation, New York, pp 11–13

Killonitsch V (1970) Sucrose chemicals. A critical review of a quarter-century of research by the Sugar Research Foundation. The International Sugar Research Foundation, Washington DC

Lieberman HR, Wurtman RJ (eds) (1982) Research strategies for assessing the behavioral effects of foods and nutrients. Proceedings of a conference at Massachusetts Institute of Technology. Pergamon Press, Oxford

Mann JI and the Oxford Dietetic Group (1982) The diabetic diet book. Martin Duntiz, London

Ministry of Agriculture, Fisheries and Food (1982) Food additives and contaminants.Committee Report on the Review of Sweeteners in Food. Her Majesty's Stationery Office, London

National Institutes of Health (1985) Health implications of obesity. NIH consensus development conference. National Institutes of Health, Bethesda, Maryland

National Research Council (1980) Toward healthful diets. National Academy of Sciences, Washington DC

National Research Council (1982) Diet, nutrition and cancer. Committee on diet, nutrition and cancer. National Research Council. National Academy Press, Washington DC

Nitzel AE (1981) Nutrition in preventive dentistry, 2nd edn. Saunders, Philadelphia, Pennsylvania

Pancoast HM, Junk WR (1980) Handbook of sugars, 2nd edn. AVI Publishing, Westport, Connecticutt

Rechcigl M Jr (ed) (1978) CRC handbook series in nutrition and food. Section E: nutritional disorders. I. Effect of nutrient excesses and toxicities in animals and man. CRC Press, Boca Raton, Florida

Reddy BS, Cohen LA (ed) (1986) Diet, nutrition and cancer: a critical evaluation. CRC Press, Boca Raton, Florida

Reiser S (1978) Effect of nutrient excess in animals and man: carbohydrates. In: CRC Handbook series on nutrition and foods. CRC Press, Boca Raton, Florida, pp 409–436

Schauss AG (1980) Crime and delinquency. Parker House, Berkeley, California

Schoental R, Connors TA (1981) Dietary influences on cancer: traditional and modern. CRC Press, Boca Raton, Florida

SCOGS/FASEB (1976) Evaluation of the health aspects of sucrose as a food ingredient. Life Sciences Research Office (LSRO)/Federation of American Societies for Experimental Biology (FASEB), US Food and Drug Administration (Prepared for FDA under contract No. FDA 223–75–2004) Washington DC

Shallenberger RS, Birch GG (1975) Sugar chemistry. AVI Publishing, Westport, Connecticut, pp 46–88

Sipple HL, McNutt KW (eds) (1974) Sugars in nutrition. Academic Press, New York

University of California Food Task Force (1974) A hungry world. The challenge to agriculture. University of California Press, Berkeley, California

US Beet Sugar Association (1959) The beet sugar story, 3rd revised edn. US Beet Sugar Association, Washington DC

USDA (1983–85) Sugars and sweeteners outlook. US Department of Agriculture, Washington DC

US/FDA/STF (1986) Evaluation of health aspects of sugars contained in carbohydrate sweeteners. In: Glinsmann WH, Irausquin H, Youngmee KP (ed) Report from FDA's Sugar Task Force (Reprinted from J Nutr 116: S1–S216

West KM (1978) Epidemiology of diabetes and its vascular lesions. Elsevier, New York, pp 224–274

WHO (1984) EHC 30: Principles for evaluating health risks to progeny associated with exposure to chemicals during pregnancy. World Health Organization, Geneva, 177 pp

WHO (1987) EHC 70: Principles for the safety assessment of food additives and contaminants in food. World Health Organization, Geneva, 174 pp